Diversity in Construction—The Gamechangers

Bianca Christina Weber-Lewerenz
Editors

Diversity in Construction—The Gamechangers

Strengthening innovation through diversity

Springer

Editor
Bianca Christina Weber-Lewerenz
Aichtal, Germany

ISBN 978-3-658-45307-7 ISBN 978-3-658-45308-4 (eBook)
https://doi.org/10.1007/978-3-658-45308-4

This book is a translation of the original German edition "Diversität im Bauwesen – Die Gamechanger" by Bianca Christina Weber-Lewerenz, published by Springer Fachmedien Wiesbaden GmbH in 2024. The translation was done with the help of an artificial intelligence machine translation tool. A subsequent human revision was done primarily in terms of content, so that the book will read stylistically differently from a conventional translation. Springer Nature works continuously to further the development of tools for the production of books and on the related technologies to support the authors.

This Springer imprint is published by the registered company Springer Fachmedien Wiesbaden GmbH, part of Springer Nature.
The registered company address is: Abraham-Lincoln-Str. 46, 65189 Wiesbaden, Germany

- *Our coexistence is based on our diversity, on respect, tolerance, empathy.*
- *This book serves as a source of inspiration for even more courage to innovate in the construction industry.*

Preface and Motivation of the Editor

The construction industry is undergoing a change in values. The awareness that humans should be at the center of technical feasibility and that progress should be value-based is increasing, as is the search for orientation. Where is the journey going to? Building for everyone must become the standard, representatives of the construction industry must increasingly engage in conversation to find new ways and position themselves innovatively. Only in this way crises can be dealt with resiliently. The goal must be to appreciate, utilize, and expand the diversity that already exists. Inclusive, fair, respectful, and tolerant. A co-author gets to the point: Meet on equal terms and together.

This book advocates for more diversity in construction and for considering and using diversity as an opportunity. It is reflected in the diversity of contributions with various changes in perspective and the empathetic commitment of all co-authors, who made this book possible.

For one co-author, her first own high-rise project is her "baby", an other co-author builds a department for innovation and diversity in the company, another one finds fulfillment in walking across the facade with a facade access system, another one knows how to handle large construction projects most efficiently and productively, yet another one is passionately committed to more innovative technologies and digitized processes in construction, while the next one considers a sophisticated management culture to be the decisive success factor in the construction industry. Diversity knows no boundaries.

As the research results of the publisher and editor in preparation for the topic demonstrate, the novelty of this book lies in the open niche with such a thematic focus on the book market. The approach is critical, constructive, concrete, holistic. Examples of existing inclusive approaches in the construction industry are presented, but at the same time existing weaknesses in terms of diversity in construction are critically named.

For these inclusive approaches, the book lays the value-based foundation and appeals for even stronger value integration: *"The reference to values of the built and increasingly digital worlds should require the securing of value references to be broadened, through participatory and inclusive procedures."*[1]

The book works out the different forms and potentials of diversity in civil engineering sciences and highlights its influences on the industry, and how diversity represents a true game changer in achieving SDGs and ESG requirements. Action strategies for promoting a diversely positioned construction industry are provided.

Representatives from politics and business as well as science and research, representatives of academic and vocational training, craft businesses, representatives of engineering and craft chambers as well as initiatives and networks that strengthen diversity and bring it into the public light have their say.

Specifically selected individual profiles serve to illustrate the diversity lived in practice as well as often very winding life and career paths, thereby creating a special closeness and authentic access. As role models, they encourage young people to have more courage to go their own way.

The book provides a basis for further in-depth research in the individual areas of diversity such as "Gender & Innovation". The explanations are intended to raise awareness of the importance of inclusive approaches and to support their integration into existing research projects. Ongoing projects and initiatives for more diversity in the construction industry are presented and combined with the latest findings from research.

The book shows for the first time the importance of New Work in crafts, a topic that has been hotly debated and researched in other industries for years, but still leads a shadowy existence in the construction industry. It is time to change.

A brief excursion into early childhood education with the "STEM approach in childhood" and the "workshop pedagogy" concept highlights that the enthusiasm for working with hands and coming into contact with manual creation and action begins in infancy. Often grandparents are the inspiration who are themselves active in manual professions or simply like to work with their hands. This early approach is crucial to gain access to manual work and thus to professions in the construction industry. This is the only way we can ensure that the existing professional profiles as well as those that are constantly being added as a result of technological progress, are filled with appropriately qualified personnel.

For companies, the economy, and politics, this book offers a guide to put their own handling of diversity to the test.

Last but not least, the book's concept which in its diversity and variety, contributes to the promotion scientific excellence and visibility of women and other underrepresented groups. The book is a call to abandon long-held attitudes and develop an awareness of the added value of innovations and a new culture of thinking and learning.

---

[1] Grunwald, Armin (2022). Technology, Ethics and the Construction Industry: a Preface. Value Accents in Construction 4.0—Ethical Observations in Dealing with Digitalization and AI. Springer Vieweg Publishing.

Imagine a world free of prejudice, stereotypes, and discrimination. A world that is diverse, fair, and inclusive. A world where differences are appreciated and celebrated. A world where the many different needs of people are considered from the start, making many life situations easier. Participation for all: This means sustainable, innovative, confident building and designing of our living and working worlds. This is what this book stands for.

The book recommendations listed at the end of the book invite you to deepen further.

Two Dancers 1949. Salvador Dalí (Spanish, 1904–1989)

Bianca Christina Weber-Lewerenz

Foreword

Dear reader,

It is an honor for me to guest foreword a book on "Diversity in Construction" whose ambition is to open realms of opportunities and potential in Civil Engineering and Architecture and, is a call for actions.

The book you have in your hands is a call for a global systemic re-thinking, commitment, and actions of all stakeholders to design, plan, build, implement, deploy, and manage innovation in the field of Engineering and Architecture bringing social value for the common good and individuals. Thus, it requires thinking ahead of the ethical, legal, and social implications and dealing with still increasing inequalities, the multiplications of conflicts and human rights violations.

This book sets strong signal for a collective responsibility to act at our level of action to be smarter and more focused on our impact and our outcomes. Societal challenges are complex and challenging thus calling for smarter approaches and smarter businesses, which involves an 'ethical and human rights by design' approach. Such approach would significantly benefit the people, the planet, prosperity, peace, and partnerships, as well as it would foster diversity and inclusivity through empathetic and enriching exchange towards building trustworthy technological innovation in line with our human values and the sustainable development goals.

The intense variety of the book chapters recognizes both unique individual paths of female career, potential and impacts of working environments grounded on gender-equal strengths. Moreover, new technologies in the Construction sector and the smart built environment are considered as key components and a key driver of social peace or inequalities and discriminations, of ethical concerns and trade-offs, of international law and human rights violations or enhancement. Thus, this book gives you, dear reader, a rich overview of the diversity of challenges to be faced to provide social value.

This book is to be praised for triggering a larger conversation about the strengths, weaknesses, ambitions and challenges in Civil Engineering and Architecture enabling all individuals to play their part in engaging into a strategic rethinking to build smarter 'smart environments' respecting international law and human rights standards, positively contributing to the environment, and bringing value to different social and political contexts. That is how important your roles are.

Paris
March 21st 2024

Virginie Martins de Nobrega

Foreword "Diversity in construction 4.0"

Martha Schneider-Bürger, Thea Luchterhand, Paula Sperle, Dr. Lydia Wladimirowa. Four names that are largely forgotten today. Four names that represent the almost 100-year history of female engineers in Germany, without whom many developments in our modern construction industry would not be conceivable and whose effects still shape our built environment today. Whether formwork for tunnels or significant research work on the development of lightweight concrete. Whether it's about patents, research contributions or—in recent decades—the promotion of sustainable construction methods: Women have long been influencing the fate of the construction industry.

This was not always the case: Historical restrictions on the participation of women in different areas of construction continue to this day. In addition to social conventions about the role of women as primarily based in the family, whose access or even leadership position in natural and technical sciences was long frowned upon, legal regulations excluding women from main construction activities were in place until 1994 (!).

The exclusion of the construction ecosystem from half of the population has effects due to a lack of family normalities, a lack of social presence of 'woman in construction' and a comparatively smaller number of role models also on the life decisions of today's generation. Breaking this path dependency and shedding old role patterns is a goal that the CONSTRUCTION INDUSTRY has consciously set for itself.

Because the industry is attractive, but still equipped with many clichés. For example, that our work is primarily about physically heavy work that can only be done by men. This outdated notion overlooks the fact that planning and process control make up a large part of the work and that automation and technologization as well as the research and use of new methods in the construction industry are already commonplace.

By 2030 alone, the industry will be short of up to 100,000 skilled workers. For this reason alone, it would be a mistake to ignore a large part of the well educated female workforce. We can't afford to do that. Because construction benefits from different perspectives, from different work and approaches, from diversity on the construction site. We could be further along than the numbers suggest today: A proportion of less than two percent of female employees in construction trades is absolutely unsatisfactory. And we can't rest on the roughly 30% of female civil engineers or the 57% of female

Fig. 1 Jutta Beeke at the "Women's Network Construction" kick-off in Berlin, HDB, 2023.[2] *Source* HDB Strehlau

draftswomen either. The same applies to the underrepresentation of women in management positions or the wage gap that statistics still show us.

That's exactly why the CONSTRUCTION INDUSTRY is involved in networks like the "Initiative Clichés-free", with the help of which outdated notions of professions and gender roles are to be dissolved. In addition, the founding of our own Women's Network Construction under the patronage of Federal Minister of Construction Klara Geywitz was a logical and overdue step for us (Fig. 1). With this network we want to provide a platform for female specialists and executives to Building exchange, knowledge transfer, and mentoring. In doing so, we consciously go beyond the boundaries of association membership and create an industry-wide network, with the help of which women from the entire construction value chain can find contacts who support them in entering and advancing in the industry. In the best case, we also prevent women from eventually leaving the industry. Thus, the network is a way to combat the shortage of skilled workers, which poses great challenges to us as an industry. At the same time, it is our response to stereotypes and serves to strengthen a female perspective on the challenges of modern construction and the construction labor market.

Why do positions in management functions have to be advertised primarily full-time? What possibilities are there to reconcile construction management activities and family responsibilities for mothers and fathers? Where are the practical hurdles for the advancement of women in taking on management functions in large projects? These and other questions and tasks are the focus of the network, as well as learning from each other and strengthening the visibility of women in our wonderful industry.

It is important to us that the concrete contents are determined by the women themselves. This is a central point of the network. We want to find out together with the women from the industry: What are your topics, what do you wish for, where do you

[2]https://www.bauindustrie.de/media/veranstaltungen/kick-off-veranstaltung-frauennetzwerk-bau (Accessed 15. Januar 2024).

hope for more support? We want to learn what female identity in the construction industry is like, what paths and goals women in our industry pursue, what hurdles they have to overcome. Based on these stories and experiences, we want to learn how we as an industry can motivate even more women to find their professional path with us and to identify with construction. This benefits not only the directly involved, but we also change the image of the industry, elevate 'Woman in Construction' to the new normal and, in the best case, make the network superfluous.

Construction is one of the most important industries for Germany with very good future prospects and so diverse that it would be wrong if this diversity were not also increasingly reflected in a normal diversity of employees and executives. It is the construction industry that is building the future in this country. If that's not an exciting challenge for people of all genders, what is?

I am pleased that this view of construction is told in the book at hand. I am convinced: It comes at the right time.

Berlin
21st of March 2024

Jutta Beeke
Managing partner of the Echterhoff Bau
Group Vice President of Social Policy
Main Association of the German Construction Industry, and Co-Initiator of the
Women's Network for Construction

Foreword

Gender Dynamics in Male-Dominated Industries

In industries where women encounter particularly harsh conditions, two central characteristics emerge. Firstly, women are often in the clear minority. Secondly, the competencies crucial for the industry are often attributed gender-specifically—with a tendency to attribute these skills more to men than to women. These two aspects are closely linked, because: Who would want to work day after day in an environment where one's own competencies are constantly being tested? Especially when one observes that the very same skills in male colleagues require no questioning due to their gender but are taken for granted. This double burden—on the one hand, the constant questioning of one's own suitability, on the other hand, the observation of seemingly effortless recognition in men—is deeply discouraging for many women and poses a significant barrier to engaging or remaining in such fields.

No company or industry can afford to ignore the potential of women. The importance of diversity as a driver of innovation and creativity is now undisputed. Teams that dispose of a diversity of perspectives often prove to be more capable of developing innovative ideas and solutions.

My research work is at the intersection of gender studies and information technology. In this context, numerous studies in the field of "Women in Tech" have uncovered systemic barriers that women have to overcome in technology-oriented professions. Given the numerous parallels between the challenges faced by women in the technology industry and those in the construction industry, I would like to share some insights in this guest foreword. These concern both the situation of women as a professional minority and the gendered attribution of competencies.

Women as a Minority: Impact on Professional Daily Life

In male-dominated industries, the minority status of women often becomes a defining feature of their professional experience. In such environments, women are often the only representative of their gender in a team or department. This singularity makes the category of gender salient, i.e., it stands out disproportionately—making it a defining characteristic. Such an overemphasis leads to these women being seen less as individuals with their own abilities and personalities, and more as representatives of the social category "woman". This reduction to gender is a core problem that fundamentally distinguishes the experiences of women in male-dominated fields from those of their male colleagues.

The desire to be seen as an individual is a universal human need. For this very reason, people like teams or groups: Only in a smaller group is it possible to be perceived with personal characteristics and individual peculiarities. In an environment where individuality is valued and promoted, we often find deeper satisfaction. In a work environment where women are isolated, they tend to be reduced to their "womanhood", while men are not seen as representatives of a category, but in comparison to other men with their unique characteristics. In addition to the fact that women are often denied recognition of their individual identity, this also puts them under additional pressure. The pressure to not only meet their own expectations, but also those placed on the entire group of women.

Such dynamic not only amplifies existing inequalities, but also makes it difficult for women to fully develop in their professional field and to be perceived as equals. To achieve true equality in the workplace, we must recognize these structural barriers and actively take measures to overcome them. Recognizing and appreciating the individual contributions of each person, regardless of their gender, is a crucial step.

Role Models and Identification Figures: The Lack of Female Role Models

The deficit of female role models in the construction industry poses a significant hurdle, especially for young women and girls who aspire to a career in this field. The sparse presence of women in leadership positions makes it difficult for them to visualize themselves in such roles and develop their own professional goals. The importance of role models who have made it to the top is essential. Equally important is the visibility of women at every level within an organization. However, they should not merely serve as tokens for diversity, but their contributions be sincerely recognized and appreciated, and they should have real power and influence.

An often overlooked but crucial role is played by role models in the immediate environment. It is essential to have examples that demonstrate: "Here is a person who is in a similar situation and whose life path I can imagine for myself"—not only in distant executive floors, but tangible and comprehensible.

In certain industries, the presence of female role models, especially in male-dominated areas, is invaluable in combating stereotypes and promoting diversity. Research findings underline that the lack of such role models is a significant barrier for women and girls considering a career in the IT sector. There is a strong need of these role models to give women who are just starting their careers confidence and a sense of belonging.

Role models act as key figures to make the professional field more attractive and accessible both for women and other underrepresented groups. They alleviate the pressure of isolation by demonstrating that success is achievable, thus paving the way for subsequent women. Through their own successes, they inspire others and help to make diversity and uniqueness of female life and career paths visible.

Division of Labor at the Workplace: A Matter of Gender

The gender-specific division of work tasks manifests itself in the distinction between invisible activities and glamour work. This central theme in gender research on workplace equality sheds light on the subtle ways in which gender-related prejudices influence the assignment of work tasks and career opportunities.

Invisible activities or non-promotable tasks are necessary, but often overlooked or perceived as undervalued tasks. They are similar to housework in the private sphere, are essential for smooth operation, but rarely receive due recognition or contribute to professional advancement. Typical examples include organizing meetings and minutes, maintaining corporate culture, or planning social events. Research findings show that women significantly more often take on these types of tasks, which ties up their resources and leaves them less time for career-promoting projects.

Glamour work, on the other hand, involves tasks that are highly regarded, create visibility, and provide direct paths to recognition and promotions. Such tasks usually involve creative, strategic or result-oriented projects with immediate impact on corporate success and growth. Men are more likely to be offered glamour work and they take on these tasks with a higher sense of entitlement, which allows them to become visible with their skills and accelerate their rise.

This dichotomy between invisible tasks and glamour work perpetuates existing gender inequalities by keeping women in positions that offer less recognition and fewer opportunities for advancement. To promote gender equality in the workplace, organizations must take targeted measures to ensure a fairer distribution of tasks. This can be achieved through transparent project assignments, the upgrading of invisible activities as valuable contributions to team success, or the rotation of tasks. Such approaches ensure that everyone participates in the necessary, but not promotion-worthy tasks. This way, career paths can be paved for all genders and an inclusive work culture can be created.

Gender-Specific Competence Attributions and Their Consequences

The gendering of competences, i.e., the assignment of certain skills and professional roles based on gender, leaves deep traces in career paths, work cultures and the innovation potential of an industry. These dynamics, fueled by outdated stereotypes and social norms, carry the assumption that men and women are fundamentally differently predisposed for certain tasks due to their gender—especially in technical and scientific fields.

The core problem is the often unquestioned attribution of higher technical competence to men, while women—regardless of their actual qualifications and professional experience—are perceived by many as less suitable for technical challenges or leadership positions.

Thus, the gender-specific dynamic is not only reflected in the division of labor and the assigned roles, but also manifests itself in the structural processes and everyday interactions. Biased recruitment and promotion practices favor men for technical and leadership positions, even when women are equally or better qualified. These practices not only perpetuate the underrepresentation of women in technical roles and leadership positions, but also cement the stereotypical image that women do not "belong" in these areas.

Women continuously face the challenge of having to prove their competence and expertise in order to be taken seriously. This phenomenon is known as "Prove-it-Again-Bias": Again and again, the burden of proof lies with the women, to demonstrate their abilities and competences, more so than with their male colleagues.

Another obstacle is the "Tightrope-Bias": A narrower range of behaviors is accepted from women than from men. This leads to a difficult balancing act that women often have to perform: They must appear competent and assertive on the one hand, but not too aggressive or emotional on the other, to avoid being judged negatively. This forces female leaders to constantly adapt their leadership style in order to gain both respect and to appear likable.

The daily work of women is further complicated by phenomena such as "Mansplaining", "Bropropriating" and "Manterrupting". "Mansplaining" occurs when male colleagues explain technical concepts to women in a condescending manner, based on the assumption that women would not be familiar with these—even if they have equivalent or superior knowledge. "Bropropriating" refers to situations where men claim women's ideas as their own, and "Manterrupting" describes the phenomenon that women are more often interrupted in conversations than men, which undermines their chance to be heard and recognized massively.

To effectively counter these phenomena, it is essential that leaders take their role in creating an inclusive and respectful work culture seriously and actively involve male colleagues as "male allies". It is important to pay close attention to how team members talk about their colleagues and how they behave in group constellations. If a few individuals dominate the conversation in meetings, this should be addressed directly. It makes sense to create and enforce a clear policy against verbal interruptions. People who tend to overshadow others should be addressed face to face about this behavior. Situations in which men appropriate ideas from women or other minorities or receive recognition for them also require the commitment from like-minded people to intervene. If a man tried to pass off a woman's idea as his own, the allied person would say, "Yes, I liked Sara's point and I'm glad you do too." By consistently addressing such behaviors, leaders and "male allies" can significantly contribute to promoting a culture of equality and respect in the workplace.

The Future of Construction: Gender Equality as a Key Factor

The dynamics and challenges observed in the technology industry—particularly the gender-specific attribution of competencies and the minority status of women—provide an informative framework for examining similar patterns in other male-dominated fields. The construction industry, with its long history of gender-specific role distribution and the underrepresentation of women, is no exception. When we turn to the topic of the construction industry, it becomes clear that the prevailing structures and cultures there reflect the discussions and insights from the technology sector in many ways.

By bridging the gap from the technology to the construction industry, the necessity and potentials of a more inclusive design of these traditionally male-dominated work fields become apparent. The book "Diversity in Construction—The Gamechangers. Strengthen

Innovation by Diversity" opens up a rich perspective on the specific challenges and opportunities for women in construction in this context. By integrating personal experiences and scientific insights, it offers a deep insight into the current conditions and necessary changes within the industry.

The book at hand underlines the importance of diversity for innovation and creativity in the construction industry. In particular, the stories of women in the construction trade, which are presented in the fourth chapter of the book, are a central element to illuminate the complexity of the challenges and opportunities in this sector. These personal experiences not only serve as inspiration and motivation, but also as an important learning source for all participants in the industry. They illustrate how essential the recognition and integration of diversity is in shaping an inclusive and innovative future for the construction industry.

A key element of the book is the depiction of how digital transformation—through technologies like BIM, AI, and Metaverse—can serve as a catalyst for promoting diversity and overcoming gender stereotypes. These technologies not only offer new work opportunities and environments, but also enable a redesign of the work culture, away from traditional gender roles towards a more inclusive and diversified industry.

The contributions in this book make it clear that the construction industry is at a turning point. The promotion of diversity and the active involvement of women in all areas of construction are not only mandates of justice, but crucial for the future competitiveness and innovative strength of the industry. "Diversity in Construction—The Gamechangers" makes an important contribution by highlighting the challenges, but also the diverse opportunities that arise from a stronger gender diversity.

For a deeper understanding of the topic of her guest foreword, the following are recommended:

- Her book with Karen Holtzblatt on women in the tech industry: Holtzblatt, Karen, & Marsden, Nicola. (2022). Retaining Women in Tech—Shifting the Paradigm. Springer Publishing.
- Her knowledge nuggets on "Socially Responsible AI Design" on the free learning platform AI-Campus: https://ki-campus.org/courses/socially_responsible_ai_design

Prof. Dr. Nicola Marsden

Acknowledgements

- I would like to thank all co-authors, all contributors, all supporters—both in front of and behind the scenes—for their openness, empathy, and insights into their highly personal life and work environments! These make the book what it is: A celebration of diversity.
- A very special thanks goes to my two editors: Sandy Lunau, whose expertise was a great support and who always drove the success of this book project. Frieder Kumm, now editor emeritus, who passionately carried the book idea successfully to the editorial review, because he never doubted the power and passion behind it.

Book Greetings

Diversity, different perspectives, and collaboration are not just buzzwords, but important anchor points when we think about shaping the future and want to implement it. The editor Bianca Christina Weber-Lewerenz has wonderfully managed to let a large number of authors have their say, who have very different perspectives on the subject of construction and yet carry in their hearts the common basic attitude that building is the future and people and organizations in the construction industry must work together to create a livable, sustainable, and innovative living environment.

Change and innovation always start with and through people. The same applies here. We learn why people have chosen craftsmanship, construction, and design and see this as their vocation. Thanks to their convictions and abilities, things change. They give construction a new, diverse face and show with their stories how change is possible but also what is still needed. The second major design impulse is technology. Not only since the public appearance of generative AI has the question arisen not only for the construction industry, but for all actors in the construction ecosystem, which technologies we need, how we use them in a human-centered and supportive way, and how it succeeds in the interplay of human-technology to first make utopias digital and then physical reality. The third element is the organizations, which economically, politically, and socially shape the debate around construction, but are also responsible for implementation. Like the other two elements, they are subject to major changes and are in a constant process of change. This creates resistance that must be actively addressed, but also creates the momentum needed to move into the future together.

In this sense, I wish all readers of this anthology diverse impulses for change and all of us the courage and confidence to tackle them together!

Stuttgart
April 16, 2024

Prof. Dr. Katharina Hölzle
Professor and Head of the Institute for Work Science and Technology Management IAT at the University of Stuttgart
Institute Director of the Fraunhofer Institute for Industrial Engineering and Organization IAO.
Technology Commissioner of the Minister of Economic Affairs of Baden-Württemberg

Greetings on the Publication of the Book "Diversity in Construction—The Gamechangers"

I would like to express my sincere appreciation and gratitude for our strong collaboration on several scientific fronts, but especially regarding diverse, inclusive, and innovative work environments in various disciplines and industry verticals, not only in construction.

As I reflect on our editorial collaborations, they encompassed a broad variety of domains, ranging from advocacy for political global initiatives, strategic decision-making, subject matter expert knowledge-transfer, or latest research trends. However, most importantly we shared a consistent dedication to advancing responsible, equitable digital innovation and digital transformation across industries that lack diversity.

It is with great pleasure that I recommend this book that lays new foundations.

As a pioneer in the field, the author has crafted a compelling narrative that examines the long overdue diversity gaps in a niche-industry: Civil Engineering and Construction.

The book also offers a comprehensive discussion of technological advances, new approaches to corporate culture, a new culture of thinking and learning. The author masterfully captures unique stories of individual career paths in the construction industry. While offering a critical examination of the intersection of law, ethics, and culture, it also highlights the need for alignment with the United Nations 2030 Sustainable Goals Agenda.

I strongly recommend this book and hope it will be inspirational to readers around the globe.

Miami, USA, as of 23rd of March 2024

Prof. Dr. Ingrid Vasiliu-Feltes
Founder & CEO, Institute for Science,
Entrepreneurship and Investments

Thoughts on the book by…

"Diversity is a Strength."
United Nations 2024

"Cities are growing again. Recognizing gender and diversity can create a higher quality of life for everyone. Wishing the book much success."
Dr. Brigitte Wotha,[3] Expert Commission Federal Foundation for Equality, Berlin; Professorship for Urban and Regional Development with a focus on Diversity Management, Ostfalia University of Applied Sciences Salzgitter; Professorship for Spatial Planning with a focus on Urban Planning and Regional Planning, University of Applied Sciences Kiel.

"Decision-Making Needs Diversity in Exchange. Wishing Your Exciting Book Project Much Success."
Waltraud Dahs,[4] Head of the Department for European and International Equality Policy European Commission; Head of Unit for European and International Equality Policy at the Federal Ministry for Family, Seniors, Women and Youth

"We build on diversity. The basis of a cultural and societal serving architectural discourse is the open and unbiased confrontation with new ideas and suggestions."
Dr. Thomas Welter,[5] Federal Managing Director Association of German Architects BDA

"Gender equality is not only a fundamental human right, but a necessary foundation for a peaceful, prosperous and sustainable world."
SDGs by United Nations 2024

"Strengthening diversity: Inclusive education as a catalyst for changes."
UNESCO 2023

---

[3]https://regionalheute.de/wolfenbuettel/ostfalia-vortrag-zur-zukuenftigen-lebensqualitaet/ (Accessed 12. February 2024).

[4]European Commission (2007). Decision-making: exchange of good practices. Directorate-General for Employment, Social Affairs and Equal Opportunities Unit G.1. EU Commission Printing and Publishing.

[5]Dr. Thomas Welter, Managing Director of the BDA (2024). We build on diversity. LinkedIn Post in January 2024. (Accessed 12. February 2024).

Summary for Quick Readers

When the editor and the co-authors in this book speak of DIVERSITY, it encompasses the variety of strengths, skills, knowledge, thoughts, creativity, innovation, technical possibilities, learning, thinking and corporate cultures—across disciplines, regardless of gender, skin color and nationality.

Over 300 pages, with over 40 contributing co-authors, a work has been created that holistically illuminates the construction industry from the diversity aspect and awakens curiosity about the many untapped potentials and critically focuses on yet insufficiently considered fields of innovation. The book provides a new and important basis for in-depth scientific research in order to provide long overdue new approaches for more sustainability and value creation in construction in research and practice.

The topics are a diverse mix, offering the reader variety and adds to the fun of reading. Each co-author comes from a different area of work and contributes to the reflection of the topic with their very individual personal and professional background.

The book is a must-read for anyone who wants to gain a better zunderstanding of new challenges in Construction in the context of digital transformation and the many still unexploited opportunities, existing gaps, needs for action and concrete solutions through a wide range of topics.

The book sends readers on a journey with personal, often humorous, winking and critical reports of their experiences. To simplify the reading flow and handling of the book, "*Strengths*", "*Skills*", "*Innovation and Technologies*", "*Knowledge*" and "*Communication*" form the main structure of this book. Short reading trips to "*AI meets Art*" and "*La Champagne*" stir up the mental cells and inspire the thinking through of the cross-connections between diversity, knowledge, technological innovation and corporate culture.

The book combines VALUE creation in and VALUE appreciation of craftsmanship.

"*If you don't move, you can't move anything*"—this is exactly where the book starts: it wants to "take people along", provide trust and orientation, show feasible new paths, share experiences and strengthen the will to innovate. In short: help, weigh up in a well-balanced and sensible way how W E want to shape the human and digital transformation today, now and many tomorrows after,, but especially our society and our environment, in which not only W E, but all subsequent generations want to (can) live.

The book exposes the societal, human, technological tensions and is equally an exciting guide in the construction industry for technological progress with high speed, recruitment of skilled workers, new knowledge, new job profiles and highlights, that could not be better suited as a role model and offer more motivation. The fact that family is THE success and happiness factor in crafts and one of the game changers, becomes even clearer in the course of the book.

The book is published in both German and English languages.

Summary for Quick Readers (German)

Wenn die Herausgeberin und die Co-AutorInnen in diesem Buch von DIVERSITÄT sprechen, dann umfasst das die Vielfalt an Stärken, Können, Wissen, Gedanken, Kreativität, Innovation, technischen Möglichkeiten, Lern-, Denk- und Unternehmenskulturen—fachübergreifend, unabhängig von Geschlecht, Hautfarbe und Nationalzugehörigkeit.

Auf über 300 Buchseiten, mit über 40 mitwirkenden Co-Autoren ist ein Werk gelungen, das die Baubranche ganzheitlich unter dem Diversitätsaspekt beleuchtet und die Neugier auf viele noch unausgeschöpfte Potenziale weckt und noch zu wenig bedachte Innovationsfelder kritisch in den Fokus stellt. Das Buch bietet eine neue und wichtige Grundlage zur vertiefenden wissenschaftlichen Forschung, um längst überfällige neue Lösungsansätze für mehr Nachhaltigkeit und Wertschöpfung am Bau in Forschung und Praxis zur Hand zu geben.

Die Themen sind ein vielfältiger Mix, die der/m LeserIn Abwechslung bietet und den Lese-Spaß zusätzlich beflügelt. So kommt jede/r Co-AutorIn aus einem anderen Arbeitsbereich und trägt mit ihrem/seinem ganz unterschiedlichen persönlichen und beruflichen Hintergrund zum Reflektieren bei.

Das Buch ist ein Muss für alle, die durch breit aufgestellte Beitragsthemen neue Herausforderungen im Bauwesen im Zusammenhang der digitalen Transformation und die vielen noch ungenutzen Chancen, vorhandenen Lücken, Handlungsbedürfnisse und konkreten Lösungsansätze entdecken wollen.

Das Buch schickt die LeserInnen auf die Reise mit persönlichen, oft humorvollen, augenzwinkernden und kritischen Erfahrungsberichten. Um den Lesefluss und Umgang mit dem Buch zu vereinfachen, bilden „*Stärken*“, „*Können*“, „*Innovation und Technologien*“, „*Wissen*“ und „*Kommunikation*“ die Hauptgliederung dieses Buches. Kurze Lesereisen zu „*KI trifft Kunst*“ und „*La Champagne*“ quirlen die geistigen Zellen durch und beflügeln das Durchdenken der Querverbindungen zwischen Diversität, Wissen, technologischer Innovation und Unternehmenskultur.

Das Buch verbindet WERTschöpfung im und WERTschätzung des Handwerks.

„*Wer sich nicht bewegt, kann nichts bewegen*“ – genau hier setzt das Buch an: möchte Menschen „mitnehmen“, Vertrauen und Orientierung schenken, gehbare neue Wege aufzeigen, dabei Erfahrungen teilen und den Innovationswillen bestärken. Kurz: helfen, wohlbalanciert und sinnvoll abwägen, wie W I R heute, jetzt und viele morgen danach die humane und digitale Transformation gestalten wollen, aber insbesondere unsere Gesellschaft und unsere Umwelt, in der nicht nur W I R, sondern alle folgenden Generationen leben möchten (können).

Das Buch enttarnt das gesellschaftliche, menschliche, technologische Spannungsfeld und ist gleichermaßen spannender Richtungsgeber im Bauwesen für technologischen Fortschritt mit High Speed, Fachkräftegewinnung, neues Wissen, neue Aufgabenprofile und Highlights, die nicht besser als Vorbild geeignet sein und mehr Motivation schenken könnten. Dass Familie DER Erfolgs- und Glücksfaktor im Handwerk eine der Gamechanger ist, wird im Laufe des Buches noch deutlicher klar.

Das Buch erscheint sowohl in deutscher als auch in englischer Sprache.

Contents

Contributors

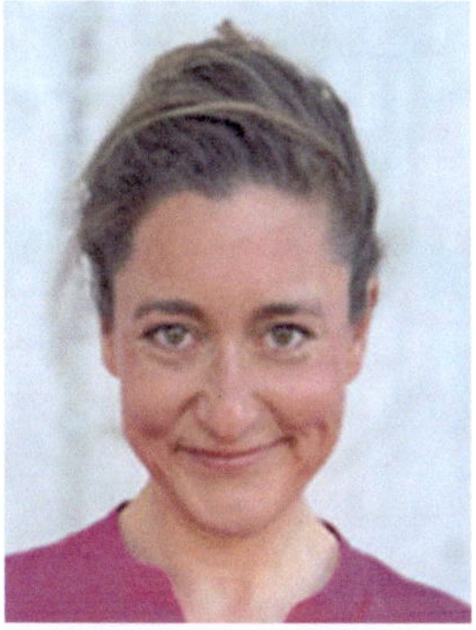

Dr. Heide Ackerbauer born in 1983, is a civil engineer in the field of building physics and energy efficiency. After graduating in 2010, she was part of an engineering office in the field of building physics and expert reports until 2020. Since 2011, she has been a research assistant at Leibniz University Hannover, where she completed her doctoral thesis from 2014-2019. Since 2020, she has been employed at the Material Testing Institute Hannover and has been heading the Building Physics department there since 2021.

Laura Attenberger Born in 2003. Originally from the district of Cham in Bavaria. After her 1.1-er high school A-Level graduation, she completed a masonry apprenticeship in her uncle's construction company. She is considering studying civil engineering.

Jutta Beeke is the managing partner of the Echterhoff Bau-Group based in Westerkappeln. In the Main Association of the German Construction Industry, she is Vice President of Social Policy and co-initiator of the Women's Network Construction. The graduate businesswoman studied business administration in Passau, Parma, and Munich before joining her family's construction company in the fifth generation.

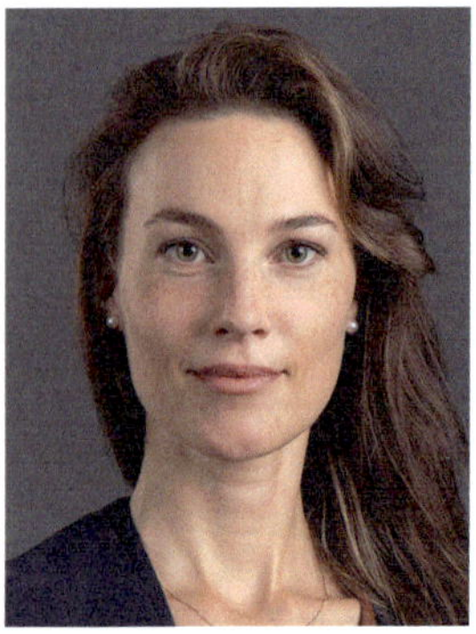

Dipl.-Ing. Natalia Bienkowski, LL.B. studied civil engineering and law at the TU Dresden. Since 2016, she has been a research assistant at the Institute for Construction Operations at the TU Dresden. Natalia Bienkowski designed and coordinated the BMBF-funded research project "Vision2028" (FKZ 01FP21011), in which the contributions of innovative female civil engineers in Germany since 1928 were researched for the first time.

Eva Bork After her high school graduation, Eva Bork completed a carpentry apprenticeship and has had many great experiences in this profession throughout her life. She constantly benefits in daily life and in many projects around the house and yard! With her family and her animals she lives near Bonn.

Dr.-Ing. Steffen Braun has been the deputy director of the Fraunhofer Institute for Industrial Engineering and Organization IAO in Stuttgart since 2024, and since 2016 he has been heading the research area "City System Design" at the institute. His research focuses on the question of how city systems, infrastructures, and the built environment can be designed to be climate-just, technology-open, and adaptable. He completed his doctorate in 2023 after a diploma in architecture/urban planning in Stuttgart's mechanical engineering at the interface of technology management and urban development.

René Breyer A business economist specializing in business informatics and a certified expert in the field of "Digital Business Management". With over 20 years of experience in the industry, he drives digital transformation in the construction industry as Chief Digital Officer (CDO). He advises both young and established companies on topics of artificial intelligence and digital innovation. In addition, as a member of the "Handelsblatt AI-Circle", he enriches the discourse on AI trends and their influence on economy and society.

M. Sc. Diego Cisterna is an expert in Lean Construction and digitization in construction. He is a co-founder of Valoon GmbH and is doing his doctorate at the Karlsruhe Institute of Technology on the use of artificial intelligence in the construction industry.

Dipl.-Ing. (FH) Sonja Ebling Since 2022, she has been Head of the Planning Department South Hesse, Hessen Mobil, in road and traffic management in Darmstadt. Previously, she was head of a technical planning department and led a planning team as a graduate engineer in the planning team. She learned the job in construction from scratch with training as a road construction technician. In 2003, she graduated in International Civil Engineering from FH Mainz with a study abroad at the University of Dundee (Scotland) with a bachelor's degree.

Dipl.-Ing. Martin Ehret is a speaker in the Department of Urban Development of the Federal Ministry for Housing, Urban Development and Construction. After studying urban and regional planning at the BTU Cottbus, he completed a two-year urban planning traineeship in Baden-Württemberg. After graduating as a government architect (building assessor), he worked as a speaker in the Ministry of Economics and Housing in Baden-Württemberg, primarily in the area of sustainable land management. Since 2020, he has been responsible for implementing the National Urban Development Policy in the Federal Ministry of Construction.

Dipl.-Ing. M.-Eng. LL. M. Michael Frahm Born in 1979. Trained in Stuttgart, Kaiserslautern, Saarbrücken in engineering and business law. Management courses at HEC Paris, HHL and Northwestern University. In 2010, he won the Young Project Management Award from GPM. Certified as a Senior PM by IAPM and as an Advanced System Thinking Practitioner by SCiO. Author of construction project management books with a system/cybernetics perspective. Numerous articles and lectures on engineering and management. Lecturer for the management of major projects and Systems Practice.

M.A. Michaela Friedrich is a research assistant in the "Service Engineering" team at the Fraunhofer IAO and deals with service development and Smart Services. As the Equal Opportunities Officer at the institute, she is also responsible for issues relating to the professional equality of men and women and the compatibility of work and private life.

Andreas Grosz MBA, connecting people, getting stuck thinking moving, forming networks from different disciplines, shaping future topics is the passion of Andreas Grosz; studied economics and political science; Dipl.-Betriebswirt.—Corporate developer and cultural manager, advises renowned companies and institutions on issues of strategic positioning and communication. In 1985, he founded and published the culture magazine LIVING; founding managing director of the World Exhibition Company EXPO 2000 GmbH; from 2000 to 2015 program director, later managing partner of the ROTONDA-Business Club in Cologne; 2003 co-founder and head of the KAP Forum for Architecture & Urban Development; together with Inken Herzig initiator of RheinDesign 2007; Economic Ambassador of the City of Cologne; publications include THE CULTURE AG. New alliances between business and culture, Hanser-Verlag; LIVING AT WORK. Trend book life and work in the future, Hanser-Verlag; Board memberships: OFW/World Business Dialogue at the University of Cologne; European Music Academy Montepulciano/University of Music Cologne. Married to the cultural journalist Inken Grosz-Herzig, father of three sons.

Barbara Hagedorn is the founder of FutureWE (www.future-we.de). For more than 25 years she was the managing director of the Hagedorn Group, the all-round service provider in the areas of demolition, disposal, civil engineering, heavy load logistics, site revitalization and digitization. She is the initiator of the "Woman in Construction" campaign, with the aim of inspiring women for construction professions and changing entrenched structures.

Susanne Haus Painter and varnisher master, restorer and business economist (HwO), has been president of the Frankfurt-Rhein-Main Chamber of Crafts since 2020 and president of the Working Group of Hessian Chambers of Crafts since 2021. Susanne Haus runs the family business with a focus on monument preservation, which was founded by her grandfather in 1933. For ten years she has been a publicly appointed and sworn expert in the painting and varnishing trade, she is also a member of various regional and national committees.

Leonie Häuser Born in 1987. First female site foreman in Baden-Württemberg, 2022. 2023 Master's degree in road construction trade. Previously commercial training as a retail saleswoman and several years of professional experience. Since 2023 she is engaged in construction and project management, followed by real estate investments.

Prof. Dr. Katharina Hölzle has been a professor and head of the Institute for Work Science and Technology Management IAT at the University of Stuttgart and director of the Fraunhofer Institute for Work Economics and Organization IAO since April 2022. Since April 2023, she has been the technology commissioner of the Minister of Economic Affairs of Baden-Württemberg. Her research areas are technology and innovation management, work science and digital transformation. A focus is always the interaction between human—technology—organization.

Dipl.-Ing. Sandor Horvath has been advocating for over 20 years for the He has been involved in the digitization of building planning and has been an enthusiastic blockchain enthusiast since 2017. Therefore, at the Technical University of Munich, he is the Head of External Relations at the TUM FinTech Center for Web3 and DeFi and runs his own educational channel with the podcast Web3Welt. In his current position, he is responsible for Web3 data economy and Industrial Metaverse at the Faculty of Construction at the OTH Eastern Bavarian Technical University Regensburg and will move to the Board of Directors of the Bauform eG data cooperative for the construction and real estate industry in the summer of 2024.

Dipl.-Ing. Jule Janson Born in 2000. Concrete and reinforced concrete builder and civil engineer from Mühlacker, Baden-Württemberg. Campaign ambassador for the craft. Winner of the Practical Performance Competition 2019 at the Performance Competition of the German Crafts.

M. Sc. Niki Karatza Her enthusiasm for timber construction led her to her role as co-founder and consultant of a service company for timber construction companies, Holzbau.tech GmbH in Rosenheim. Her motivation is to contribute to advancing timber construction and making it more efficient through technologies and process optimizations. She has completed the B.Eng. Wood Technology, as well as the M.Sc. Applied Research and Development.

M. Sc. Agnes Kelm is a researcher and head of the BIM lab at the BIM Institute of the Bergische University Wuppertal. In addition to her work in research in the field of BIM, digitization and the digital transformation of the construction and real estate industry, she is actively involved in working groups of DIN and VDI as well as at BuildingSMART.

Dipl.-Ing. Anne Keßler studied architecture in Weimar, Berlin and Gothenburg and completed the building construction traineeship with the state of Mecklenburg-Vorpommern. In the federal construction department, she worked in the areas of procurement law, general matters of construction and social urban development; ESF. She currently heads the Department of Basic Urban Development, Building Culture, Research and Coordination.

Univ.-Prof. Dr.-Ing. Dipl.-Wirt.-Ing. Katharina Kleinschrot has been a professor of construction process technology and circular value creation at the Institute for Construction Operations at TU Dresden since April 2023. Her research focuses are circular construction and construction in existing buildings. Previously, she was head of sustainability development at Zech Group SE and worked in traditional construction.

Dipl.-Wirt.-Ing. Heike Kling Diploma in Industrial Engineering, 58 years young, Head of Engineering and authorized signatory at Hilti Deutschland AG.

My mission: "Efficient construction through value engineering and digitization", my basic understanding: "There is no such thing as impossible"

M. Sc. Nils Koch to Krax works as a researcher at the BIM Institute of the Bergische University Wuppertal. His academic journey initially led him through a dual study program, which included training as a road builder parallel to his bachelor's degree. Subsequently, he successfully completed a constructive master's program in the field of civil engineering. His research focuses particularly on the fields of digitization and digital transformation, with his focus on the executing processes in the construction and real estate industry. Here, his special attention is on the integration of construction and craft businesses.

Christine Lin Graduate of Shanxi University of Finance and Economics and Beijing Foreign Studies University. Worked at CBRE Beijing Company,
Yicheng Commercial Real Estate Co., LTD.
Today, she is a freelancer and investor.

Dr. rer. nat. Moritz Julian Maier works as a Senior Researcher at the Center for Responsible Research and Innovation of the Fraunhofer IAO. Here, he particularly deals with participatory research processes and innovative methods for stakeholder involvement and foresight.

Prof. Dr. Nicola Marsden is a research professor for socio-informatics with a focus on gender issues in the Faculty of IT at Heilbronn University. She is an expert in Gender and IT, e.g., for the 3rd Equality Report of the Federal Government, at EU level for ERAC SWG-GRI and for the German UNESCO Commission for AI and Gender. She heads the Social, Behavioral and Economic Sciences research unit in the Baden-Württemberg doctoral association, and volunteers, e.g., as a founding member of the Participation working group at the Society for Informatics (GI) or as deputy chairwoman of the nationwide competence center Technology-Diversity-Equal Opportunities (kompetenzz.de). She has long-standing expertise in the project management of inter- and transdisciplinary processes, researches the development and implementation of participatory and user-centered methods of designing innovations, e.g., as a real-world laboratory, and has a broad network in the economy.

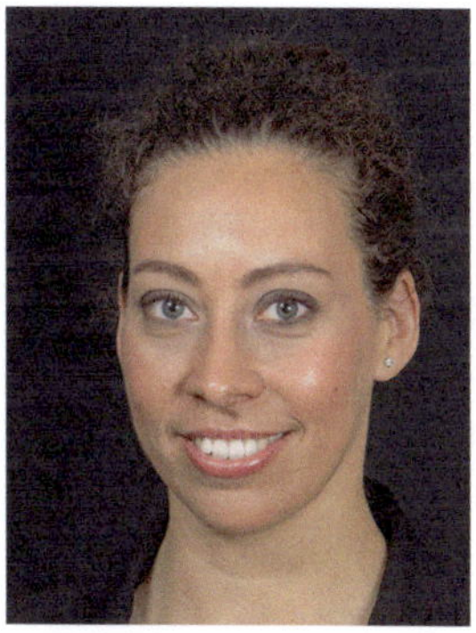

Virginie Martins de Nobrega Master of Law is a lawyer and consultant for international law and human rights. Since 2018, she has also been working as a specialist in ethics, international law and human rights as well as societal AI. She holds an Executive Master in Management of International Organizations (SDA Bocconi) and two Masters in international and European law (Paris X Nanterre). Since 2004 she has been committed to sustainability and value creation.

Prof. Dr. Anica Meins-Becker heads the BIM Institute at the Bergische University Wuppertal and holds a professorship for the digital transformation of the construction and real estate industry. Her current research focuses are in the area of process management, the creation of digital product, supply chain, and building twins, as well as the continuous further development of the Building Information Modeling method.

B.A. Antonella Menrath Bachelor's degree in Romance Studies and Economics. As of March 2024, she is a newly qualified plant mechanic for SHK and winner of her year and chamber. She works in the heating, ventilation, and plumbing department at SHK Friedrich Morsch GmbH & Co. KG. Since February 2024, she has been a master student in the plumbing and heating trade. She speaks fluent Spanish, is active as an influencer for skilled worker recruitment, and as a role model for more women in construction.

M.Sc. Annemarie Mulder After 20 years in the construction industry in Europe and China, her desire to manage large projects led her to the German civil engineering business 5 years ago. She holds the role of overall project manager for the construction consortium for a PPP project in Thuringia for VINCI Construction—as the first woman in Germany. Her management style is shaped by her Dutch roots, international experience, and a Master's degree in Public Policy & Management. Her focus is on contract, stakeholder, change management, and process optimization.

Dr. Manuel Mühlbauer conducted research in the field of computer vision and artificial intelligence with a focus on generative building design. His current work at the Nuremberg University of Technology involves the development and evaluation of software to enhance creativity in the context of building data modeling.

As an entrepreneur, Dr. Mühlbauer is interested in the automated assessment of feasibility and profitability of residential projects in the real estate industry. Earlybuild also offers the construction and real estate industry features for climate-positive building planning and green investments.

Kateryna Pylypchuk Master of Law is from Ukraine, an internationally acclaimed expert in real estate, strategic marketing, and AI-based business management, with experience spanning over 15 years. Renowned for her troubleshooter prowess, business acumen, social investment initiatives, and entrepreneurial spirit, she serves as the Strategic Business Advisor at "Modulina Straw Panels LLC," a premier Lithuanian manufacturer specializing in prefabricated straw panels for carbon-sink construction. Kateryna Pylypchuk is the driving force behind the "Rebuilding Ukrainian Cities

Green" initiative in FIABCI-Ukraine, championing sustainable urban development in the postwar reconstruction era under the Global Charter "Building Better Lives."

Dr.-Ing. Katharina Ritzer works as a senior engineer at the ISEM—Institute for Smart Development and Machine Elements at the Hamburg University of Technology (TUHH). There, she investigates how New Work can be implemented in product development using industry-related projects, focusing on methods to improve distributed collaboration.

Karolin Röhring After graduating from high school, she initially began studying business administration in 2016, but soon decided to against it and finally began her training as a metalworker in the fall of 2017. After completing her training, she switched to her parents' company and has been working there ever since. In 2022, she successfully passed her master's examination and subsequently completed further training as a welding specialist/iws (international welding specialist). She then started a further training as a certified business economist according to HwO, which she is expected to complete in the fall of 2025.

Dr.-Ing. Kathrin Rüschenschmidt The Münster native and doctorate physicist began her professional career in 2006 in project management and technology development of power semiconductor components in Austria. After a several-year "interlude" from 2009 to 2013 in the product development of power semiconductor components in southern Germany, she has been living and working in Hanover / Laatzen since 2014. There, she has been passionately shaping the topics of project management and innovation for the Discrete Automation business area of Emerson since 2023. As an ambassador and neurodiverse pioneer for change and inclusion, she draws strength from her partnership, networks, and personal values.

Dipl.-Ing. Julia Schneider Born in 1995, project manager in construction management at Drees & Sommer.

Silke Stark has been head of department at the German Chamber of Crafts (DHKT) since 2023, responsible for the nationwide image campaign for crafts. Previously, she was responsible for press and public relations for various companies and non-profit organizations, mainly in the fields of education and health. From 2012 to 2022, she headed the corporate communications department at the Annedore-Leber Vocational Training Centre Berlin, a supra-company training institution for young people with disabilities.

Prof. Dr. Ingrid Vasiliu-Feltes Doctor, University of Medicine and Pharmacy, Carol Davila.
Executive Master in Business Administration – Health Policy and Management. Chief Ethics Officer, Government Blockchain Association
Faculty, University of Miami Business School. Executive MBA, Business Technology. Bioethics Certification, Harvard University.
Certification for AI and Business Strategy, MIT Sloan School of Management. Certification for Blockchain Technology and Business
Innovation, MIT Sloan School of Management Innovation & Entrepreneurship, Stanford Graduate School of Business.
Financial Certification, Harvard University, Business School.
Negotiation Certification, Harvard University, Law School.
Consultant, UN-Legal and Economic Empowerment Network.
Consultant, EU Blockchain Observatory and Forum.
WBAF World Excellence In Entrepreneurship Award.

Award for Outstanding Education.
Award for Outstanding Research Achievements.

Dr.-Ing. Dipl.-Ing.(FH) is a civil engineer, studied in Konstanz, Mainz and Stellenbosch, South Africa. Since 2007, she works as a freelance consultant. From 2008 to 2014, she lived and worked in Beijing, China. Since 2019, she has been researching and shaping the corporate responsible digitization (CDR) in construction. With the Excellence Initiative that she founded in 2020 for sustainable, human-led AI in construction, she gets the construction branch involved into the general debate on digital transformation and a trustworthy, sustainable AI. In 2024 she completed her doctoral thesis in this field (www.bwl-engineering.com/iethics) at RWTH Aachen University. She is an author and committed as a speaker to advancing digital change in construction.

Prof. Dr.-Ing. Burghilde Wieneke-Toutaoui is a mechanical engineer, was a professor and university leader. She has been active in the Association of German Engineers (VDI) for decades, including in the network "Women in Engineering", and works as an advisory board member and evaluator. After retirement, she completed a BA degree in Area Studies with a focus on Japan.

Dr. Gabriele Wydra-Somaggio has been a research associate at the Institute for Employment Research (IAB) since 2009. Her research focus includes the transition from vocational training to the labor market. Since 2021, she has been leading the Vocational Training Panel for Crafts and is investigating the labor market entries of graduates after training in crafts.

1 Introduction

Stereotypes still exist in the twenty-first century and in the age of Industry 5.0—not only, but to a large extent in construction. They are still promoted and cultivated; just like timidity, prejudices, discouragement, fear, apprehension, and discouragement. Once again, it demands a disruptive, new thinking and learning culture, especially in a male-dominated, traditionally conservative industry like architecture, engineering, and construction. Towards more diversity, more passion, and openness for the diversity of individual ways of thinking, more female role models and motivators, who offer students and young adults enthusiasm and inspiration for STEM subjects and trades professions. With their refreshing "My Story" in this book craftsmen and experts share their daily work, their very personal enthusiasm and their journey to construction. They not only undermine gender stereotypes, but make enthusiasm tangible. They know where the hammer hangs and defy the hurdles brick by brick. Their passion for the craft and the pride on the result of their daily work is deeply moving when reading. Here the decision between nail polish and drill is simple: both! Go for both!

And so at the beginning of a book there is always a special idea. In the present case, it has fully unfolded in view of technological change and with regard to the interaction between humans and technology in construction: At the center of the consideration is the power and possibilities diversity offers for shaping a truly innovative and contemporary construction industry. The diversity caused by new technologies in companies and society open up new potentials and opportunities. There are already diverse approaches and first practical experiences, even role models for responsible handling. The key is to critically weigh innovations, define and follow sensibly new strategic paths, consciously use potentials, and design the construction industry with this diversity openly and consciously innovative.

When the editor and the co-authors in this book talk about DIVERSITY, it includes the diversity of strengths, skills, knowledge, thoughts, creativity, innovation, technical

B. C. Weber-Lewerenz (ed.), *Diversity in Construction—The Gamechangers*, https://doi.org/10.1007/978-3-658-45308-4_1

possibilities—interdisciplinary, regardless of gender, skin color, and nationality. Diversity also means different ethnic affiliations, diverse levels of education, gender balance in AI training and in job placements with tasks around AI (keyword Gender Bias). The time is ripe for a critical inventory of the conditions for an inclusive work culture. Learning, thinking, and corporate cultures explosively collide and reveal the previously underestimated, unrecognized complexity and diversity of untapped potentials. Diversity becomes the game changer and innovation engine in construction.

The book looks at the history, highlights entrepreneurial paths towards more diversity, but also sheds light on problem areas that hinder the innovation process, suggests new ways that let this flow freely, and investigates the conditions yet to be created.

There are certainly enough points of reference in concrete examples for lived diversity in construction or exactly the opposite of it: The prevention of innovation by excluding different perspectives. All the more it is the editor's concern, to emphasize how diversity can strengthen innovation, create new opportunities for even more innovation in all trades, and significantly contribute to the design of societal sustainability.

A thorough examination of the many facets of diversity on over 400 book pages, a sophisticated selection of more than 40 contributors and experts, who from the editor's close-knit, timed author care and proofreading over 6 months ensure motivation—on the publisher's side, Co-authors and not least the editor. The power to adhere to the ambitious timeline and the goal of using the book, among other things, for in-depth scientific research were constant motivators for the editor and co-authors. And perhaps the book's theme can even be used as a flagship in the publishing department of construction and the niche "diversity" discovered there.

Stylistic guidelines, such as writing from the first-person perspective or in a neutral form, were not set. For some, an interview survey was easier due to time constraints, and so a diverse mix was created here as well. The goal was at all times to preserve the authenticity and individual form of each author's text and thus the diversity. Only then could the kaleidoscope that lies in your hands, dear readers, be created, which offers the reader variety and may additionally fuel the joy of reading.

The breadth of expertise and the individual backgrounds of the authors, the age and the associated wealth of experience that speaks from it, could not be more diverse: The oldest contributor was born in 1948, the youngest in 2004. Some look back on decades of professional experience, have—retired—embarked on a completely new path and are now committed to a very personal concern in society with a network that has grown over the years, others are in the starting blocks or bravely switch to another area or field of activity. While some contributors were writing their essay, they were working on their second vocational training or part-time master's degree, had just successfully completed a doctoral thesis, were on their way there, or had gotten married, reoriented professionally, made a fresh start, or had just become a mother or father.

One of the biggest challenges for the editor, it must be said at this point, was to select from the plethora of ideas and contributions submitted. Thus, chapters were created as building blocks of lived diversity, which give the book its structure: "*Diversity in Con-*

struction—My Story", "*Intercultural Bridge Building*", "*Start in the Early Stages—Do What Your Heart Beats For*", "*New Work in Crafts*", "*Human Transformation*", "*International Law and Human Rights: European Politics Including Diversity and Inclusion*", "*New Thinking and Learning Culture*", "*Degree Programs and Training*", "*What Makes Good Communication in Architecture and Construction?*" and "*Initiatives and Networks*". Sharing individual experiences, viewpoints, and opinions creates personal and professional closeness to the readers, creates credibility, allows empathy with different perspectives, and thus promotes participation and access of the readers in various areas.

The chapters, following one another, form the book's structure. Preface, guest prefaces, very personal impulses formulated in book greeting words, and the introduction tune the readers into the topic and underline the necessity to call things by their name, to discuss critically, and to provide decision-makers with arguments to shape necessary framework conditions and to bring speed into the implementation.

To simplify the flow of reading and handling of the book, the cornerstones with which the editor approaches the concept of diversity, such as "*Strengths*", "*Skills*", "*Innovation and Technologies*", "*Knowledge*" and "*Communication*" are the main divisions of this book. The individual chapters assigned to these sections have emerged from a natural sequence of thoughts, following the question of how and where diversity in construction arises, what it is based on, how it constantly evolves, where it is particularly noticeable, what framework it needs for development. The flow of thought also includes the limits diversity often encounters and which new paths need to be defined and taken to overcome these limits. In this critical observation, diversity appears as THE key factor for sustainability and value creation.

It is the teamwork lived in the book creation process together with the authors that makes the uniqueness of this book. It was and remains the editor's obligation and wish to let diversity speak beyond professional boundaries. Because a holistic approach to this complex range of topics can only be implemented together, value-based and with new, interdisciplinary knowledge. At this point, a big thank you with a heartfelt hug is expressed to all authors. The many long joint exchange talks helped in finalizing the range of topics of the book.

With one's own history and often surprising career choice, it is possible to be a role model, to strengthen innovation through diversity and to provide inspiration to attract professionals.

But is that enough? No, because it depends on what W E now make of it, how W E shape the world of life and work, how W E use new technologies like AI, Metaverse, Augmented Reality sensibly and live our potentials. Today, now and many tomorrows after, W E shape the humane and digital transformation, but especially our society and our environment, in which not only W E, but all subsequent generations want and can live.

There are women who courageously go their own way, because "*Who does not move, cannot move anything.*" But still, gaps in the system become visible and noticeable. They lead to questions like:

- What is behind CVs in which a completed study is followed by a vocational training, and sometimes in unrelated fields?
- Where does the devotion to/the passion for the craft lie?
- What is wrong with our advice and support in career orientation if studying is considered the logical step after high school? Why are possible training professions not mentioned here (too often) at all, but only the choice of the subject of study is expected? Shouldn't the thoroughly thought-out study and career choice be at the forefront of preparing for life?
- What changes come with an award, a recognition?
- How openly do women speak in management positions?
- What career steps move women to even hold back statements about the "conditions" and prejudice-laden obstacles such as the glass ceiling and further to mobbing in construction?
- Where does the construction industry actually stand when we women are still sometimes shouted at, but we dismiss this as an additional motivational factor and still want to be better? Is shouting actually louder? Or does it rather speak of fear not only of the self-confident female counterpart but above all of everything new, as it does the digital transformation also demand from all of us?

Diversity goes beyond the pursuit of gender justice. To keep up with the rapid progress in technological feasibility, people need to be "included". Also, possible unintended consequences of technologies on society and the environment need to be assessed and dealt with. We are therefore in a societal, human, technological tension field.

Added to this is a growing shortage of skilled workers, demographic change, and ongoing business and craft business transfers to the next generation. Here, the pride of the "old hands" in decades of experience and success, young innovative ideas, societal expectations in the context of the 17 UN sustainability development goals (SDGs), the broadly lack of will and courage to take new paths, the lack of adaptation of working time models, and the widespread lack of digital infrastructure collide.

Transferred to the heavily male-dominated construction industry, it is all the more important to drive progress and innovation through more diversity, not only in the fields mentioned before, but also to overcome prejudices against women and all innovations as well as traditional attitudes. The rejection of everything new (still paper instead of digital), the dust in some views (work as always, no innovation, everything is running) represent obstacles in the minds. These need to be dismantled. The lack of visibility of women in the industry and in science and research are facts. As is the frustration of researching due to lack of appreciation and temporary employment contracts. Also, the precarious conditions in the evaluation of scientific contributions, as well as sexism and discrimination do not suit us well. Added to this is the lack of a clear definition of excellence in education and science.

The chapter "*New Work in Crafts*" focuses on the diversity of possibilities within a lived new work culture to meet the very individual needs of those working there: for example, with flexible working time models adapted to different family situations and care responsibilities and other forms of family support. The formula is: Family = Success (and happiness) factor! A diverse workplace, where inclusion is at the center, means that the treasure of different abilities, creative visions, unbiased approaches, and experiences that a heterogeneous workforce brings with it, is specifically lifted. More balanced decisions, more innovative products and services, and more sustainable technological progress are the fruits that can be harvested. Conversely, a lack of diversity leads to a lack of innovation.

Inclusion also means enabling people with disabilities or physical or mental impairments to contribute professionally without obstacles and to unfold their potential.

Awareness of the need for inclusive measures is also increasing in the crafts sector: The Crafts Code was first sealed on 17 Sept 1953 with the HwO[1]. The last change was on 14 Feb 2020 with the reintroduction of the master craftsman's obligation[2]. For the first time in the Crafts Code (§§ 42p ff.) and in the Vocational Training Act (§§ 64 ff.) it is regulated how the training of people with disabilities should be implemented.

The chapter "*State of the Art—What is the Status Quo?*" shows which positive developments towards more diversity in construction we can already look back on and which milestones we have already achieved. Looking back at history shows that technological progress must be accompanied by a debate about the human-technology relationship, if trust in technology is to be maintained, that diversity is the key to success, and this realization is important in order to set the course for a sustainable digital and humane transformation now. From the question "Where do we want (need) to go?" various calls for action are derived.

The chapter "*Start in the early stages*" highlights that children grasp with their hands and workshop educators, educational activists and neurobiologists know: Passion for STEM subjects begins in the crib. "*Enthusiasm is fertilizer for the brain*", says Prof. Dr. Gerald Hüther.

The fields of activity are diverse and new profiles, qualification requirements and fields of action in society, politics, education, science, research and user practice are emerging in the course of the digital transformation. The increasingly complex diversity of technical possibilities forces societal discourse to consider risks and ethical questions associated with the technology change, because the normative uncertainties increase with technical feasibility. In this value debate, disciplines should contribute their expertise across disciplines.

Companies undergo a change in values. They look for definition of their own digital strategy, in the selection and application of meaningful technical support, in the design of the digital change and human transformation, as well as in the employee retention and repositioning as an attractive employer for orientation and experience values, perhaps

[1] www.zdh.de/data-and-facts/crafts-ordinance (Accessed 27. November 2023).

[2] www.laws-on-the-internet.de/hwo/index.html (Accessed 27. November 2023).

even role models. It counts among the critical success factors to secure the intangible values in the company such as human factor, knowledge, innovation, company affiliation and to make them reliably assessable as a trust-building factor and to communicate transparently. Especially in the construction and architecture world, the way these values are communicated plays a major role. One chapter addresses the question: "*What makes good communication in architecture and construction?*" What formats are there or are new ones needed? One author affirms: "*Architecture is built communication.*"

Where and how networks and communicative exchange work is examined by the chapter "Initiatives and Networks".

The chapter "*Human Transformation- Human and Technology*" shows how much human and technology depend on each other and without them there can be no sustainable digital change.

Finally, a chapter acts as "*Intercultural Bridge Building*" in construction. It presents perspectives and experiences from countries like Ukraine, China, Japan.

For understanding and deepening, literature sources, book reviews, a comprehensive glossary with term definitions must not be missing. Small eye-catchers are two short reading trips to "*AI meets Art*" and "*La Champagne*". They encourage to perceive cross-connections between diversity, knowledge, technological innovation and corporate culture. The two topics stem from the love of art and music and the affinity for Champagne of the editor.

The book identifies the key factors and core competencies, which are crucial for diversity and innovation. Such a role model of the construction industry could lead to a pioneering role of the construction industry at home and abroad. Crucial for this are the preservation and the protection of rights, which the chapter "*International Law and Human Rights: European Politics including Diversity and Inclusion*" goes into more detail.

Awareness of this is shaped by lifelong learning and expanded. It serves as an important pillar for success in the construction industry. The chapter "*Degree Programs and Training*" presents, various paths of education and further training and defines them as a central task of the industry. With new technologies such as AI, digital twin, metaverse, new job profiles and new professional fields are added. New ways are opening up to become active in the crafts sector, so that the industry will become even more diverse. But these new profiles need to be filled with the right candidates.

In the past, numerous efforts have been made to strengthen the crafts sector. These include various innovation, STEM and image campaigns, such as those by the Central Association of German Crafts (ZDH). This image campaign for the crafts sector is present on Facebook, Instagram, YouTube, and Twitter. The goal of the initiative is to bring the crafts sector more into the focus of public attention and to convey a contemporary and modern image of crafts—especially to young people. The image campaign presents the different professional fields of crafts, shows the many opportunities for development and career opportunities and wants to encourage young people to train in one of over 130 craft professions.

The crafts podcast "Who Makes Tomorrow?"[3] is primarily aimed at those who significantly influence the career choice of young people: parents and grandparents, teachers as well as opinion leaders and multipliers. It speaks for itself, when a grandchild is inspired by early childhood participation in grandpa's carpentry studio, by getting a taste of the construction site or by helping out in the bakery, later chooses a craft profession.

Short facts for lovers of numbers, data, facts about the order of crafts:

The Central Association of German Crafts (ZDH) provides comprehensive reports on the subject area "Women in Crafts"[4]. Women are key performers in various positions in the crafts sector. Three facts stand out: Successors are sought, mixed teams are economically successful and women bring momentum to the climate and energy transition. With a joint action plan, the Federal Ministry for Economic Affairs and Climate Protection (BMWK) strengthens female entrepreneurship[5]. As an independent entrepreneur, master, founder, in a dual leadership role with the partner or as a successor in the family business, women have claimed their place in leadership positions. Almost every fifth successful master's examination was passed by a woman in 2022 (17.9%). Every fourth craft business (24.2%) is led by a woman. In addition, over 74.3% of craft businesses are family businesses, which are jointly managed by a (married) couple. However, the development potential of women is far from being exhausted[6].

Three figures on women in SMEs, crafts, as founders and in start-ups send signals:

- 1.2 million self-employed women, of which 293,000 have an immigration backgroundin 757,000 companies,
- 257,000 female founders (42%),

37.8% mixed start-up teams and

20.3% Female Teams.

Promoting diversity, seeking exchange, engaging in conversations, sharing one's skills and knowledge, shaping the value change with one's own strengths—these represent fields of responsibility of each and every one.

We agree that the course content, i.e., the curricula, and the teaching qualifications should be reviewed and updated to meet current standards, so that employers receive the professionals trained for the new job profiles.

We, you, dear reader, and we authors are a team full of power. With our knowledge, we are a secret weapon.

---

[3] www.handwerk.de/podcast (Accessed 27. November 2023).

[4] www.zdh.de/about-us/department-social-security/women-in-crafts/#c4680 (Accessed 27. November 2023).

[5] www.bmwk.de/Redaktion/EN/Dossier/more-businesswomen-for-the-middle-class.html (Accessed 27. November 2023).

[6] www.zdh.de/about-us/social-security-department/women-in-crafts/#c4680 (Accessed 27. November 2023).

Sometimes it is surprising what is possible in here and now, thanks to new technologies, and how diverse the tasks and interactions in construction are: One person works at a highly technological workplace with the digital twin and makes his project virtually walkable, seeing the performance promise as a digital model in front of him and working in it. Another uses AI for cutting wood, in stonemasonry work, or at the management level for data processing, to make strategic business decisions based on it. One communicates from Germany with partners around the globe and in different languages for the joint project in the desert, using BIM as a common unified data platform. One designs a Smart City for a tender and shows in a virtual world how buildings intelligently communicate with each other, visualizes energy consumption, makes traffic and transport processes transparent, and virtually represents leisure behavior. For another, managing large projects without implementing an innovative thinking culture and daily access to a digital and AI-supported infrastructure in the company is no longer conceivable. And another finds her passion in global, intercultural real estate projects and builds bridges between countries. One researches and develops AI-based chats with the aim of facilitating and improving communication and data exchange in everyday construction. Yet another team researches and tests a variant of GPT specifically tailored to construction law, to have the legal procedure suitable for the problem at hand immediately available in everyday construction. All technological research and approaches for new solutions have one thing in common: The human being is at the center.

Therefore, value creation is only possible if the human being is appreciated in his diversity and technology is further developed in such a way that it meaningfully supports and makes human work even safer. This requires bite, passion, and perseverance.

Together we can do this.

Now, dear readers, let the following enchanting personalities and the diversity of topics work their magic on you.

State of the Art—What is the Status Quo? 2

2.1 Definition of Terms

To approach the "State of the Art", or the Status Quo regarding "Diversity in Construction", the first step is to define the term diversity. The approach to the conceptual definition is not easy given the range of existing diversity and variety in construction. Because it is only in conversation, in mental reflections and playing through the range that the complexity becomes visible—especially for those outside of the construction profession. This is not just about, as is often assumed, the buzzword "women's quota", but about difference, diversity and variety in all areas and at all levels:

It includes people, countries, languages, cultures of origin, corporate cultures, leadership and management methods, professional profiles, fields of activity, competence levels, knowledge, skills, strengths, knowledge and abilities, innovations, salaries, performances, educational paths, training, trades and sub-trades, preliminary and follow-up work, work techniques—today and yesterday, machines, aids, materials, construction products, raw materials and their extraction, workflows, sizes, tables, measured values, regulations, requirements, standards, norms, testing procedures, methods, proofs, documentations, project ideas, planning and procedures, types of buildings, construction environments such as city, country and nature, used energies like sun, wind and water, infrastructures, recycling possibilities of construction and recycling materials,[1] material cycles, environmental influences, energy intensities, classifications, procedures, supply chains, life cycles of built environment, weather conditions according to which the work outside is directed, initiatives, networks, associations, chambers.

[1] www.buildingforum.com/knowledge/sustainable-building-and-renovation/circular-building/zero-waste/ (Accessed 27. November 2023).

B. C. Weber-Lewerenz (ed.), *Diversity in Construction—The Gamechangers*,
https://doi.org/10.1007/978-3-658-45308-4_2

Due to this high complexity, which lies in the terms of diversity and variety related to construction, the Status Quo obviously cannot extend to all mentioned areas. It would simply exceed the scope of the book. Each individual field would deserve its own State of the Art.

Therefore, the editor's book bases on the dimensions "strengths", "skills", "innovation and technologies", "knowledge" and "communication". The corresponding introduction provides a brief overview of the historical, societal, socio-cultural background of the topic spectrum "Diversity in Construction", which contributes to the holistic understanding of the readers.

2.2 History "Woman in Construction and Technical Professions"—Visibility and Aspects of Social Cohesion

Let's start with the history "Woman in construction and technical professions", women in the sciences, their visibility and the aspects of social cohesion derived from diversity.

"Without friction, no heat", it takes disruptive thinking, beyond professional boundaries, to engage in dialogue, break down traditional barriers, shake up previously purely male-dominated industries and professions and recognize and utilize the strengths, knowledge and abilities of women as innovation drivers. Emily Warren Roebling (1843–1903)[2] was a pioneer, and it is worth sharing her story. She was an outstanding civil engineer (and one of the very first, Fig. 2.1) and played a crucial role in the construction of the Brooklyn Bridge. Particularly impressive is that she took over the management of the project after her husband's sudden withdrawal due to illness, thus becoming a pioneer for women in engineering.[3] Furthermore, she remains an example of the fascinating careers in our industry to this day.

Besides Emily Roebling, there are personalities in civil engineering who bring a wide variety of educational backgrounds and were predominantly men at the time: Fritz Leonhardt[4]; Jörg Schlaich[5]; Gustave Eiffel[6] and Heinrich Gerber.[7] Joseph Monier[8]

[2] https://www.ardalpha.de/knowledge/history/historical-personalities/emily-warren-roebling-brooklyn-bridge-new-york-woman-history-womens-history-100.html (Accessed 27. November 2023).

[3] https://queens-of-structure.org/pioneers (Accessed 27. November 2023).

[4] https://www.dbz.de/news/dbz_The_Art_of_Constructing_Fritz_Leonhardt-Exhibition_in_Stuttgart._From-144392.html (Accessed 27. November 2023).

[5] https://www.ilek.uni-stuttgart.de/team/Schlaich-00002/ (Accessed 27. November 2023).

[6] https://structurae.net/en/persons/alexandre-gustave-eiffel (Accessed 27. November 2023).

[7] https://www.german-digital-library.de/person/gnd/118690620 (Accessed 27. November 2023).

[8] https://www.britannica.com/biography/Joseph-Monier (Accessed 27. November 2023).

Fig. 2.1 Emily Roebling. *Source* https://www.frauenbauenstadt.at/exhibition/emily-warren-roebling

invented "reinforced concrete" as a gardener. Leonardo da Vinci[9] was a brilliant builder, Ulrich Finsterwalder[10] and Franz Dischinger[11] were the inventors and pioneers of prestressed concrete construction.

The outstanding women selected and presented in this book and from around the globe have achieved highly regarded careers in construction with role model effect. Herta-Maria Witzemann[12] was an interior architect and furniture designer. After an apprenticeship as a carpenter—unusual for a woman at the time—she began her studies at the School of Applied Arts in Vienna in 1938 and completed them at the Academy of Applied Arts in Munich in 1942.[13] She worked in well-known Stuttgart architecture firms and set up her own business as an interior architect and furniture designer in 1948.

[9] https://www.ad-magazine.com/article/leonardo-da-vinci-artist-portrait (Accessed 27. November 2023).

[10] https://www.db-construction-newspaper.de/knowledge/technology/engineer-portrait-ulrich-finsterwalder/ (Accessed 27. November 2023).

[11] https://www.db-construction-newspaper.de/knowledge/technology/engineer-portrait-franz-dischinger/ (Accessed 27. November 2023).

[12] https://www.saai.kit.edu/inventory/herta-maria-witzemann.php (Accessed 27. November 2023).

[13] https://www.akbw.de/service/forum-house-of-architects-and-architects-hda/the-halls/hall-witzemann/vita-witzemann (Accessed 27. November 2023).

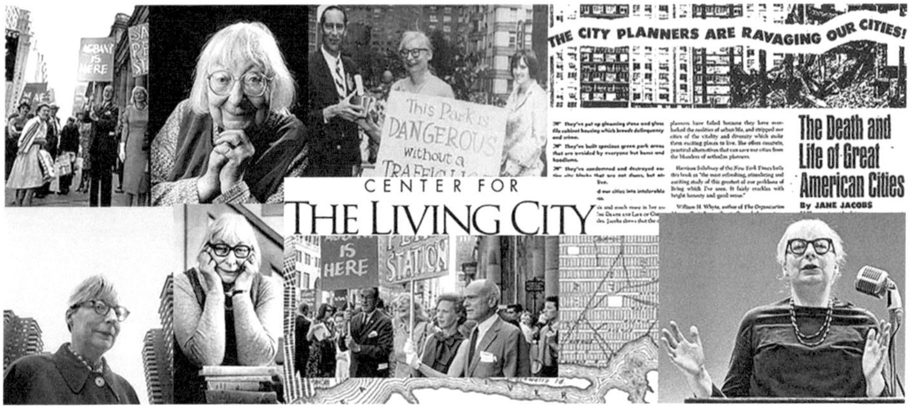

Fig. 2.2 Jane Jacobs. *Source* https://centerforthelivingcity.org/janejacobs

She was a university lecturer for interior architecture and furniture construction at the State Academy of Fine Arts Stuttgart, state chairwoman of the Association of German Interior Architects (BDIA) and president of the Association of German Interior Architects. Through her work as a professor, she supported female students in developing strong self-confidence. Looking into the second half of the twentieth century, she was far ahead of her time: "*Security, reflection, peace … and a conscious feeling of life are some of the essences that from the irrational world in that want to be incorporated into our rationally comprehensible environment.*"[14]

In New York, Jane Jacobs (1916–2006) fought against radical clear-cutting in the years 1950 and 1960. As an urbanist, city and architecture critic (Fig. 2.2), she investigated the renovation measures for the elimination of slums in entire New York neighborhoods (Jacobs 1961). The Lower East Side and Greenwich Village neighborhoods were the result of large new developments, and newly founded citizens' movements prevented the destruction of some neighborhoods. Her goal was to stand up for neighborhoods where social capital, i.e., networks through neighborhood, good social mixing, resilience, mutual trust, and vitality, are paramount. With the Jane's Walks named after her,[15] a series of neighborhood walks have been created in more than 100 cities worldwide. The current mayor of Paris since 2014, Anne Hidalgo, similarly advocates for better quality of urban life. Hidalgo plays a leading role in urban climate innovations, setting signals with new standards as the deputy chair for Europe of the "C40 Cities" initiative.[16]

---

[14] https://archive.state-announcer.de/moments/categories/country-people/herta-maria-witzemann/ (Accessed 27. November 2023).

[15] https://janeswalk.org/cities/ (Accessed 27. November 2023).

[16] https://www.c40.org/en/steering-committee/mayor-anne-hidalgo/ (Accessed 27. November 2023).

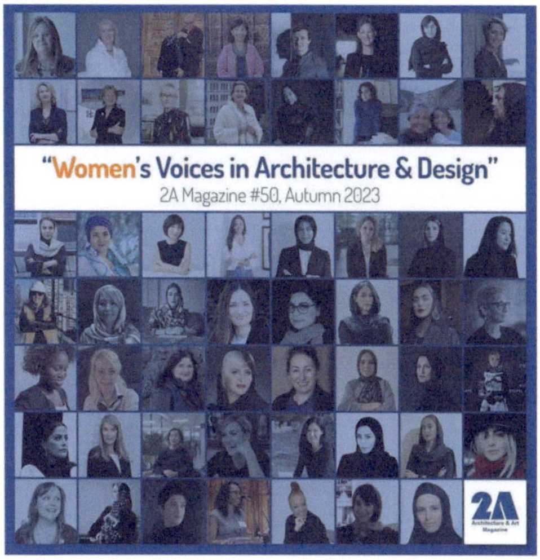

Fig. 2.3 Women's voices in architecture and design 2023. *Source* 2A Magazine, Issue 50, Autumn 2023

The Urban Land Institute (ULI) honors her commitment with the "Award for Visionaries in Urban Development" in 2023.[17]

Since Emily Warren Roebling, the arc has spanned widely over the last century and the pioneers drive the wheel of technological innovation, digitization, and transformation,[18] (Jaeger 2023; Koelbl 2020) and represent loud voices in construction, architecture, and design (Fig. 2.3). The research blog "bauingenieurinnen.de" particularly highlights women in science and research, their successes, and their significant contribution to scientific progress (Fig. 2.4).

2.3 Aspects of Societal Cohesion

The question of how visible women and their co-creation of architecture, built environment, and living and working worlds are, has relevance for and impacts on societal cohesion. Women communicate differently than men, they bring in their perspective, design approaches and experiences. The resulting design approaches in construction play a role model, because women are visible in these areas. They have an impact and radiate to the young female generation, they inspire. "Zia" comes from Arabic and means "the shining one", "the brilliant one". The initiative of the same name by DIE ZEIT "ZIA—Visible Women in Science" aims to make women in research more visible. Women have a strong

[17] https://stadtundgruen.de/artikel/visionen-fuer-paris-anne-hidalgo-erhaelt-auszeichnung-19539 (Accessed 27. November 2023).

[18] www.womenincivilengineering.de (Accessed 27. November 2023).

Fig. 2.4 Making female civil engineers visible. *Source* Research blog www.bauingenieurinnen.de

impact on the young female generation, who are facing the choice of a craft apprenticeship or a technical degree or are looking for their way in technical professions. They offer orientation for career development paths, yes, even when breaking through "glass" ceilings.[19] Where one has already walked the path, gained experience, hesitated, and then successfully embarked on her path, one likes to come together, exchange in networks and initiatives, strengthen each other, search and share new solutions and approaches, encourage a change of perspective. The enormous number of women-founded networks, initiatives and campaigns (see chapter "Networks") is evidence. Not only at the top management level like in boards are social media campaigns for more women in construction are no longer unthinkable, but more and more craftswomen, master craftswomen, managing directors of construction companies, as well as chamber presidents increasingly use their own profiles to inspire for careers in construction, to inform about job profiles and career paths and to give deeper insights into daily working routines. They ensure visibility of what is possible, visibility of and for more women in construction, visibility for the craft and its appreciation, visibility for sharing successes and mutual reinforcement.

The biochemist and Nobel laureate Jennifer A. Doudna[20] answered the question "Is it important for you to be a role model, especially for young scientists?" as follows: "*I have always done what interested me in my scientific career. And it's very exciting what has come out of it. This has little to do with the fact that I am a woman. But of course, it is an important signal*". Researching and developing means going through highs as well as lows. Often unwanted, forced.

[19] https://www.zeit.de/wirtschaft/unternehmen/2024-01/studie-frauen-vorstand-verdoppelt-paritaet (Accessed 27. November 2023).

[20] Jennifer Doudna and Emmanuelle Charpentier, the two "Inventors" of the CRISPR/Cas procedure, received the 2020 Nobel Prize for Chemistry. Article in Spiegel No. 51 from December 16, 2023. Interview with Jennifer A. Doudna.

The social media campaign "#IchbinHanna"[21] was born out of the need of many people working in science. Their wish was to give a face to the practice of temporary contracts in science (WissZeitVG) with the hashtag and thus to ensure visibility. An explanatory video illustrates this practice.[22] Unreasonable, human-targeted working conditions have become daily routine and normal. But thousands of scientists are resisting (Bahr et al. 2022) against the special regulation of the WissZeitVG, to chain fixed-term employment contracts endlessly.

However, it would be too short-sighted to allocate unbearable conditions in science and research only in the area of contract practice. These range from insults, sexual assaults and violence to sexism, discrimination as well as power and physical abuse. These phenomena appear across genders and regardless of professional position but are often found in relationships between employees and supervisors, where one or the other confuses professional superiority with power. Abuse of power is also evident in the salary practice, where performance-related payments are neither fair nor equal. The so-called gender pay gap also affects female professors. The problem has been thoroughly researched (Burghardt 2021).[23] (Lipinsky et al. 2022; Frohn et al. 2017; Lipinsky and Schredl 2023)[24] And asks the question: Who rocks the boat? Who rocks the whole thing and questions given conditions and long-established ways?

The in-depth engagement with diversity, especially the low proportion of women in executive boards, reveals the particularly large gap and associated, urgently needed actions and makes them visible and raises awareness. To achieve a balanced ratio on the boards requires a strategic decision for a healthy mix of women and men on the board. The clear positioning of the supervisory board is part of its responsibility. Women are steadily gaining, but much too slowly and unevenly in terms of professional opportunities, salary, status, and decision-making authority at the workplace (Charter of Diversity 2016). This concerns especially the still male-dominated construction industry due to gender-specific bias, whether conscious or unconscious.

Apart from the need to address fundamental inequalities, there are many reasons to close the gender gap in corporate leadership. The greatest benefit lies in the exploitation of economic advantages. Innovation processes require the ideas of smart minds that need to be brought to the market in order to remain competitive and resilient. Smart ideas only arise where—and only in a sustainable way—diverse exchange exists, where diverse perspectives are welcome and encouraged, beyond disciplinary boundaries, where for-

[21] https://iamhanna.wordpress.com/ (Accessed 27. November 2023).

[22] https://web.archive.org/web/20210611145015/https://www.bmbf.de/en/media-video-16944.html (Accessed 27. November 2023).

[23] https://www.gesis.org/fileadmin/cews/www/CEWSjournal/CEWS-journal121.pdf (Accessed 27. November 2023).

[24] OECD, the World Health Organization, UN organizations and other published indices on economic empowerment.

ward-looking discussions are maintained, and new solution approaches achieved. This inclusive practice provides inspiration for the development of new technologies, ensures business growth through balanced strategic decisions, and thus favors an innovative, economic development. To implement it requires interdisciplinary networking, mentoring of young talent, and support in the technical-infrastructure adaptation of work environments. A rethinking may lay sustainable foundation for the entrepreneurial thinking and learning culture adapted to our times. Female success stories serve as inspiration and provide role models.

2.4 Charter of Diversity

The next step of investigating the Status Quo leads us directly to the Charter of Diversity, an employer initiative to promote diversity in companies and institutions. It states: "*We can only be economically successful if we recognize and utilize the existing diversity*" (Charter of Diversity 2016).[25] Why is diversity important in construction? A diversely structured company allows for a culture of thought that allows, respects, and incorporates diverse perspectives and viewpoints into considerations and decisions. A company thus becomes a place where different opinions, ideas, and backgrounds come together—for the benefit of the company and ideally for the benefit of the community and society. Companies should be places where job satisfaction and employee retention are increased. However, the opportunities and enrichment that diversity brings require active, constructive action and require that challenges are willingly accepted, and hurdles are creatively overcome. Diversity means corporate self-commitment[26] to lived diversity and appreciation of this diversity.

For example, to improve communication with clients, diversity in the sense of a balanced gender ratio in companies is a crucial factor, as female social communication is one of the female strengths. Embracing diversity usually also means that one can look forward to an increase in empathy and understanding in the team and external exchange. In the daily interaction of colleagues, communication is a key factor. The public discourse on diversity of opinion[27] shows how difficult a balanced conversation exchange is today. Diversity of opinion is desired, but not on every topic. Therefore, the State of the Art of diversity includes dialogue about communication, maintaining and protecting values, and showing attitude.

---

[25] https://www.ide-journal.org/article/2017-volume-4-number-1-warum-vielfalt-eine-enorme-herausforderung-darstellt/#:~:text=der%20unterschiedlichen%20F%C3%A4higkeiten%20und%20T-alente,und%20entsprechend%20nutzen%20zu%20k%C3%B6nnen. (Accessed 27. November 2023).

[26] https://www.charta-der-vielfalt.de/ueber-uns/ueber-die-initiative/die-urkunde-im-wortlaut/ (Accessed 27. November 2023).

[27] https://www.zeit.de/gesellschaft/2024-01/harvard-universitaet-usa-campus-claudine-gay (Accessed 27. November 2023).

2.5 Dialogue About Communication, Maintaining and Protecting Values, and Showing Attitude

The academic world has been deeply divided since the attack by the militant Islamist Hamas on Israel on October 7, 2023, and the subsequent start of Israeli bombardments of targets in the Gaza Strip. Anti-Semitic incidents and hate crimes against Jews, but also Muslims, have greatly increased in the US. In particular, universities with anti-Semitic and Muslim-hostile actions came into focus:elite universities like Harvard University, Yale, Columbia, Cornell, University of Pennsylvania and the Massachusetts Institute of Technology (MIT). Students call for "genocide of Jews" on campus, university presidents have to answer in court[28] and face the accusation of not having done enough against anti-Semitism. Hate-filled, reckless, offensive actions are contrasted with the Harvard president's reference to freedom of speech.

Not only in the US, but also in western countries, clear statements from academic institutions are missing. Especially since the expectation is close that these places of learning, knowledge, thinking, personal-human development, which often carry "elite" or "excellence" in front of their university name, should not tolerate such behavior, here empathy, attitude, demarcation from hostility, openness, tolerance determine the value guide. Universities are challenged to provide a safe environment. To ensure that all students feel accepted with their diverse backgrounds. Students or teachers should no longer have to fear physical and emotional attacks, even attacks on their lives, at their places of study or teaching.

We need people who take a clear stand, enforce limits, set boundaries where human, democratic and societal values are attacked and where they need to be protected. Because it should not be normal and accepted when people are afraid, feel fear and have to say: "I have felt like a stranger here for a few weeks".[29] After October 7, many Muslim and Arab people report an increase in racism: They do not want to be under general suspicion. Since the terrorist attack by the Islamist Hamas on October 7 and the subsequent war, the atmosphere in Germany has also become colder and more irreconcilable. On the streets, in parts of the cultural scene, in front of synagogues, there have been shocking anti-Semitic intimidations and violence. Jews report feeling unsafe in Germany, the country of the Holocaust.

Is this in line with the Basic Law and with the value of diversity? Article 3 of the Basic Law states:

---

[28] https://www.zeit.de/politik/ausland/2023-12/usa-kongress-universitaeten-antisemitismus-islamophobie-nahost (Accessed 27. November 2023).

[29] Contribution by Tülay Karakuş, DIE ZEIT December 30, 2023, www.zeit.de/zett/politik/2023-12/antimuslimischer-rassismus-nahostkonflikt-anstieg-deutschland

(3) No one may be disadvantaged or preferred because of their sex, their descent, their race, their language, their homeland and origin, their faith, their religious or political views.[30]

2.6 Obstacles in the Traditionally Conservative, Male-Dominated Industry

The Status Quo of diversity in the construction industry reveals the obstacles in the traditionally conservative, male-dominated industry, still present in people's minds. "*Not only moving mountains on the construction sites, but also in people's minds*," is what Barbara Hagedorn calls for.[31]

Since discrimination within the construction industry has not been in focus so far, the Berlin Chamber of Architects founded the working group "Equality and Diversity" in November 2022.[32] The circle of initiators is formed by Hille Bekic, Vice President of the Berlin Chamber of Architects as well as the architects Inken Bühring and Sudipti Roychoudhury. They say: "*We need a transformation and the embedding of different perspectives and ways of life*." How does this work?

It requires raising awareness. From the very beginning, models of future design should be considered along the process of shaping such design and completely thought through in order to meet societal expectations of living and working spaces, to achieve sustainable and climate-just construction, and to use new technologies for more efficient, productive work. Close cooperation with interdisciplinary experts provides solutions beyond one's own field. They bring diverse experiences, knowledge and strengths. Disruptions, i.e., completely new perspectives, must also be allowed. Perhaps one should even take a step back and rethink. This approach can be seen as an extremely positive catalyst for human transformation towards balanced decisions. The rapid, complex changes and interactions require well-considered steps as well as calmness and transparency in decision-making. Courage to break new ground and the will to innovate lay important foundations.

And it is precisely these innovative, rapidly growing technologies, whose potential applications, but also risks and unintended effects need to be understood, require new approaches in science and practice. These are border areas at which we move: the overlapping fields between technology, natural and social sciences, ethics, politics and society. Coming from these different perspectives to address the same topic—this combination is needed.

[30] Art. 3 GG.

[31] www.baublatt.de/not-only-moving-mountains-on-construction-sites-but-in-minds (Accessed 27. November 2023).

[32] Newsletter of the German Chamber of Architects Berlin DAB January-February 2024, p. 5 ff.

2.7 Innovative Power, Knowledge, Experiences, Technologies, Skills and Strengths

The State of the Art of diversity in the construction industry illuminates equally innovative power, knowledge, experiences, technologies, skills and strengths. These aspects increasingly merge. "Take-Make-Waste" is outdated. It is important to develop and protect the valuable things that our world offers. Good examples serve as role models for this purpose. Sustainability in the social, societal and environmental context must be promoted by architects and engineers. The guiding vision should be to create spaces in which we feel comfortable and safe, to build in a socially sustainable and climate-neutral way, to implement new construction processes—supported by digital methods and AI technologies.

The team of the Fraunhofer IAO sharpen the focus in their book contribution: "*future construction and living worlds require a significantly more differentiated and versatile offer, which should be shaped by targeted diversity research and knowledge transfer from science. ... Especially in the design of technology and transformation, diversity dimensions play an increasingly important role ... In addition to cultural diversity and the increasing mobility of the young generation worldwide, accessibility or social integration through the creation of living forms that bring people of different ethnic, religious and social backgrounds together, is a strategic location factor of attractive cities.*"[33]

The economist Ulrike Malmendier, political advisor, professor of financial market economics and member of the Economic Advisory Council,[34] is convinced, "*AI is an incredible opportunity for more productivity, especially in small and medium-sized businesses. Everywhere in the world, smart minds and good engineers are needed, of course also reliable machines. And who could do that better than us? Let's get on with it*".[35] She not only sees growth opportunities in AI and green transformation, but the need for action in politics. This is also the view of researchers and developers, for example in companies in the construction industry, who are working on sophisticated technology to make construction processes even more productive and efficient and to relieve people in their work.

The construction industry is undergoing a major, all-encompassing change. Academic teaching is addressing the issue of sustainability more strongly. The following four examples are highlight the look at the present, here and now, and into the future:

---

[33] Cf. Chap. Human Transformation—Man and Technology: Contribution by Katharina Hölzle, Steffen Braun, Moritz Maier and Michaela Friedrich (2024). Human-centered technology design and societal perspectives in the transformation of built environment.

[34] https://www.sachverstaendigenrat-wirtschaft.de/about-us/council-members/prof-dr-dr-ulrike-malmendier.html (Accessed 27. November 2023).

[35] In an interview for an article in SPIEGEL, issue No. 9 from February 24, 2024.

To meet challenges such as the energy and climate crisis, extreme weather events, the digital revolution and the shortage of skilled workers in the engineering sector, the Faculty of Civil Engineering at the University of Architecture and Construction (HAB), today Bauhaus University Weimar, will be strategically realigned under a changed name from April 2024: "Faculty of Construction and Environmental Engineering Sciences", in short: Construction and Environment. The faculty was founded 70 years ago, and the focus on innovative teaching content and sustainable and resource-saving construction, development and research of green technologies, new construction methods with renewable raw materials and recyclable material, conversion of existing buildings, optimized load-bearing structures is becoming increasingly visible.

The research of digital transformation, construction robotics with its own learning factory, lifecycle-oriented building construction and circular construction are the focal points at the Chair and Institute for Construction Management, Digital Construction and Robotics in Construction (ICoM) at RWTH Aachen University. Enabled by interdisciplinary knowledge transfer and cross-disciplinary collaborations, academic teaching, research and development are oriented towards the entire value chain of construction.

The state-of-the-art digital laboratory "building.lab" of OTH Regensburg represents the ecosystem of digital construction and aims to make human-centered digital innovation in construction tangible and experiential. The field of Digital Construction is therefore firmly anchored in the study of civil engineering: Parametric planning, BIM2SIM, BIM2Field, industrialized and additive manufacturing processes, data analysis, AI and robotics, etc. form core fields. The laboratory interacts cross-disciplinary with scientific research and teaching in construction informatics, construction operations, Digital Sciences, and the Centre of Artificial Intelligence.

The world's largest hall for construction robotics dedicated to research is the Robotic Fabrication Laboratory of the Institute for Technology in Architecture (ITA) at ETH Zurich. It houses robots for all automation purposes with a focus on additive manufacturing in construction on a large scale and the automation of the entire production process. The seven professorships and 200 members of the ITA are involved in leading global research projects such as the Singapore-ETH Future Cities Lab, the EU Climate KIC and the Swiss Competence Centre of Energy Research (SCCER). The Robotic Fabrication Laboratory (RFL), the 3D Printing Lab, the Concrete Materials Lab and the Adaptive Building Systems Lab form a network of interconnected laboratories. They enable large-scale tests of building components and systems on a large scale and across multiple areas. To test applicability under realistic conditions, ITA researchers often shift lab-based research to Living Labs outside of ETH, such as the NEST Living Labs DFAB House and HiLo and the UWCSEA Office Living Lab in Singapore. Living Labs enable the transfer from research to innovative practice and involve leading industry partners from design, engineering and construction.

In addition to advancing knowledge through research in their respective areas, the examples mentioned, like increasingly other academic educational institutions, focus on the education of the next generation of architects, civil engineers and designers.

2.8 Assessment of the Sensible, Responsible Use of Technologies

An assessment of their responsible use of technologies, of sensible areas of application, considerations of how the unintended consequences can be anticipated and taken into account in the long term to protect human and societal values, is equally part of the inventory of the diversity of technologies in the construction industry.

These considerations are increasingly filling public societal, political, scientific and educational policy discourses. Because when technologies such as AI, Virtual Reality, digital twins, chatbots or GPT meet human interaction, ethical and moral questions arise. The increasing technical feasibility leads to an increase in ethical social responsibility (Kohnke 2017). Such a new approach in civil engineering offers a new field in scientific research, which the editor of this book is advancing with her "Excellence Initiative for Human-Led, Sustainable AI in Civil Engineering", her scientific research, and the concept of "Corporate Digital Responsibility (CDR) in Civil Engineering".[36,37,38] (Weber-Lewerenz and Traverso 2023) Only since 2020/2021 the steady interest in researching practical, value-oriented, engineering and scientific publications in this field significantly gained momentum.

The digital transformation is one of the main drivers for technical innovation and dynamics of all work processes. This requires a will to innovate, suitable guidance for users, and a legal framework that not only provides a protective space for the trustworthy, responsible handling of new technologies but rather catalyzes openness and the will to innovate, to drive meaningful technological progress and to exploit potentials to generate new values. Architects and engineers share the pursuit of progress as well as a common understanding of values, accordingly, increasing the sustainability and efficiency of buildings and cities. CDR represents in these areas a binding approach that inspires companies, researchers and practitioners of all disciplines in defining their own digital strategy (Weber-Lewerenz 2021). This requires close interdisciplinary interface work across disciplinary boundaries to define the necessary key factors.

The Main Association of the German Construction Industry (HDB) has initiated a comprehensive reform process with its own restructuring and alignment of its activities on Digital Transformation, Sustainability and Securing Skilled Workers. This approach starts where it is most important and urgent for the members of the association, namely in operational business.[39]

---

[36] https://www.bwl-engineering.com/iethics (Accessed 27. November 2023).

[37] https://sustainabilitylandscape.rwth-aachen.de/activities/doctoral-project-innovation-promotion-in-construction-4-0-through-the-corporate-digital-responsibility-cdr-approach-A-new-scientific-research-field-for-the-digital-breakthrough (Accessed 27. November 2023).

[38] https://reser.net/expertises/innovation-ecosystem-open/ (Accessed 27. November 2023).

[39] General Construction Newspaper ABZ from 15.12.2023: "Reform process completed".

The architect Werner Sobek,[40] Stuttgart, knows, the causes themselves must be discussed. These include demographic changes, unsustainable resource consumption, increased waste generation, soil sealing, and changes in land use law.

The architect Francis Kerré[41] from Burkina-Faso puts the fight for spaces for people at the center of the discussion, because for him there is "*no realization of our dreams without architecture*".

The President of the Federal Chamber of Architects, Andrea Gebhard, gave a guest speech at the New Year's reception 2023 of the city of Mannheim[42] a summary: Time is short (quote: *New York Times*), 70% of sustainability goals must be implemented in cities, i.e., it can only be achieved through the collaboration of cities. Europe has a particularly high responsibility to maintain the global standard of living in the north. This can only be achieved with the help of resource-saving approaches as well as flexible thinking and working. It is necessary to send the message: We, Europe, are taking the lead! The European Green Deal is important for resilience, the ability to—even in times of crisis—agilely adjust, adapt, be resilient and forge new paths. It is important to focus the entrepreneurial on frugality and optimization, and to keep in mind the limits of growth in strategic decisions (1972, Club of Rome). Because climate is a common good for all and by all (Encyclica of the Pope). The 20s are now decisive, whether we meet these challenges. Technical solutions must be reconsidered. Everyone involved in construction is responsible for the emission of 40% CO_2 and 45% of total waste; we are part of the cause but also part of the solution.

Thus, interdisciplinary discussion is essential and forms the basis of assuming responsibility.

2.9 Diversity of Possible Innovative Approaches to Sustainable Building

The State of the Art of diversity in construction is the basis for the diversity of possible innovative approaches to sustainable, resource-saving, even more efficient, even more productive, even safer construction. The core of such considerations is based on the awareness of our human values and ideals. Innovative approaches and technologies are only meaningful if their research and application take into account social compatibility and sustainability and are in harmony with nature and the environment. Not only demolition, but above all the "departure" of a new mindset is what the Hagedorn group stands for. A best practice company that has committed itself to diversity, as well as sustainable work processes. This

[40] https://www.wernersobek.com/en/ (Accessed 27. November 2023).

[41] https://www.kerearchitecture.com/ (Accessed 27. November 2023).

[42] https://www.youtube.com/watch?v=1M63VSatVR8 (Accessed 27. November 2023).

is how the family business, since its foundation in 1997, succeeds in linking demolition, material flow management and recycling, urban mining and circular economy.[43]

Increasingly complex technical possibilities to work more sensibly, efficiently, and productively, bring equally complex areas of responsibility with them: the high consumption of raw materials and the much too high waste output of the construction sector demand the reduction of CO_2 emissions and construction and demolition waste.[44]

"Zero Waste" is a sustainability effort not only in the European Union. Green Deal[45] is the key word. A new EU action plan is supposed to significantly improve the circular economy[46] and switch to circular systems for production and consumption, so that the EU's climate goals of climate neutrality by 2050 can be achieved.

Examples of Zero-Waste certification are the "Net Zero Waste Rating System" of the IGBC[47] for buildings and built environment, which is based on the 3R principle (Reduce, Reuse, Recycle), as well as the TRUE certification,[48] a program for zero-waste certification of physical facilities, their operation and maintenance.

Since January 2024, Switzerland has had a coordination office for circular construction.[49] With the goal of normalizing circular construction by 2033, the "Circular Construction Catalyst 2033" (C33), officially known as the "Association C33—Swiss Coordination Office for Circular Construction",[50] is intended to serve all actors in civil and structural engineering as a neutral point of contact for information exchange, as well as facilitate the connection of existing activities and encourage the sharing of knowledge. The strategy was developed with representatives from the entire value chain in the construction sector.

The original designers of the Expo 2000 Görlitz consider the world as a garden. The goal is to bring this garden to bloom, without wasting resources. A good gardener is characterized by the fact that he weighs: what grows well, what do I have to remove, so that other things have more space and thrive better. Transferred to urban design, this involves the holistic consideration of the entire site with its surroundings and future—

[43] https://www.baugewerbe-magazin.de/demolition-recycling/urban-mining-and-circular-economy--new-future-location-of-hagedorn.htm (Accessed 27. November 2023).

[44] www.bbsr.bund.de/BBSR/EN/publications/bbsr-online/2020/bbsr-online-17-2020-dl.pdf?__blob=publicationFile&v=3 (Accessed 27. November 2023).

[45] https://commission.europa.eu/strategy-and-policy/priorities-2019-2024/european-green-deal_en (Accessed 27. November 2023).

[46] https://www.consilium.europa.eu/en/policies/circular-economy/ (Accessed 27. November 2023).

[47] https://igbc.in/igbc-net-zero-waste-rating-systems.php (Accessed 27. November 2023).

[48] https://www.greenindustryplatform.org/en/platforms/true-program-zero-waste-certification (Accessed 27. November 2023).

[49] https://punkt4.info/social-news/news/switzerland-gets-coordination-center-for-circular-construction.html (Accessed 27. November 2023).

[50] www.circularconstructioncatalyst.ch (Accessed 27. November 2023).

breathing, growing, green. Transferred into politics it represents providing the appropriate framework for the development of building type E, setting conditions and at the same time facilitating implementation.

The call is addressed to everyone, because the challenges of our time are great and complex.

What connects us as a society? One answer might be that it is the common fundamental ideas about law, justice and fairness. These are based on moral principles and values that have been shared by all cultures since time immemorial and are supported by common practical experience. This is the basis for, for example, the **Global Economic Ethos**.[51] This is a manifesto, issued by the World Ethos Foundation, which sets out the basic principles and values of a global economy. These in turn arise from the declaration of the Parliament of the World's Religions on the World Ethos.[52]

Further International Book Recommendations

- "Women Leadership in Digital Era". International Journal of Responsible Leadership and Ethical Decision-Making, IGI Global Publishing (DOI: https://doi.org/10.4018/IJRLEDM.300803)
- "Empowering Digital Innovation by Diverse Leadership in ICT—A Roadmap to a better value system in computer algorithms". Humanistic Management Journal (DOI: https://doi.org/10.1007/s41463-022-00123-7)
- "Impact of Women Empowerment on SDGs in the Digital Era". Book, IGI Global Publishing (DOI: https://doi.org/10.4018/978-1-6684-3637-0)
- "Impact of Digital Twins in Smart Cities Development". Book. IGI Global Publishing (DOI: https://doi.org/10.4018/978-1-6684-3833-6)

References

Bahr, A., Eichhorn, K. and Kubon, S. (2022). #IAmHanna. Precarious Science in Germany. Suhrkamp Publishing. https://www.suhrkamp.de/buch/ichbinhanna-t-9783518029756.

Burghardt, J. (2021). Sexual Harassment, Sexualized Avoidance and Sexism. In: Workplace Science. Springer, Berlin, Heidelberg. https://doi.org/10.1007/978-3-662-63893-4_12.

Charter of Diversity (Ed.) (2016). The Charter in its wording. Diversity as an Opportunity—The Charter of Diversity of Companies in Germany. http://www.charta-der-vielfalt.de/charta-der-vielfalt/die-charta-im-wortlaut.html.

Frohn, D., Meinhold, F. and Schmidt, C. (2017). "Out in the Office?!" Sexual Identity and Gender Identity, (Anti-) Discrimination and Diversity at the Workplace. Cologne: IDA | e. V. Institute for Diversity & Anti-Discrimination Research.

[51] www.global-ethic.org/wp-content/uploads/2022/10/Global-Ethic_Economy_Global-Economic-Ethic.pdf (Accessed 27. November 2023).

[52] The Global Ethic is the formulation of a basic set of ethical norms and values, which can be derived from religious, cultural and partly also from philosophical traditions of human history.

Jacobs, J. (1961). The Death and Life of Great American Cities (Eng. Death and Life of Great American Cities).

Jaeger, L. (2023). Brilliant Women in Science. Hidden Contributions that have Changed the World. Springer Publishing. https://link.springer.com/book/10.1007/978-3-662-66528-2.

Koelbl, H. (2020). Fascination Science—60 Encounters with Pioneering Researchers of Our Time. Knesebeck Publishing.

Kohnke, O. (2017). It's not just about technology: The people side of digitization. In: G. Oswald and M. Kleinemeier (edt.). Shaping the digital enterprise. Springer Publishing, 69–91. https://doi.org/10.1007/978-3-319-40967-2_3.

Lipinsky, A. and Schredl, C. (2023). International comparative research on forms of gender-related violence in scientific organizations. In: Pantelmann, H., Blackmore, S. (eds) Sexual Harassment, Discrimination and Violence in the University Context. Springer Gabler, Wiesbaden. https://doi.org/10.1007/978-3-658-40467-3_4.

Lipinsky, A. et al. (2022). Gender-based violence and its consequences in European Academia. Summary results from the UniSAFE survey. Report, November 2022. UniSAFE project no.101006261.

Weber-Lewerenz, B. (2021). Corporate Digital Responsibility in Construction Engineering. In: International Journal of Responsible Leadership and Ethical Decision-Making (IJRLEDM), Vol. 2, Iss. 1, Art. 3. (140920–010117). https://doi.org/10.4018/ijrledm.2020010103.

Weber-Lewerenz, B. and Traverso, M. (2023). Innovation Empowerment in Construction 4.0 by the CDR-Approach. A new field of scientific Research for the digital breakthrough. Journal of Architectural Environment & Structural Engineering Research (JAESER), Vol. 6, No. 3, p. 41–64. https://doi.org/10.30564/jaeser.v6i3.5915.

Part I

STRENGTHS

3 Diversity in Construction

3.1 Diversity in Construction—Paths into the Craft: It doesn't Matter Where You Come from, but Where You Want to Go

Susanne Haus, President of the Frankfurt-Rhein-Main Chamber of Crafts

After my first day of vocational training, I would have liked to throw everything away and hide at home. This is not how an apprenticeship should start. It was in August 1991. A very hot day. With my first own car, I was full of energy and with a pounding heart on my way to my first construction site in the Taunus: A new construction. Attic conversion with insulation between the rafters. I wanted to make a good impression and literally knelt down to work. What I didn't know is that you shouldn't kneel too eagerly into mineral wool, because at the end of the day it itches considerably and persistently on the skin. The fact that no one had pointed this out to me and surely had a great laugh behind my back about the high school graduate with little knowledge, I don't want to add to the other "great" experiences of that day. Thus, I joined the long line of new apprentices who were thoroughly teased in the first days of their training.

I come from a traditional craftsman family and as the daughter of a master painter and his working wife, I grew up between overalls, brushes, paint buckets and sales representatives, who also liked to sit at the breakfast table. It was obvious for me to plunge into this familiar area in the professional world.

But there was another very significant and formative event that may have also influenced my final decision for the crafts. My four years older brother Markus, who was diagnosed with leukemia at the age of 19, died in December 1989 after two years of struggle. He had learned the profession of painter and varnisher and worked as a journeyman in the family business. The clear expectation was that he would one day follow in his father's footsteps and take over the business.

B. C. Weber-Lewerenz (ed.), *Diversity in Construction—The Gamechangers*,
https://doi.org/10.1007/978-3-658-45308-4_3

His death left an empty space, and I think my parents besides the deep grief for their son, suddenly also questioned the meaningfulness of their professional activity. "Who are we doing this for?" "How is this going to continue?" These questions arose now, and I remember a very depressed and "woolly" mood.

After passing my high school diploma, I initially informed myself about study opportunities and quickly realized that I could not find the right course of study for me and that all of this seemed much too theoretical to me. But I also can't remember that we dealt intensively with our future possibilities and the diversity of academic professions at high school, let alone information about vocational, dual training or career paths outside of a degree.

Eventually, I realized: I wanted to work with my hands and know how crafts work. I had always been interested in my grandpa's work (Fig. 3.1), and so I decided to pursue a vocational training in crafts. My parents never pushed me in a specific professional direction. Deciding for a craft apprenticeship was—and I still feel this way—a free and interest-led decision, which I have not regretted for a second to this day. The day I told my parents that I wanted to go into training, I will remember as one of the happiest days in our lives.

After completing my apprenticeship as a painter and varnisher, I worked for a while in the family business and gained experience. But I was always looking for challenges and wanted to learn more. The next logical step was therefore the master's examination. I remember a very beautiful, carefree (… at least in the beginning, when no one was thinking about exams yet) year at the master school in Lahr in the Black Forest. We had the opportunity to try things out, to learn and to spend time together with like-minded people in the crafts. Having time for these things is a great good, and unfortunately, it often falls short in everyday life.

As a vegetarian, this time particularly "hardened" me, because since then I know all (really all) jokes at the expense of meatless nutrition and its followers. If you have sur-

Fig. 3.1 Author at the age of about 4 or 5 years with her grandpa. *Source* Susanne Haus

vived that, you can stay cool at the buffet during lunch breaks in later seminars, smile mildly and continue to enjoy your side dishes, because the fact that there were vegetarian alternatives to choose from as a matter of course, as is the case today, was not common in the 90s and certainly not for breakfast on the construction site.

With the master's examination in my pocket, I went a step further and expanded my knowledge of the world of craft restoration. As a restorer in the painting and varnishing trade, I was then able to implement my passion for old things, buildings and historical techniques well in my own company, and even today we are often involved in craft monument preservation.

In preparation for my entry into the management of the family business, I wanted to learn more about business basics and be well equipped to be able to independently manage the company's fortunes one day. Therefore, I completed part-time training as a business economist in the crafts, which fed me with many practical contents and helped me a lot, for example to understand what the tax consultant had made of our numbers and whether that was going in the right direction.

My last professional qualification was the successful examination for publicly appointed and sworn expert in the painting and varnishing trade. A very interesting and demanding activity, which due to the necessary intensive research and the occupation with the diverse topics of the expert opinions to be prepared contributes to the actuality and depth of one's own professional horizon.

Once socialized in our family, you cannot avoid voluntary commitment. Following the example of my father, I got involved in the guild early on and took over the chairmanship of the apprenticeship examination committee. A short time later, I was offered the position of the master of our painting guild—another new challenge. I agreed and was then master for 13 years. I was also put on the candidate list for the full assembly of the Chamber of Crafts by the district craft guild and elected as a member of this supreme body. The meetings of the full assembly twice a year were then also my first contacts in the chamber.

One of my personal passions, which initially had little to do with my profession, was acting. In an amateur group I played theater intensively for many years and always felt very comfortable and in my element on stage. At that time I did not yet suspect that this stage experience would come in very handy would. First, I was allowed to moderate the graduation ceremonies for the journeymen and journeymen of our district craft guild. On these evenings there were always many prominent people from politics, society and also from our Chamber of Crafts present to show their respect and recognition to the young people. Also my predecessor in the office of the President of the Chamber of Crafts was often present and always brought a festive speech, which I was then allowed to announce. They were always very pleasant encounters, and over time you got to know each other.

At one of the next full assemblies, which are always led by the president, he approached me and asked if I would be interested in working on the board of the Chamber of Crafts. There it was, the next challenge. It was also mentioned that the incumbent vice president did not want to run again, and with a for me still not quite comprehensible, spontaneous self-confidence I heard myself say that I was also ready to run for the office of Vice President. On the way home, I slowly became aware of what I had just said…

Indeed, I was then elected Vice President, and the five years of this first term brought me great joy. I have learned a lot and observed intensively how the craft and its organizations function. At this point, for all those who are surprised by these structures again and again and discover new things, let it be said that I too regularly have to expand my knowledge about the craft organizations and what and who else there is, and continue to learn to this day.

If you now understand a little about how I tick, you already know how it continued.

My predecessor did not run for election after 15 years in office. I felt another challenge, threw my hat in the ring—how could it be otherwise—and was elected President of the Chamber of Crafts Frankfurt Rhein-Main (Fig. 3.2). I consider it a great honor to be able to represent the crafts in the region, in Hesse, nationwide and even internationally in this voluntary position. The tasks of a president are very diverse, and it is not easy to name and describe them all, because there are many different levels at which a president operates, and ultimately it always depends on the person acting how they shape the office and which focuses and topics are close to their heart.

Formally, the tasks of the board of the chamber are regulated in the Crafts Code (HWO) and in the statutes and therefore also the tasks of the president and the full-time managing director. Everyone contributes to these tasks with their actions.

The chamber is entrusted with sovereign tasks, such as the management of the crafts and apprenticeship roll, the appointment and swearing in of experts and the regulation of examination regulations, their implementation and much more. These tasks are usually carried out and monitored by the full-time staff, as well as the management of the professional and technology centers, where the inter-company courses and other training and further education courses are offered and carried out. The highest body in the chamber, namely the general assembly and the board that forms from it with the presidency, however, is filled by volunteers. In addition to the president, there are two vice presidents,

Fig. 3.2 Author in front of the HWK Frankfurt Rhein-Main, which she leads as President, 2023. *Source* Susanne Haus, HWK Frankfurt Rhein-Main

one from the employer and one from the employee side. The members of the general assembly are two-thirds from the employer side and one-third from the employee side. Together we represent the interests of the crafts.

The omnipresent shortage of skilled workers is of course also a concern for our chamber. At the center of our actions are the task and the idea of telling young people about crafts and arousing interest in our diverse professions. In the context of our excellent image campaign, we initiate events, are present at trade fairs, go to schools with our new "Crafts Experience" bus and give the students with many trial stations the opportunity to get a personal impression of the crafts. Because only if I know something or have heard about it can I consider taking up this profession. This particularly affects young women, who may therefore be less likely to choose technical careers because they often do not find out about them beforehand.

Career orientation, as I experienced it in my school career—or rather, did not experience—is an extremely important instrument to know after school which professional direction I want to take. Unfortunately, in the various types of schools, these measures to get to know as many options as possible are very implemented differently. Many high schools still assume that it is sufficient to inform about the possibilities of academic education paths. Thus, young people are deprived of a large part of the information about possible career paths.

We should rethink and not speak of higher and lower degrees, but simply of different qualifications, all of which are urgently needed. A master carpenter has a different expertise than a lawyer. The lawyer would struggle just as much in the execution of carpentry work as the master carpenter would in a job as a lawyer. Both are excellently positioned in their field, but not exchangeable and certainly not dispensable. Nevertheless, society assigns different values to these professions. Surveys repeatedly show that respondents consider craftsmanship indispensable but do not consider it for their own career path or that of their children.

Craftsmanship has functioned for decades and has done its performance. From my point of view, it has always done this somewhat detached from the perception of society. It was there, kept many things running and yet acted "under the radar" of public and also media attention. However, when craftsmanship appeared in the media, it was usually portrayed as not very competent or reliable. I remember, for example, scenes in the "Tatort" where the fictional police station was being renovated and the craftsmen acting as extras, working in the background, always appeared male, older and not very agile. This annoyed me back then, and there would certainly be many more examples to mention where a staging often portrays craftsmen as almost stupid and in no case on par with an also portrayed academic. We know that the media portrayal contributes to shaping the image of people and thus also to press certain professional and other groups into stereotypes that are not appropriate. Fortunately, we are currently experiencing a rethink and attempts are being made to show people in their diversity and equality, which leads to closer scrutiny. Actors increasingly portray people from different population groups, origins, professions or also with various restrictions, just as life is in reality.

Craftsmanship recognized early on that it needs to take measures for its own image, but also for the self-understanding of its own guild. The Central Association of German Crafts launched an image campaign in 2010 with the vote of the entire craft sector. The visibility and achievements of the crafts were to be carried into the attention of society as "The economic power from next door" in a sympathetic, somewhat cheeky, confident, willing to perform and powerful way. The image campaign has been filled with content year after year, designed with different focuses and discussed internally with commitment and controversy. In any case, the campaign under the magenta umbrella brand "The Craft" is a success story, because the attention for the professions of the craft has increased, and the realization that there is an equally attractive and fulfilling career path in the craft besides the academic career is becoming more and more established.

The visibility of women in crafts, especially in "non-female typical" crafts, has also increased, which is an important message for the career orientation of young women. "Role models" provide inspiration in a simple way and expand the horizon for one's own ideas. This way, young women perceive that there is a large and interesting selection of professions available to them, apart from the known professions.

Digitization and thus also the diverse formats of social media help in directly addressing young people. Videos and entertaining information appeal to the generation that we need as young talent in the craft.

The Hessian craft sector has developed a VR cardboard glasses in times of Corona and therefore no career orientation taking place at schools, which allows you to watch information films about approx. 35 craft professions on your own mobile phone using an app. We sent these 'cardboard glasses' to the school classes. The 360-degree view films show trainees who present their own profession very passionately, vividly, and most importantly, on an equal footing with the students. With these glasses, girls can learn about traditionally male-dominated professions in a safe space, something they might not do publicly in front of other people.

We are fighting for the recognition of craft professions on an equal footing with other degrees, and it cannot be mentioned often enough that society would not function without the craft industry and its daily achievements.

Let's look back to the beginning of the "first" refugee crisis in 2015. The craft industry was the part of society that immediately took action. The logical and practically oriented view of the situation inherent to the craft industry meant that many young people who had arrived in Germany were in training and employment in the craft industry shortly afterwards. Because the integration of "foreign" people works best through personal contact, through learning the language in daily use, and through a task that conveys satisfaction and belonging at the end of the day. To complicate matters, not only German but also Hessian must be learned here, which does not necessarily simplify things. Teaching my journeymen High German also doesn't work. So our Afghan intern arrived on the yard on the third day with a cheerful "Morsche" and was visibly proud of it.

Openness and cultural diversity have always characterized the craft industry from my point of view, because our motto is: "It doesn't matter where you come from, but where you want to go." If the will is there to learn craft skills and to get involved, it is not a question that attempts will be made to overcome all other hurdles together. This is not easy, not quick, and not without conflict in many cases, but it is clear to everyone that the effort is worth it.

With my voluntary work as president, I try to contribute to the multitude of topics that move the craft industry and help us to position ourselves for the future. This requires addressing many different groups and levels. Politics is an important addressee when it comes to placing the interests of the craft industry so that suitable framework conditions can be created or maintained. This involves personal contact with the various political parties and especially addressing ministries that are important to us. The craft industry must draw attention to itself and its needs so that we are seen and our concerns are included in laws and decisions. This is important not only at the regional level, but also at the state and federal levels. In 2015, for example, the intervention and protest of the German craft industry played a decisive role in ensuring that the master's title could be retained as a qualification for access. Together, European politics could be convinced to preserve the "master", which was critically classified by other states as a restriction for the free labor market, as a valuable and quality-assuring professional degree.

A postcard campaign underscored the demands of the craft industry at the time: Thousands of postcards were brought to Brussels in large sacks in a public-effective manner and placed at the feet of the responsible decision-makers. This shows that every individual can contribute to change or protect things with their voice. In sum, our voices have a powerful and decisive effect.

At this point, I would like to talk about one of the most important forms of organization in the craft industry: volunteering.

The craft industry, in its form as self-administration, relies on the volunteer women and men from the various professions. I have already mentioned the general assembly of the chamber as a body of volunteers. Many of the elected members come from the so-called "voluntary track"; this term refers to the guilds and the state and federal guild associations formed from them. Membership in the professional guilds, which have emerged from the historical guilds, is voluntary. The company consciously decides to join the regional guild of its profession and to support the work of the guild for its profession through its financial contribution. This usually automatically includes membership in the state and federal guild association. The guilds regionally join together to form the cross-professional district craft associations. Often, the necessary committees are made up of members of these guilds, for example, the examination committees, which have been assigned the sovereign task of conducting and accepting examinations. These can be journeyman and master examinations, but also the examination to become an expert. Volunteering is therefore indispensable for a functioning self-administration and represents a central part of the DNA of the crafts. In this area too, in my view, women in the crafts are underrepresented. I would like to see more women who enrich these committees and the work in guilds and the chamber with their expertise and their actions. The female perspective on things, the leadership style of women as well as the approaches to solving tasks and also the willingness to cooperate are of course motivated from a different perspective and are therefore indispensable in order to make the crafts future-proof and flexible. We must succeed in attracting more women and girls to the crafts and building or expanding the company cultures necessary from a female perspective as entrepreneurs or in leadership positions, which are important so that the female offspring feel welcome. It is not always easy to break up the old habits in existing workforces and make it clear to everyone what correct behavior should look like and which jokes and allusions may be hurtful and inappropriate. Not because this was done and expressed on purpose before. In many places, it is crucial for a mutual understanding to bring about a change of perspective among colleagues and, for example, to put oneself in the situation of a young woman who spends the first day in a foreign environment with a completely new daily routine and certainly does not yet bring the self-confidence to correctly classify and cope with a supposedly harmless comment. This applies equally to young men who also need such a welcoming culture in order to start a professional life well and positively. Young men from a different cultural background may need to be made even more aware of the basics of flawless cooperation and communication with the opposite sex. When I look at myself, I have to admit that this has not always been as successful in my company as I am now demanding and presenting here. Admitting this is bitter, especially when you set high standards for yourself as president, but giving up is not an option. Learning from mistakes is important, as is constantly initiating changes and involving the workforce, because such processes only succeed together. As a seasoned woman in the crafts, my stepdaughter told me that she was always very appreciated and supported by the men on her construction sites. Quote—and this is absolutely meant positively: "They strut quite a bit." She works as a locksmith on large machines in scrap yards and thus in an absolute male domain, at least as far as West German scrap yards are concerned. In the East, there are many more women who quite naturally work in these supposed "men's jobs". I consider it my heartfelt task to connect women who have already decided on the crafts and to create opportunities for exchange and the establishment of a network. Together with two committed colleagues in the crafts, Sonja and Viktoria, we founded the women's network **WerkH**. The big H stands for Handwerk (crafts) and not for Herren (men). Finding a name was not easy, as many names are already taken or have negative connotations. It was important to us that our network has its own name in order to enable clear identification and communication.

The network is being excellently received by the women and more and more are joining, who learn about it in various ways. This makes me very happy, because it means that the women are talking about it and perceive it as a value-add to attend our meetings.

We have dealt very intensively with how we approach the women with our first invitation in order to arouse interest in becoming part of this network. The decision was made for an analogue invitation, which we sent out thousands of times. The response was good, but considering the number of shipments, there was still room for improvement. Whether we would have achieved a greater response with an email, I dare to doubt; our invitation would certainly have been lost in the flood of digital information that lands in our accounts every day. The question now arises for me: How do we manage to generate attention in the digital world and really reach people? This not only affects our invitations, but especially our diverse efforts to reach young people via social media and the internet. Unusual ideas and information and formats tailored to the target groups are certainly part of the solution. If we manage to address people personally and reach them emotionally, the content will also go viral. Equally important is the consistency of these formats, because it takes a long time for things to penetrate the breadth of society and make it into the sustainable perception of society. This can be seen excellently in the example of our image campaign. The "economic power next door" is now a fixed term, not only in political speeches, but also in society as a whole. "The craft" has arrived in people's consciousness.

The digitization does not exist. Some are overwhelmed by this term and the willingness to deal with it decreases rather than being understood as an opportunity. I believe that breaking down this large, initially intangible topic to a manageable size is the right way, which I also want to take on the mega-topic of AI. I don't have to serve digitization, but it has to serve me. Artificial intelligence is only of value to me if it can relieve and enrich me in my everyday life. To do this, it is important to define concrete application possibilities and to present them, for example, in our women's network. I always find innovations interesting when they are tangible and applicable. It is important to work out these helpful tools and to offer them in a manageable way in the diversity that is already possible, thus reducing the fear of contact with new technologies, conveying security in their use and thereby lowering the hurdles for dealing with more complex applications. This is one of our tasks as a chamber and craft organization: to take our members by the hand, identify concrete and helpful things and support the companies. In Corona times, for example, the format of online offers has established itself and continues to be used as a supplementary and well-frequented means. Many online meetings and seminars save time, money and also considerable CO_2 emissions, which brings us to the topic of sustainability.

Sustainability has always been firmly anchored in the DNA of the crafts. Not only in the area of building restoration and the so-called "climate crafts", which embodies this topic very plastically, but also in the durability of the things that are made in the craft. They are durable and repairable, whether it is a solid dining table, a custom-made shoe or a musical instrument. This serves the ecological pillar of sustainability, but also the economic one, because although a handmade piece may seem expensive at first, the investment made is relativized over the course of its use, and these pieces are often passed on to the next generation. Right now, I'm looking in our kitchen at the kitchen cabinet that my husband's grandfather had a carpenter build in the twenties. This cabinet is 100 years old, and no door is jammed or crooked. And even if it is, it can be straightened again, and—this is now my specialty—the color of the furniture can be refreshed at any time.

This brings us to the mega-topic of "refurbished". In principle, it is exactly what I just described; give it an English designation, and it is already "up to date". But that would be too short sighted. It is more than overdue that we as a society throw away less and thus contribute less to the pollution of the seas with microplastics and reduce other burdens on the environment. I also find it unbearable that we outsource the disposal of our problematic waste abroad and have it done by others. Out of sight, out of mind.

It is indispensable to further advance sustainability in the sense of ecological aspects in the craft and to extensively promote our skills, which we have in this area. In terms of content, it is **the** craft topic par excellence. Things are repaired and revised and resold. So it may well be that old kitchen cabinets will soon be absolutely hip again; maybe you should secure one already.

The third pillar of sustainability, the social one, is still missing. I like to speak of the craft family in my speeches, which I am allowed to give on various occasions. Most of our companies have up to five employees and are family businesses. These small structures enable short distances and great reliability: The boss knows everyone personally and is also always the contact person. Often the companies are regionally involved in volunteering beyond the craft. At my last event, the mayor reported that during his company visits in the city area he sometimes sees half the fire brigade in the workshop of a company. Craftsmen and craftswomen are very often active for various associations, for the church, for the youth, for the local fair and other festivals. This is social commitment and is quite naturally at home in the craft.

You see, or rather you read, there is still a lot to do in craftsmanship. Unfortunately, like many industries, we have major concerns about the next generation. In order to be able to face the often mentioned and multifaceted challenges that society faces, and to be able to gradually transform our way of life and work, we need people who want to drive things forward. We need those, who are willing to put in an extra effort and contribute to the whole with personal performance. I am in favor of working time models that meet individual needs and make it easier for both women and men to combine family tasks and professional life. We still have a lot of development and potential for change.

What I don't quite understand is that work itself—at least that's how I perceive it—is often no longer seen as a good and satisfying part of life, but as a burden that should be reduced as much as possible. Is that so or am I mistaken? What do we need in the companies so that people can again identify with their work and be proud of their performance? What is missing so that work is perceived as fulfilling and not as a "thief of leisure time"?

I wish for many young people in the craft who recognize the potential of these diverse professions and above all find their own talents and ideally implement, live and expand these in "their" craft.

In the end, it is only important that each person finds his or her personal vocation. That he or she finds what ideally carries a lifetime, is expandable and feeds personal interests. So satisfying and fulfilling, the profession becomes a good part of life and makes people into personalities. Craft shapes.

The chambers, the craft guilds and associations do a lot to draw attention to the craft, the opportunities and possibilities, the good permeability and the many further education offers for the craft family. I have mentioned some things, much more we report regularly on our channels, whether it's the social media, our websites or our craft newspapers. We always try to meet the needs of our target groups. To address, and we are constantly trying out new formats. Digitization is making its way into the chamber and into our businesses and associations in various ways, from small tools, such as the digital evaluation of a seminar via a mobile app, to a committee portal that provides meeting documents, and ChatGPT, all the way to BIM (Building Information Modeling), which will change our way of building in the future.

The craft industry is always on the move and takes advantage of the innovations of the time. I describe the craft industry as very pragmatic, solution-oriented, and personally binding. In the whirl of a fast-paced and global environment, the craft industry manages to remain down-to-earth, but not old-fashioned, modern, but not short-lived, large, but still approachable, sometimes edgy, but not heartless, and also very committed beyond the professional horizon. This is what I appreciate and love about the craft industry, and I am a craftswoman with great joy and all my heart.

3.2 Flexibel, digital, weiblich. Die neue Baubranche bricht mit alten Klischees

Barbara Hagedorn

Women can't operate large machines, they are too emotional for the coarse language on construction sites, they lack muscle strength and a sense for technology. These prejudices are part of a long list of clichés that exist when it comes to the professional role of women in the construction sector. But clinging to clichés leads us women away from our goal. We don't move anything, we don't change any structures, if we focus on what is still stuck in some people's minds.

I know exactly what it feels like when others don't take you seriously at first because widespread clichés stand in the way like a wall. After my commercial training in wholesale and foreign trade, I became self-employed in 1992 by founding a tire and car specialty store. I was 23 years old at the time and was never afraid of the male-dominated business. But I quickly had to learn that my abilities and achievements were often tested more than those of my male competitors—and that just because I am a woman. For example, for a man, it was enough to say that a tire is round and black. As a woman, I had to explain the tire in the smallest detail, disassemble and reassemble it to prove myself. This is not fair, but in the end, I would always say that such situations have led me to not let myself be easily defeated and to know exactly how strong I am.

The tire business was going well, but when my future husband Thomas rolled into the yard because he needed tires, everything changed. In 1997, he had become self-employed with a demolition company. Two businesses are one too many, we quickly agreed on that. In 1999, I decided to give up my company and join his (Fig. 3.3). A good deci-

Fig. 3.3 Barbara Hagedorn with her employees at a construction site in Bielefeld. *Source* Hagedorn

sion, as it quickly turned out. While we initially only demolished single-family homes, later it was large industrial buildings, wind turbines, and power plants. Our business quickly expanded beyond just demolition, and our service portfolio gradually expanded to include waste disposal, civil engineering, site revitalization, and later also heavy-duty logistics and digitization. Meanwhile the Hagedorn Group has become the fourth largest demolition company in the world, having a uniquely strong process chain at disposal, and employ more than 2000 employees. Sometimes I have to pinch myself to realize what has grown in the last 26 years. This success has cost us, me, a lot of sweat and energy, but at the same time, every step up was also proof of doing many things right. And I was brave—our team was brave. Only those who have the courage to try new things, to leave their comfort zone, and to take risks will be successful—I am convinced of that. If I hadn't been brave, I wouldn't have founded the tire business or given it up for my husband's company. And I wouldn't now, after I left the Hagedorn Group, become again self-employed, but more details later. We think too much instead of just doing. Those who are too cerebral to make quick decisions will ultimately find it harder to be successful.

Specifically for women, there are additional hurdles: Many industries are male-dominated and there is a lack of female role models. We need women who are visible. Women who support women. This means not that we want to displace men—on the contrary. I encourage to focus on the positive, to come together and recognize the profitable interplay that arises when genders combine their strengths. According to my own long-term experience the result of the collaboration is unbeatable.

Nevertheless, I do not close my eyes facing facts. And the fact is: More than 90% of all excavator operators are male. Women on the construction sites are rare, because in no other industry is the proportion of women as low as in the construction industry. The proportion of women here is just 13%, in main construction trades it is even only 1.6%. Although the number of female civil engineers is now increasing, still: Too few women choose the industry. Here is a strong need to change. When I say that we can no longer afford to be unattractive to women, I am not just referring to the economic aspect. If a company refuses to allow women, diversity and diversity, this will lead in the long run to this company no longer being able to operate successfully. This means: Anyone who wants to be successful, convince the next generations and gain a lead in the market, should and must be open to allowing changes.

It is becoming increasingly difficult to fill vacancies. In addition, in the next ten years a quarter of the German construction workers will retire due to age. We must now start changing structures, otherwise we will face even bigger problems in the coming years.

Of course, I am aware that we cannot change the world from one day to the next, but we are laying a first foundation and making small steps. Because the crucial thing is that you start and change things. You can do this in a quiet room or you talk about problems and find new supporters and inspire others along the way. As part of the campaign "Woman in Construction", initiated in 2020, we at Hagedorn Group have founded a network that includes more than 35 companies from the construction sector and the crafts. Together, we had set ourselves the goal of inspiring more women for the industry and beyond, encouraging them to follow their self-chosen path (Fig. 3.4). All members shared the conviction that women can do the same job as men and many are eager to enter male-dominated industries, but ultimately do not dare to take the step. In regular meetings we exchanged ideas, openly addressed problems and developed concepts to make a difference.

Fig. 3.4 Barbara Hagedorn at the 2nd Hagedorn network meeting with Speaker Tijen Onaran, Hagedorn headquarters in Gütersloh, 2024. *Source* Hagedorn

The goal of supporting women and changing structures went across all channels: we have put up posters in our hometown, given many interviews, shared social media posts, produced our own video series, started an industry-wide survey and much more. Each step involves our own employees and employees. Because fundamental changes must come from within a company and many were directly on board and have contributed ideas. Because advertising, being visible, is one thing, but to change framework conditions and structures, it is additionally indispensable to openly address problems within the company and, step by step, create a new basis. It is important that everyone pulls together—from the management level to the intern. But this does not happen overnight, so first the will and motivation must be there to change something.

The branch offers various jobs with a future, jobs with purpose. Especially in view of meeting climate targets, renovating dilapidated infrastructure, and the need for housing. Those who work in construction create opportunities, perspectives. The Hagedorn Group tackles disused industrial wastelands and makes them valuable again—thus avoiding the sealing of green spaces. This work is an important building block for structural change in Germany. The meaningfulness of action, especially in a professional context, is of considerable importance for the coming generations. As is the corporate culture of a company. Hagedorn Group therefore relies on flat hierarchies, a "you" culture and the values "honest, fair, binding". Every employee benefits from a company pension scheme, private supplementary health insurance, offers for health and fitness, and an academy for personal and professional training and further education. With access to a company-owned polarity therapist and barber shops, as well as the possibility of a sabbatical, Hagedorn Group is the living proof that the physical and mental health and well-being of employees come first. Sustainable handling of the human resource is crucial and brings about positive effects for every company.

With the project "MIThelfen", employees can get involved socially during their working hours—this strengthens the sense of community and makes a further contribution to society. Today, it is no longer enough to just offer home office and free coffee. As female entrepreneurs, we must do more to attract future applicants. With the children's university, for example, Hagedorn Group arouses interest in the industry at an early stage. Children between 10 and 14 years learn in five two-hour lectures how strong our process chain is. And the learning material is not only taught theoretically, but also in a practical way: the participants are allowed to lay water pipes, drive excavators and wheel loaders, and test our excavator simulator. And in the end, there is a diploma and the guaranteed promise of an interview. The children's university started for the third time this year and there are almost as many girls as boys among the 30 participants. These are amazing, positive advances! Even though the path to equality is still long and it will take some time before we no longer have to talk about equal pay, we are setting small mosaic stones with such steps before a solid foundation is created. But this foundation needs many supporters.

The starting position for the construction branch is generally good. So the conditions today are better than a few years ago. In the past, work on the construction site was different, for example, there were no quick changers on the machines. That's different today—to modify attached tools, no one has to get out of the excavator anymore. This is just one example of many. Due to digitization and the associated latest technology, it mainly requires brains and a delicate touch, not muscle power. Digitization enables flexible working times and opens up new working fields in favour of female personel. When I hear that it fails because of small things like a second toilet on the construction site, I can only shake my head. We must not resist going new ways. We as female employers must start early. This means that if a young girl decides to start an apprenticeship in construction and the parents are worried about the prejudices about the industry, we invite them to the interview and answer all questions. We should promote the girls and women who are interested in the industry. This implies to find solutions to reconcile job and family.

Predictable, flexible working hours are, for example, a first step to promote a timely working culture. And also for jobs with responsibility. Because many women work in jobs that are underpaid, poorly socially secured and associated with fewer professional development opportunities due to their family situation. Thus, many women with children depend on a partner. Working hours should therefore not become an access barrier for women for certain professions and hierarchy levels. In addition, companies should try to make it possible for women who are part-time employed to increase to full-time if they wish—or vice versa. This should also rule for males.

Politics can and must do much more to overall improve the situation of women. Due to a lack of childcare places, many mothers are forced to cut back at work. For many women, therefore, working more is not a question of wanting to, but of being able to. Often they cannot find a nursery or kindergarten place for their offspring, or the facility closes so early that they cannot take a full-time job. More all-day schools or after-school care places would make it easier for women to extend their working hours.

However, it must be noted: If there were not the large salary difference between men and women, there would also be fewer gender-specific challenges in the workplace—because the decision of whether the man or the woman works more hours would then no longer be just a question of salary. As long as women, for example, earn little in many areas and have few opportunities for development, they also more easily—whether they want to or not—fit into the traditional division of labor model.

If the salary allows, there is the possibility that man takes longer parental leave. This has become common practice at Hagedorn Group—although I must admit that I was a bit taken aback when a project manager told me a few years ago that he would be out for a year. I was taken aback only because it was the first case for us at the time—the development is a positive one and a relief for many working mothers. As a mom of two daughters, I know what I'm talking about, and it wasn't always easy to juggle everything. And I can honestly say: I couldn't have done it without support, for which I am very grateful and I am aware of how fortunate I was to be able to reconcile business and family. But still, it was not easy.

In general, we need more women. Not only in our industry, but also especially in politics. Because even over 100 years after the introduction of women's suffrage, women are underrepresented in parliaments at the municipal, state, and federal levels. After all, a democracy needs both women and men and equal participation of women in political decision-making processes.

Moving towards this goal, we need more female role models. The fact that there is a lack of female role models is no secret and the result of an industry-wide survey we initiated. More than 92% of women want more female role models in the industry and stronger professional promotion. To change this condition, I show our face and encourage women to simply dare (Fig. 3.5).

I am very proud of what we at Hagedorn Group have already achieved. When the campaign was launched at the end of 2020, we hoped, but never expected, that it would make such big waves. The response of the large media led us back on the table, talking more about it and finding solutions to support women. This means a lot to me personally. At Hagedorn Group together with colleagues we have not only founded a network, but for example, we have been able to win former professional boxer Regina Halmich as a prominent supporter and started a cooperation with Zeppelin Construction Machinery for the promotion of women. The best part is: Meanwhile there are 59 trainees, 17 of whom are female and six of them work directly on the construction site—as prospective construction equipment operators or deep construction workers in road construction. In the industry, this is an absolute rarity. The "Woman in Construction" campaign and the associated marketing support has helped to attract young women to the commercial sector. One of the employees started her training as a construction equipment operator in 2021 and therefore had to endure a comment or two in her circle of friends and family. She previously worked in a veterinary practice and wanted to go into construction. She was brave and took this step despite the comments—today she works as a project management assistant after successfully completing her training. These life stories make me proud.

Fig. 3.5 Barbara Hagedorn on stage at the Coreum, Future Conference "Bits and Machines", February 2024. *Source* Hagedorn

Women still represent the minority. Changing, this is the challenge. But with the interplay of all genders and generations, strong teamwork, and the will to change something, we will master these challenges. And it is precisely because of this conviction that, after 25 years, I left Hagedorn Group and founded a new company again: Future WE. We support and accompany companies in their transformation. It is important to break through everyday life, To create space for common thinking, exchange and the implementation, practical and hands-on. We are convinced that The future is a team sport that can only be solved through dialogue between generations can be!

3.3 Feminist Urban Development Policy—A Gender-Diverse and Care-Oriented City is a Better City for Everyone

Dipl.-Ing. Anne Keßler and Dipl.-Ing. Martin Ehret

Beautiful weather: After work, my neighbor at the daycare loads her toddler into the child seat and cycles to go shopping—while doing so, some cars dangerously overtake her, there is a bike path, but it is not physically separated and the road is very narrow. The way to the nearest shopping opportunity is relatively far. Finally, at the checkout, she calms the whining child with a chocolate bar: 'too many sweets' she thinks, but she doesn't want to disturb other shoppers. Finally, she heads to the next playground, but wait, she needs to go to the toilet. And that's not so easy, because public toilets are scarce in her spatial environment. After what seems like an eternity, she finds the nearest toilet. However, she has to pay for using it, it's free for men. And how is she supposed to relieve herself now and at the same time hold the toddler in her arms so that it doesn't run out or crawl on the dirty floor? For my neighbor, the weather is suddenly not so nice anymore!

This situation reflects a multitude of everyday experiences, especially for those who perform a large part of the care work—the so-called care work. In Germany, these are still predominantly women: On average, they spend 43.8% more time on unpaid care work than men. Converted, this corresponds to a difference of 77 min per day (VCD 2024). The design of our cities and municipalities, with their individual buildings, public spaces, and transport infrastructures, has a significant influence on the quality of life of all people and their advantages or disadvantages.

The professional discourse on feminist urban planning is not new. In the German-speaking world, there has been increasing criticism since the 1970s of spatial planning that originated almost exclusively from a male perspective. Although attempts are increasingly being made in urban planning to take into account gender-equitable, intersectional needs, among other things through gender-sensitive planning. However, the reality often looks different: The infrastructures developed under the guiding principle of the car-friendly city in the 1950s and 60s are still manifested in the urban environment today. They often meet the needs of working people to get from home to work and back as quickly and comfortably as possible by car. These structures primarily favor the mobility behavior of men, who travel 13 km more per day than women and drive more cars (VCD 2024). Women, on the other hand, more often use public transport, bicycles, or walk, and their travel relationships are more complex due to a larger daily proportion of care work. They often combine chains of routes to do shopping, take children to kindergarten or school, or run other errands.

Closely related to this, the design of public space also plays a significant role. Due to lack of lighting, narrow alleys, or non-existent sidewalks, "fear spaces" can arise, which especially force women, but also people from marginalized groups, to take detours to avoid them. Sidewalks that are too narrow or traffic islands that are too short pose obstacles for strollers or people with impairments.

These are only a few examples that show how complex and demanding planning for a fair city are, and that gender equality manifests itself in our built environment, in architecture and urban development. The International Women's Day, which took place for the 113th time on March 8, 2024, under the motto "InspireInclusion", offered a good opportunity to focus on issues related to the design of our cities and communities. One of the central action areas of International Women's Day highlighted how planning and construction of infrastructures should meet the needs of women and girls (action area "Designing and building infrastructure meeting the needs of women and girls") (International Women's Day 2024).

Whether in the design of public spaces, mobility or housing—fair planning must take into account the interests of all population groups. This means equally: Planners have a great responsibility to create sustainable structures—which usually exist over decades—for all population groups. So how is it possible to advance gender-just, intersectional and care-oriented urban planning? What tools, levers and programs does municipal practice require? What are good examples that can serve as role models and be transferred to other places?

3.3.1 The Strategic Basis: The New Leipzig Charter and the "just City"

The strategic compass and basis of modern urban development policy in Europe and in Germany is the New Leipzig Charter, which was adopted on November 30, 2020, by the ministers responsible for urban development in each member state under the German EU Council Presidency (BMWSB 2024). The Charter was created in an intensive two-year participation process at national and European level. It describes three overarching dimensions and associated guiding objectives that need to be thought of in an integrated and related way: The fair, the green and the productive city.

The postulate of the fair city mentioned in the New Leipzig Charter aims at the development of a city for all. To achieve this aim, a stronger focus must be placed on feminist planning in the future. Feminist urban development is an important guiding objective, not only to advance gender equality, but to enable all groups and social milieus to participate equally in society in the sense of integrated planning. Thus, the feminist perspective uses an intersectional view, i.e., gender justice in connection with age, migration background, education, disabilities, income, household type and sexual orientation, as well as life phase orientation with a special focus on the care work performed by those affected.

In Germany, the goals and contents of the New Leipzig Charter (Fig. 3.6) are implemented with the National Urban Development Policy as a joint initiative of the federal government, the states and the municipalities. The cross-level approach of the joint initiative makes it possible to address current challenges of urban development flexibly and quickly and to work together on solution approaches and framework conditions for municipal practice. This is achieved, i.e., through the promotion of innovative projects, the development of strategic framework documents and guidelines, the initiation of congresses and university days, etc. The insights gained from municipal practice flow into the work of the federal government in order to adjust the legal instruments and funding instruments in the area of Continually modernizing urban development (Weigel & Schulze 2022).

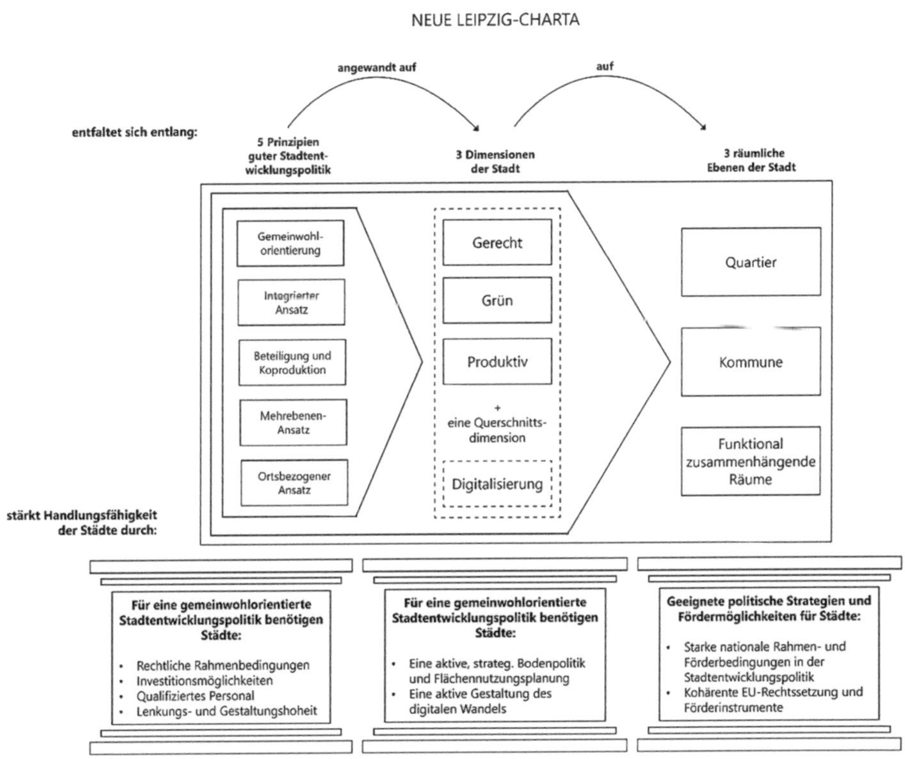

Fig. 3.6 Structure of the New Leipzig Charter with principles and the three dimensions of the city. *Source* Federal Ministry of the Interior

3.3.2 The National Urban Development Policy: Lever for Feminist Urban Development

The Federal Ministry for Housing, Urban Development and Building (BMWSB) has initiated a broad process within the National Urban Development Policy during the 20th legislative period of the federal government, which for the first time explicitly addresses feminist urban planning and development. The instruments and formats of the National Urban Development Policy can and should be used to provide a platform for the discourse on a gender-equitable, intersectional, and care-oriented city. Exchange formats offer opportunities at different levels to connect knowledge from municipal practice with the state of scientific research, to integrate data collections as a basis for argumentation, and to make exemplary examples known.

To initiate the discourse on feminist urban planning on a broader stage, a side event on the topic of feminist urban development policy took place at the 16th Federal Congress of National Urban Development Policy on September 12 and 13, 2023, in Jena. Experts discussed with an interested and committed audience about the feminist focus in planning, which should not be neglected due to current challenges, such as climate change, but should be brought into harmony. Because especially in the effects of crises, it is true that women are disproportionately affected. Examples include the fallback into traditional roles during the Corona restrictions or the strong impact of heat stress on women, as they are disproportionately dependent on affordable housing and therefore often live in disadvantaged neighborhoods and also spend time in heat-stressed public spaces due to care work.

Achieving participation in planning processes, requires to significantly increase the attractiveness for women, but also other marginalized groups, and to work with outreach formats. Another demand was the goal of the solidary city, in which all those involved in urban development recognize their own privileges and consciously share resources and power positions with marginalized groups and jointly shape the city with them.

The participants called on the federal government to actively name the issues and to recognize in its own publications, legislation, and activities. They expressly encouraged the Federal Ministry of Construction to develop guidelines for feminist urban development as a strategic basic paper.

The BMWSB started the process of formulating guidelines for feminist urban development with a roundtable in the spring of 2024 (Fig. 3.7). With the roundtables, the National Urban Development Policy has set up a thematically focused format since 2017 to initiate an open scientific discussion on the positioning of current issues and future topics of urban development. In the past, for example, topics on resilience, health, or mobility in the context of urban development were discussed, leading to recommendations for action or guiding principles.

Experts from the field of feminist urban planning discussed and deliberated within the roundtable which strategic approaches should be further developed for future action in the

Fig. 3.7 Participants of Roundtables "Feminist Urban Development" of the National Urban Development Policy, March 2024. *Source* Federal Ministry of Housing, Urban Development and Construction

BMWSB and beyond, in order to achieve a gender-equitable, intersectional, care-oriented, and thus livable city for all. The event laid basis for establishing guidelines on "Feminist Urban Development" and started a cooperation process between an advisory board of experts with different perspectives and professions, the BMWSB and the Federal Institute for Building, Urban and Spatial Research (BBSR). In addition to the roundtable, other formats of the Federal Ministry of Construction serve to identify ideas, requirements, and recommendations for action for politics and society in this complex of topics, which in turn are incorporated into the guideline process. Flow into it. It includes a research study by the BMWSB and BBSR titled "Gender-Just Urban Development: Potentials for Sustainable and Diverse Cities". The study examines how gender-just urban development can be promoted and further strengthened in practice (Nexus 2024).

3.3.3 Contents of Feminist Guidelines for Urban Development Policy

The content of the guidelines will be, on the one hand, to show positive guiding and future images of feminist urban development. They intend to serve as a recommendation for action for municipal planning practice, as a basis for argumentation for resolutions for corresponding strategies in cities and municipalities, or for the development of corresponding planning tools and methods. The focus will be on:

- safe **public spaces** and residential environments with high quality of stay and diverse green, free and play areas that meet the needs of care work, in order to take into account the various social roles, life phases, and cultural backgrounds of the residents and users from the outset in the needs analysis and planning.
- barrier-free, affordable and needs-based **offers and spaces of mobility**. In this context, a better networking of transport modes and the increased quality of use of public transport, walking and cycling should be highlighted.
- **affordable housing** embedded in **mixed-use neighborhoods**, where living, working and daily needs services are within easy reach in the spirit of the New Leipzig Charter.
- **participation processes** in planning, which should be designed to be low threshold, connected with everyday life and often also with outreach formats, in order to enable all residents to actually participate.

In addition, these guidelines formulate objectives for the working method in the BMWSB and subordinate authorities. This will involve, for example, focusing on the needs of women, people of other genders when setting up funding programs (e.g. by conducting gender checks[1]), to ensure that they have equal access to resources and opportunities. An initial impetus for this took place in parallel with the guideline process in 2023: Two gender workshops were held with the Federal Foundation for Equality in cooperation with the Gender CC association. Based on specific strategy papers, funding programs and research projects, gender checks were carried out and recommendations were developed. It is also important to take into account the intersectional and life-phase-oriented perspective in research projects. Especially in surveys and participation of experts, more attention must be paid to ensuring that the diversity of perspectives is represented.

Last but not least, a solid data basis is strongly encouraged, one that shows disadvantages of women, people of other genders and marginalized groups. BBSR data collections can be aligned with this. There was a strong vote from the experts and experts at the roundtable for the firm establishment of monitoring in the guidelines of feminist urban development.

The aim of publishing the guidelines for feminist urban development is also to contribute to the current political discourse.

3.3.4 Where Are Good Approaches?

Of course, it requires the dissemination of good examples of feminist urban development, both in their spatial impact as well as at the process level in participation pro-

[1] See working support Equality-oriented legal impact assessment according to § 2 GGO, BMFSFJ.

cedures and governance structures. For positive future images in spatial planning, planning women must be given a strong voice and the recognition of their works must be promoted. This is supported by the BMWSB for example with the "patronage" for the Woman in Architecture Festival 2025, a decentralized networking event for the strengthening of female architectural culture. Nationwide, the achievements of women in design, in planning processes, and work methods or office management as well as in teaching, research and politics are made visible in various events.

At the municipal level, there are already flagship projects, such as the project "Her-City Weimar". Starting from the Her City initiative by UN-Habitat methods, tools were developed to achieve more equality and inclusion in urban development. The students Lia Zinngrebe and Lisa Maßel conducted city walks and surveys to identify the needs of women in public space and to include these into the discourse with the city administration to sensitize for a feminist perspective in urban planning.

Another showcase project, demonstrating a good implementation of a feminist urban development at the municipal level, can be found in Vienna (Fig. 3.8). The Seestadt Aspern is one of the largest urban development areas in Europe. By 2028, about 20,000 people are expected to live here and almost as many jobs are to be created in the district—living and working take place here side by side, there is a good functional mix and the distances are short (Urban Change Academy GbR 2022). This corresponds to the guiding principle of a "city of short distances" and thus benefits especially those who perform care work. In the Seestadt Aspern, emphasis was also placed on the design of parks as well as the width and accessibility of sidewalks, the furnishing of public spaces with seating and good lighting of streets to avoid fear spaces and last but not least on street names of famous women (Aspern 2021). The planning of the district was accom-

Fig. 3.8 Seestadt Aspern in Vienna. *Source* https://intelligentmobil.de/seestadt-aspern

panied by an expertise on the master plan with the topic "Gender Mainstreaming in the Urban Development Area Flugfeld Aspern" to make optimizations of the conception from the perspective of all population groups.

3.3.5 Conclusion

What makes guiding principles of feminist urban development so important for the municipal practice, the working method of the BMWSB and the political discourse in the professional public?

A feminist perspective on the design of our cities and communities, holds the great opportunity to create a fair city for all residents and to secure and improve overall quality of life. It includes an intersectional consideration, i.e., a joint consideration of different lines of inequality, to thus enable participation for different marginalized groups (Heinrich Böll-Stiftung Bremen 2022). Thus, not only acceptance is promoted, not only created for necessary transformation processes, but democratic participation and thus democracy and its institutions are overall strengthened.

A democratic tolerant society that gives all target groups space to realize their needs and enables participation is also a prerequisite for so-called soft location advantages. In addition to affordable housing or daycare places, it requires the acceptance of people in all their diversity, in order to give them a home. These are important prerequisites for skilled worker immigration, which we urgently need. Urban development can also make an important contribution to economically attractive locations from a feminist perspective.

Links to Other Relevant Websites
https://www.bmwsb.bund.de/
http://www.nationale-stadtentwicklungspolitik.de

References

Aspern (2021). Aspern Seestadt als Vorreiterin in Sachen Gender Planning. https://www.aspern-seestadt.at/city-news/aspern_seestadt_als_vorreiterin_in_sachen_gender_planning (Accessed 01. March 2024).

BMWSB (2024). Die Neue Leipzig-Charta. https://www.bmwsb.bund.de/Webs/BMWSB/DE/themen/stadt-wohnen/stadtentwicklung/neue-leipzig-charta/neue-leipzig-charta-node.html (Accessed 01. March 2024).

Heinrich Böll-Stiftung Bremen (2022). Stadt für Alle?! Feministische Perspektiven in der Stadtentwicklung. https://www.boell-bremen.de/de/2022/10/13/stadt-fuer-alle-feministische-perspektiven-der-stadtentwicklung. (Accessed 01. March 2024).

International Women's Day (2024). Let's inspire inclusion (https://www.internationalwomensday.com/Theme (Accessed 01. March 2024).

Nexus (2024). Gendergerechte Stadtentwicklung: Potenziale für nachhaltige und vielfältige Städte. https://nexusinstitut.de/gendergerechte-stadtentwicklung-potenziale-fuer-nachhaltige-und-vielfaeltige-staedte/ (Accessed 01. March 2024).

Urban Change Academy GbR (2022). Urban Change braucht Gender Planning! p. 48. https://urbanchangeacademy.com/wp-content/uploads/2022/08/Diskussionsstoff_Urban-Change-braucht-Gender-Planning.pdf (Accessed 01. March 2024).

VCD (2024). Die Verkehrswende muss feministisch sein. https://www.vcd.org/artikel/feministische-verkehrspolitik (Accessed 01. March 2024).

Weigel & Schulze (2022). Stadtentwicklungspolitik als nachhaltiger Gestalter, in PLANERIN. 3/2022. Politik der Ampel – Zwischen Stillstand und Aufbruch.

Individual Profiles in Construction—My Story

4

4.1 Factually Constructive + Smile but remain determined = Dangerous Woman!

Annemarie Mulder M.Sc., EUROVIA, VINCI Construction Germany in an interview with Bianca Weber-Lewerenz, as of 6th of November 2023 and as of 26th of January 2024.

What is my Background?
I have a master's degree in public policy and management. Years later, I deepened the area of commercial real estate with a bachelor's degree. My current job is project management of large construction projects (Figs. 4.1, 4.2 and 4.3).

What Inspired Me to Pursue a Working Life in Construction?
I got involved in construction projects through my husband, who is already active in construction, and our start together in France in 2000. I was self-employed in the real estate sector from 2006 to 2012. I then worked in the construction department of AIRBUS for four years. One of the absolute highlights was the construction of the A330 Completion and Delivery Center with all taxiways in Tianjin, China. To do this, I traveled long distances very often, which posed a great professional and private challenge. Afterwards, I was involved in construction projects in Hamburg for two and a half years. I have been with the French construction group VINCI Construction since 2021.

When asked, what is « My Story», there are many aspects.

B. C. Weber-Lewerenz (ed.), *Diversity in Construction—The Gamechangers*,
https://doi.org/10.1007/978-3-658-45308-4_4

Fig. 4.1 Project Manager Annemarie on the construction site PPP B247 Mühlhausen-Bad Langensalza at BW303, 2023. *Source* VINCI Construction Shared Services GmbH/Dirk Mathesius

Fig. 4.2 Project Manager Annemarie on the construction site PPP B247 Mühlhausen-Bad Langensalza at BW1, 2023. *Source* VINCI Construction Shared Services GmbH/Dirk Mathesius

What Does Diversity in Construction Mean to Me?

Keywords that immediately come to mind include woman, man, diverse, talent, qualification, specialist, generalist, expert, European, Asian, international, a mixture of everything, because in my opinion that is exactly what makes an ideal construction team.

Fig. 4.3 Project Manager Annemarie on the construction site PPP B247 Mühlhausen-Bad Langensalza at BW305, 2023. *Source* VINCI Construction Shared Services GmbH/Dirk Mathesius

Would I like to See Even More Diversity, and if so, in Which Areas?

Yes. My great wish here is that civil engineers think and understand holistically. And that there are diverse teams in the construction industry that think and act across disciplines. A construction project consists of many participants. It is important that everyone accepts the "different tick" of others and can empathize with them and that everyone involved in the construction project has a conversation and is empathetic towards each other.

How Did I Get My Career in Construction?

Good question, and perhaps not understandable to everyone: the construction industry is purely coincidental. Traveling with my husband has always been a priority in my life. I have a diverse background and interests. Construction, its technical subject matter and everyday working life are anything but "dry". What absolutely fascinated me from the start was the direct interaction, the strong sense of humor, the many social contacts, with—as I experienced it—colleagues who didn't hold grudges. Every day is different. This also applies to the weather dependency of a construction site, the effects of seasons, climate and temperatures on a construction project. We are talking about extremes here, such as summer heat, constant rain, thunderstorms, strong wind conditions, sub-zero temperatures and ice. This means that a high degree of flexibility is required in construction.

What Ups and Downs Do Exist?

I remember a low point straight away because it's personal relationships that stay in one's mind. During a construction project in France, I sought a conversation with my

boss because he treated me with very little professionalism and contemptuously. But he didn't respond to my concerns at all, instead just talked about his problems. I discovered that the boss was not following his lead.

A highlight that I gave myself was the following: When I was asked whether I was interested in a project management position, I told my boss directly and with complete conviction: I want to be a boss! And do you know what that did to me and my boss? There was a lot of confidence, it increased trust between us and also the perception that I could do this job!

Which Words Describe Me Best?

Positive, Imaginative, Quick to Analyze, Motivating for Those Around Me, Resilient, Solution-Oriented, Constructive, Perfectionist. Because I Want It Exactly and not Roughly.

What Qualities Have Always Helped Me in Construction?

The laughter and the humor and joke because it reduces the tension in many situations and moments. There is an anecdote that I remember very fondly: At a New Year's reception the atmosphere was very lively. And when it came to reporting—along with champagne—about current projects and discussing professional expectations, I was approached directly by upper management about my project. When I described the facts and my personal assessment clearly, unambiguously and factually with a smile, the person I was talking to said: "Oh, you are a dangerous woman!" That's why the formula applies to me: Factually constructive + smiling but remain determined = dangerous woman!

Fig. 4.4 Construction Summit in Hamburg, April 2023. *Source* Construction Summit

Were There Mentors or Special People in My Life Who Motivated Me?

Yes, as a person, for me it was always my mother. She was a role model in every respect: When she was young, she was already completely independent, had bought her own house, owned a car, and ran her own practice as a midwife. My mother was always positive, everyone around her always approached her with enthusiasm. She is pure inspiration for me!

In my professional life there were also people who had a special impact on me, namely my first supervisor in the construction industry. He had made it to millionaire, but always remained grounded. He wasn't afraid to transport a sheep in his S-Class, even though his car needed complete cleaning afterwards. He checked the quality of the sand directly with his hand on his knee despite his expensive suit. He spoke to the cleaning staff at the same linguistic level as he did to football stars and real estate kings. He was able to deal with all types of people on all levels and find ways to communicate with others. It was a role model for me because he was so familiar with people of all "levels" and never felt or showed himself to be superior.

What Does It Take to Balance Woman—Construction—Starting a Family—Further Career?

When it comes to this question, I have to start at the very beginning and in the family relationship: Only if I clearly say in my relationship what I want, what expectations I have, what I can and cannot achieve, what needs I have, then something can actually happen and change. In the second step, I then succeed in motivating other women to achieve their goals. For me it carries responsibility to inspire other women, because it ultimately leads to added value (Fig. 4.5 and 4.6). *Especially among the younger gen-*

Fig. 4.5 Annemarie as Speaker at the Kick-off of the Main Association of the German Construction Industry as of 21st of September 2023. *Source* Hauptverband der Deutschen Bauindustrie

Fig. 4.6 Annemarie at an event for women in construction on Women's Day on March 8, 2023 at the Dutch Construction Industry Association. *Source* Bouwend Nederland

erations who must contend with increasingly complex challenges in the digital age. They also openly support it, as I often notice. I would like to explain this using the example of the GPS-controlled excavator. The younger generation is enthusiastic and sees the potential and advantages. I have male, older colleagues who say with complete conviction that people these days no longer know how to work themselves: they have become completely dependent on new technology. As if technological innovation in construction means a deterioration and a threat. I think the fact that the proportion of women is constantly increasing in all areas is a great sign for the future of construction: women are open to technological advances, as well as simplification and improvements in everyday working life. There is still a lot of room for improvement in some areas of construction.

And if We Look at Your Current Company, How is Digital Transformation Linked to the Entrepreneurial Culture There?

Like many places, we have a mix of modern young people and older people with a lot of experience. There are performance areas in which hardly anything has changed, and some colleagues have a conservative attitude that is typical of the construction industry. But the first steps toward change have taken place for two years. They remain at the top of the agenda.

I think digital transformation is an overarching theme: a mix of workforce development, training, technology and research & development. If it is to lead to success, it definitely has to go hand in hand. We work "lean" and our company management plays a role model. The balance of power and the trolls of the hierarchy change through BIM, because information and data are no longer withheld, but rather shared. Overall, this leads to more transparency. In my opinion, the connection between digital transformation and entrepreneurial culture can only work through:

- *Change Management*
- *Acceptance and enthusiasm of everyone involved in the company*
- *Philosophy and values, and diversity as key value*

What Opportunities Does Digital Transformation Offer (BIM, AI, Metaverse..)?
In terms of my field of activity, it offers great relief in daily working routines and processes. Maybe the chances are less in small projects, but still very important for practice. They are essential in large projects. Here, the priority is to keep an overview of everything. Due to the increase in complexity, the use of BIM, AI, etc. *is essential to improve handling of costs and time (e.g. a tool for analyzing all correspondence, immediate reporting of potential conflicts between project participants or with potential additional costs). Such technologies represent strong support for management decisions and strategic measures.*

When it comes to the construction industry in general, e.g. the exchange of data and information between colleagues, BIM or AI is no longer a nice-to-have. We are talking here about a broad topic that influences many areas of life and work: it extends to the problem of resources, their conservation, the supply chain, the sensible use of personnel when it is even more needed in other areas outside of construction (maintenance of older people, for example).

What General Conditions Are Needed for Construction that Have so Far Been Ignored?
There needs to be greater acceptance that women want and should take on a different role. The support of superiors is required so that they recognize and model the necessary way of thinking adapted to the new circumstances. More women in construction means more diversity in thinking, which contributes to success. Some outdated role traditions must go, and a new culture of thinking in construction must start!

How High—on a Scale of 1–10—Do You Estimate the Willingness of Construction Companies to Integrate Digitalization?
Willingness Very High (8–9), Degree of Implementation Very Low (4)

How High—on a Scale of 1–10—is the Need for Education About Digitalization and AI?
Very high! That's why it's extremely important to have platforms such as the Construction Summit Hamburg (Fig. 4.4), *which enables exchange, sharing of experiences and new knowledge, networking and the growth of professional expertise.*

When You Talk About Trends and Tendencies in the Construction Industry in the Digital Age, Where Would a New Culture of Thinking and Learning, for Example, Help?
To be specific: digitization should not lead to more interpersonal distance. If I have something to clarify or complex tasks to assign, I first seek direct conversation. With such

behavior, personal closeness, trust and reliability, the young generation will also have a lasting interest in the construction industry. This has an even greater impact on attracting skilled workers who would like to work in an industry that is "humane".

What is the Current State of Values, Ethics, in the Company and in Everyday Working Life in Construction?
I can't give a general answer to this question. There are two sides of a coin. But what changes: The undesirable behavior is addressed! Example: I was informed this week that my former high-ranking boss from a previous company had been dismissed by their Ethics & Compliance Department. He had probably placed his personal interests higher than the interests of the company. By the way, for more than twenty years, but no one dared to talk to him about it.

Where Do You See a Need for Action?
I won't change the whole world, but I can influence and stimulate the world around me. I can say what is important to me and what I expect, which rules of the game my team should (or must) adhere to.

4.2 The Architect in Construction Management

Julia Scheider (M.A.), born in 1995, Project Manager in Construction Management at Drees & Sommer

At the age of 13, I wrote in a letter to my future self "What I want to become: Maybe an architect". I received this letter 15 years later, by then I had already studied architecture and was just becoming a project manager. The professional, self-confident role model I have in my family is my father. The communicative and emotional strength I have from my mother. I love the aesthetics of my built environment. It fascinates me how architecture affects people, with its negative facets, but also all the positive influence. What it does to people, who are in a building and move or just pass by it and marvel. Be it in private, in my home, at work, in the office or in the evening, for example in a restaurant. Do I feel safe? Do I feel small? Do I feel powerful? Organized? Chaotic? Architecture creates emotions and reactions in all people, whether consciously or unconsciously. To name two architects who inspire me, the first is of course Zaha Hadid, who with each of her buildings, translated form into built figure. The other is Mariam Kamara, who with local products builds open architecture and thus in my opinion represents a role model for what we understand by sustainability.

4.2.1 Studying Despite School

As much as I can get excited about architecture, so little did I manage to do that for the "school system". The ninth grade required two attempts. After the tenth grade, there was the forced change from high school to a technical/environmental school, where I was able to get my technical baccalaureate in technical computer science. The prevailing hybrid school system had practical and theoretical components, which suited me much better. On practical days, I gained experience in the marketing of a DIY store chain. Thanks to the vocational school, it was possible despite initial difficulties to achieve the necessary degree to ultimately meet the admission requirements for a study program.

The architecture studies at Kaiserslautern University of Applied Sciences suited me significantly better. After the master's degree in 2019, the supervising professors asked me if I could imagine working as a lecturer—I could. It was a feedback that surprised me. Because this trust was placed in me. And because it made me look positively at the work done so far and the professional competencies acquired. Since then, I have been teaching students in the Department of Building and Design in computer-aided design methods as well as in representation and design. These are two subjects that are thematically related. The analog content worked out in "Representation and Design" finds its digital transformation in "Computer-aided Design Methods" and makes it easier for students to get a design entry into their architecture studies. How does design work? How can such process look like? What tools do I have to get things out of my head onto paper, into drawing programs, models and renderings as an image to visualize? So that I can later show others what it could feel like, what I had previously imagined.

4.2.2 Teaching as a Source of Learning

Teaching is a mix of lecture, conveying of theoretical content and practical exercises. I appreciate the incredibly inspiring environment. Especially during the time of my career entry, in which I was simultaneously on the construction site and in the comparatively strict management environment, teaching was a significant balance for me personally—between "free spirits" at the university and the mostly self-proclaimed "old hands" on the job. This is also the core of my motivation: I want to stay involved, hear and see the fresh ideas of the students and take these contents into my everyday professional life. Where is the future going? What are the "big construction sites" that we need to tackle, for example, to create an environment in which young people can make their own thoughts and ideas a reality, instead of passing on their own mental corset. "We've always done it this way!" It is human to fall back on one's own pool of knowledge, which one has worked out for oneself. It is difficult to understand that it is a pool and not a sea. And—to stay with the metaphor—you can achieve the same result in the other pool, you swim. Through teaching, I have learned that rarely anything is absolute. I find this incredibly motivating and want to be able to motivate others with it.

Despite my master's degree in architecture, I am not working as an architect today, but in construction management, thanks to a practical semester. A preliminary internship on a construction site is common in architecture studies. However, a deepening during the practical semester is already rarer. Anyone who plans buildings should also know how they are practically created. I lacked this knowledge until then—as do most people who study architecture. What does this dashed drawn waterproofing membrane look like later in connection with the window, the door, the facade base? I only learned all this on the construction site. Although the studies provide the theoretical basics. Nevertheless, in my view, the learning curve on the construction site is much steeper and practice is essential to understand architecture and to be able to build good architecture.

4.2.3 Between Lecture Hall and Construction Site

First, I extended my practical semester from four months to six. Then I got the chance to continue working for my former internship company parallel to my architecture studies. Ahead of me was a phase that was extremely demanding in terms of time, physically and mentally. Constantly commuting between my place of study in Kaiserslautern and the construction sites in Saarbrücken and Stuttgart was exhausting in the long run. Looking back, the changes of location, but especially the change between the different requirements, were highly stressful. For me personally, this parallelism between studies and construction site was the right way. This put me right in the middle of the action—in construction management—and I stayed there in a positive sense. However, I cannot and do not want to recommend this type of career entry across the board.

On the construction site, I already had a lot of responsibility in my early 20s. Even though this was chosen by myself and the opportunities outweighed for me, I look back more critically on this time with a few years of distance. Construction management, especially site management, with its many short-term deadlines, where quick, pragmatic action is required, attracts characters according to my experience: people who like to expose themselves to performance pressure in order to then perform at their best. Here, the employees themselves, but also the companies are obliged to provide a balance in time. Tools must be provided to work in a structured way and to recognize one's own limits in a controlled and early manner.

4.2.4 Ballet School for "Learning to Walk"

The first project was the construction of the John Cranko Ballet School in Stuttgart—a building with a museum-like character and demanding architecture. I still remember one of the first challenges: Due to the almost continuous exposed concrete requirement class 4, all empty pipes were pre-installed in the shell—including concrete core activation, which was fed by an ice storage. During this time, I became aware of the importance of a well-functioning team. Such a team is essential to find out more about yourself, how you like to work and structure yourself, and it helps to find your own rhythm.

Over the construction period of the ballet school, I developed from an intern to a working student to a permanent site manager, who completed the project together with colleagues. During this time, I also got to know and appreciate my later colleagues from

Drees & Sommer—first and foremost Joachim Thomas. We shared an understanding of team structures, site organization and execution.

4.2.5 Door as a Symbol for Interdependencies

After my switch to Drees & Sommer, it turned out that I was able to support Joachim Thomas initially in the TGA site management (TGA = Technical Building Equipment). In retrospect, an experience, which provided me with enormous professional added value. A good example with symbolic character for this were the doors. For me, it was initially difficult to understand why my colleagues insisted so much that a certain door had to be installed at a certain time—considering hundreds of doors that had to be installed in the project. In fact, there are many dependencies in a construction project. Expert inspections, for example, must take place in closed rooms. No door, no acceptance. Thus, due to a single missing door, the entire project schedule can be thrown off balance.

This understanding of the substantive work of my colleagues to sharpen, still helps me daily and has influenced my way of working early on. It helps me enormously to detect collisions in construction processes earlier and thus to work more efficiently in execution.

After some smaller intermediate construction sites, another highlight project was ahead of me—the Porsche Design Tower, also in Stuttgart. I like to refer to it as my first own "baby"—a high-rise (Fig. 4.7). This project was also very demanding in terms of construction and quality. There were strict requirements—including fire protection. Also, the logistical handling on an inner-city construction site without a construction site setup area was a challenge to deal with. This was compounded by the handling of material deliveries during the Corona time. For me, these were many new contents that had to be learned every day.

4.2.6 Meet on Equal Terms with Enough Self-Confidence

When the foundation stone was about to be laid at the Porsche Design Tower, we as site management were allowed to put something in the time capsule. We didn't skimp and had a plate engraved at the time—with the names of the then known team members. That was one of those key moments when I realized: We are really building this now. What followed was an emotional roller coaster ride: From "Yes, of course, we can do this" to "Let's see what happens" was intermittently all there. I am grateful that so much trust was placed in us as a team—despite our average age of under 30 years and without long-term experience in risk and supply management. This trust is also needed, and from all sides—be it from the client or the employer. And also a certain self-confidence. In other words, the right balance between the necessary respect for the task and the trust in the know-how and the strength of the team.

Fig. 4.7 Author on the Porsche Design Tower, Stuttgart, 2023. *Source* Julia Scheider

On the construction site itself, I love when something visibly progresses. The shell construction phase is the most visually tangible. When a 120 m high tower crane is set up, it still gives me goosebumps. I am fascinated by this height. During the construction of the Porsche Design Tower, as is usual with high-rise buildings, there was a facade access system, which later became For example, even the facade cleaners need it. In this case, we built an element and post-and-beam facade that cannot be cleaned exclusively from the inside later on. During the briefing, we took advantage of the opportunity and also went over the edge of the building. The view was fantastic. It was one of those moments that made me realize how special such a project and the job are. In addition to an action like the spectacular ride along the facade, for me, the many small social events with the teams on the construction site are also part of it. Whether eating tarte flambée with the shell construction workers, pizza with the medium voltage system fitters, breakfast with the drywall builders, or barbecuing with the crane erectors—all this contributes to understanding oneself as part of a team that is creating something big together. And meet on equal terms and together.

4.2.7 Promoting and Motivating

When I joined Drees & Sommer in 2020, I was initially a construction manager. My current position is project manager in construction management. This means that with this job profile, I am responsible for the success of a project by ensuring that it remains within the cost, schedule, and quality framework for our client and is economically viable for us as a company. At the same time, I lead the project team by supporting and motivating the individual team members.

In our company, it is possible to develop further from any field of activity and gradually take on more responsibility as an expert or manager—for topics, employees, and projects—up to partnership. For me personally, I can say that working with teams, building independent and self-determined team structures, promoting the interests of colleagues are the things that make me happy and suit me. I like sharing my enthusiasm for what we do and having the ambition to empower people and affirm them in what they can do and what they aspire to.

4.2.8 Family Planning is (not) a Big Deal

These development opportunities exist regardless of someone's gender or family situation, or plans regarding family planning. Nevertheless, in my view, women still approach the topic differently than men. Anyone who wants to have (more than) one child as a family tries to plan further: When is the right time to start a family? In a year? In two years? In five? And despite all planning—in construction management, pregnancy is associated with greater restrictions than in a classic office job. While in the office, in many cases, all activities are possible until shortly before birth, it is something else to walk around the construction site heavily pregnant. So especially in the construction industry, you need an employer who understands how to react appropriately to the situation. It would thus be possible for pregnant women in construction management to take on a higher-level, coordinating role. You don't necessarily have to go out to the construction site, but perform these tasks from the container office. Prerequisites are open communication, the establishment of representation rules, and flexible workplace design—including adequate structures for working from home. A well-structured team and on behalf of the pregnant woman the ability to let go of tasks are essential.

A company-wide regulation for mobile working, such as we have, is supportive. Specifically, if the activity allows it, employees can also work from home in consultation. The most important factor is and remains acceptance. Colleagues and superiors must all understand and accept that a (temporary) special situation requires an adequate approach. From my perspective, pregnancy is no different from many other circumstances: If I break my leg or tear my ligaments, I'm out. If I have psychological problems, I'm out. This can happen to anyone without exception. And at least in the case of the accident example, no one is later asked what it did to their own career. Taking accidents into account is part of normal risk management. And with a structured reintegration phase, there are clear processes that regulate re-entry after a longer absence.

4.2.9 Understanding Instead of Patronizing

Pregnancy, on the other hand, is often seen in relation with weakness in the minds of many people. If superiors use pregnancy as a criterion for career development, it can still become a career killer. Because in my view, there are still too many prejudices in society as a whole: Does the employee return as a "full" mother or "only" part-time? When is she planning the second or third child? How long will she be absent? Is she fully resilient afterwards? Do her priorities shift towards family?

But what about employees who have relatives in need of care at home? Or simply those who pursue time-consuming hobbies? What about those who take several months of sabbaticals? I think more sobriety, more normality is needed in this specific contect. Pregnancy and parenthood are enormously important individual changes in a person's life. But they should not be evaluated by third parties more emotionally and prejudiced than other personal changes and incisive experiences. We all have tools to organize ourselves so that we can deliver the required performance at the workplace. That's why understanding is needed, but not patronizing.

Furthermore, it helps if employers officially have the topic of parenthood on their radar. For example, we have had the option for some time that employees who go on parental leave can keep their laptop if they wish. This can help to stay involved in ongoing processes and not lose track. Of course, people on parental leave are still invited to events to be able to exchange ideas in a relaxed setting.

4.2.10 Childcare Costs as a Financial Burden

One issue that preoccupies many new parents is the question of toddler care. On the one hand, it is difficult in many places to get a place at all. And even if it succeeds, so-called "external care" is often associated with high financial expenditure. It is not acceptable that the salary, that someone earns because the child is cared for during the day, goes one to one or mostly for this care place. For example, our employees with preschool-age children receive a subsidy for childcare costs. Here, employers should promote that it is financially worthwhile for parents to work.

4.2.11 Change Management—Keyword "Digitization"

Of course, changes are not only in private, but especially in the professional environment. The construction industry is now also in the midst of digital transformation. Building Information Modeling (BIM) is just one buzzword. It's about the digital networking of all processes, products, and participants. Building Information Modeling bundles all relevant data of a construction project. Where architects or engineers each had to create their own plans and inform themselves about every small change, all information is available in a digital model with BIM.

Documentation on the construction site is also done digitally. For hard-to-reach places, for example, a robot dog can theoretically take over this job now. Artificial Intelligence is making its way into many processes—whether it's defect management, schedule controlling, or quality assurance. This will become much more apparent in the near future. Ultimately, it's also about building more efficiently and improving risk management.

4.2.12 Change in Mind and on the Screen

But there are other factors that slow down these developments or at least their spread. Not everyone who is supposed to use new tools and systems already has the necessary knowledge. Here, in addition to regular training and persuasion are necessary. There is the underlying fear of replacing oneself and making one's own job redundant through Artificial Intelligence. Even if in 2022 Thomas Zurbuchen, former Science Director of NASA in the ZEIT podcast "Alles gesagt" concludes: "The fact is, there is no scientific problem that only humans can solve but the fact is, humans are much better."

The concerns do exist. And these concerns must be taken seriously by companies. In recent years, a lot has happened in the field of digitization. This can overwhelm the people directly affected by it. Whether project manager, site manager or architect. This development affects all of us. Because everyone has to decide every day, what exactly do I look at more closely? What do I deal with more intensively now, besides my daily operational tasks, which still continue? Which new tool might be outdated by tomorrow? What will prevail in the long term?

4.2.13 Giving and Taking Time for New Things

In situations where things need to move quickly—which is in the nature of our work—we naturally resort to the tools we master and with which we feel safe. To get to know a new tool or to deal with new technologies in general, one must be given time to learn. We cannot expect everyone to understand and implement everything intuitively. Here I see the employers social responsibility for ensuring that their employees have the necessary time available. On the other hand, employees are also asked to demand these times for themselves and to seek an environment in which this is possible. Especially with regard to artificial intelligence, clear legal foundations are also needed, such as something very essential like the protection of private data and intellectual property.

In addition to technological change, sustainability in construction is becoming increasingly important. Hardly a construction project that does not contribute to one or more of the 17 UN sustainability goals. Typical are for example sustainability goals related to health and well-being, quality education, clean water and sanitation, affordable and clean energy, industry, innovation and infrastructure, sustainable cities and communities or measures to combat climate change. The Sustainable Development Goals

(SDGs) represent a comprehensive concept of sustainability. But especially in relation to the environmental and climate protection, the construction and real estate industry has a large lever, as it is responsible for 38 percent of global CO_2 emissions. In addition, in Germany, half of all waste consists of construction and demolition waste.

4.2.14 Circular Construction

There are numerous possibilities to make construction more sustainable with a view to the climate. Using renewable raw materials for building materials, whenever possible, as these bind CO_2, is just one. The Cradle-to-Cradle principle (C2C)—"from cradle to cradle"—plays a key role:I in the best case all products are recyclable. In buildings, for example, furniture or even entire buildings are designed in such a way that they can be disassembled into their components or dismantled at the end of their life cycle. Without loss of quality—as is usual in "normal" recycling. If this idea is consistently pursued, no waste is created in the first place. In construction, this presupposes that one uses chemically harmless materials and also avoids adhesive connections and instead uses plug-in systems or rotary connections. This requires more time in the pre-planning phase and a detailed examination of the **how** is built, not just **what** is built.

In this context, for buildings, neighborhoods and cities, urban mining plays an important role. This requires an understanding of building material as an asset and repeatedly utilizing these. Cities are urban mines—with real estate as material storage. Existing buildings contain many secondary raw materials that need to be extracted and reused. A building resource pass provides information about which materials are present in what quantity and quality in a building. Only when such a certificate is available, urban mining can fully come into play.

4.2.15 The Future Lies in the Existing

What significantly adds to sustainability in construction is the shift towards more renovation projects instead of new buildings. Thus, no new areas are sealed off through a renovation of existing buildings—and resources are reused, for example, when applying the urban mining principle.

Relying on renewable energy has become the standard for buildings. The repertoire includes, for example, photovoltaic systems, geothermal energy, or the use of rainwater for, among other things, toilet flushes. Also, green roofs or facades and the provision of bicycle parking spaces when constructing office buildings are common measures to make construction projects more environmentally friendly. Besides these obvious measures, much can be achieved in the background. Relying on regional products and short supply chains for building materials has a positive effect.

4.2.16 Trends and Tendencies—Today for Tomorrow

I cannot free myself from the fact that what I do as a human being every day has a negative impact on our environment and the habitat of all. One must become aware of the responsibility for one's own actions. This not only affects private life but especially what one does during one's own working hours. The construction industry can massively influence what the coming generations will see and have to deal with. This, we define with our actions today. I get the impression that (especially young) people are becoming more and more aware of their responsibility and thus find new, alternative solutions. And if there are already solutions, such as alternative building materials, then it is a must to use them, spread them more widely, and further develop them.

4.2.17 New Culture of Thinking and Learning

Whether changes in terms of sustainability, new technologies, or a change in values in society as a whole—all this goes hand in hand with a new culture of learning and thinking. "I don't need to know everything; I just need to know where it is"—that belongs to earlier times. Today the sentence is instead: "I don't need to know everything, but I need to know how to distinguish the right from the wrong." This is a massive difference. Because for everything I say, with three keywords and a search engine fed with Artificial Intelligence, within seconds, an assessment can be generated that answers my questions, confirms or refutes my statements. But is it also well-founded? What sources does it refer to? Where do the contents shared via social media, which I have read and found to be correct, come from, to which no sources are given? Of course, the algorithm repeatedly confirms what I think I already know and pushes content into my feed that, according to my digital footprint, I 'm interested in.

The confirmation bias additionally ensures that I perceive what confirms my opinion more strongly. To change this again, it takes a multitude of refuting arguments. This is dangerous because it happens so quickly and easily. We have to learn to deal with this. To come back to the topic of new thinking and learning culture: If everyone fights with the same weapons—in this case with the same tools and systems for information procurement—you quickly meet on equal terms. This equality is certainly unusual for many, but of immense value. We are then talking about the keywords competence-driven development and the possibility of flat hierarchies.

4.2.18 Shortage of Skilled Workers and New Job Profiles

In relation to the shortage of skilled workers in the field of construction management, we must we need to mentally detach ourselves from rigid systems that mean: first high school diploma, then university, then five years of professional experience. Just as I, trained architect with some professional experience in this field, can work as a construction and project manager, so can an electrician, a plumber, or a drywall installer with some professional experience as a construction manager. The required qualification is disposing of an understanding of construction, processes and procedures, and anticipatory thinking—not a degree. It is long overdue to deal with alternative development models in order to be entrepreneurially successful and especially innovative in the long term. Therefore, I advocate thinking flexibly in every respect: Open minds and open doors!

Links and QR codes for further reading and deepening:

- On the new construction of John Cranko School: New construction John Cranko School (dreso.com) www.dreso.com/de/projekte/details/neubau-john-cranko-schule
- On Porsche Design Tower: New construction Porsche Design Tower, Stuttgart (dreso.com) www.dreso.com/de/projekte/details/neubau-porsche-design-tower-stuttgart

4.3 Going to the Construction Site as a Woman Despite Having a 1.1 A-Level?—Why not!

Laura Attenberger (Mason)

First, I would like to introduce myself briefly. My name is Laura Attenberger, I am 20 years old and come from Bavaria, more precisely from a small village in the Upper Palatinate. In 2021, I passed my high school diploma with an A-Level of 1.1 and, since August 2021, work in a small construction company with 15 employees in the neighboring town to complete my apprenticeship as a mason. In a few months, Ime1) will take my graduation and can then proudly call myself a certified mason (Fig. 4.8).

4.3.1 My Professional Career—from Childhood Dream to Reality

That's all about me in a nutshell. For many the question now automatically arises why, with such a supposed "top degree", one does not pursue an academic education, but rather prefers to "bluntly" stack stones on top of each other, smeared with dirt in wind and weather. Well—on the one hand, the mason's profession involves so much more than just the physical, dirty work. Not only the diversity of activities in the professional field was what quickly convinced me. More details later. Personally, it was already clear to me subconsciously from an early age where I wanted to go professionally in the future. Even when I was in the 3rd grade and had to state my career aspiration in a friend's book, I

Fig. 4.8 Author during school training in staircase construction, Chamber of Crafts, 2023. *Source* Laura Attenberger

neatly wrote "Mason"—how unusual for a little girl. At that time, I couldn't imagine much about this profession yet (Fig. 4.9, 4.10 and 4.11).

Today I know that my wish had a deep origin: My uncle was and still is an enthusiastic mason and from an early age I was allowed to help out on Saturdays with renovation projects at home and learned early on what plaster and mortar feel like in the hands or how within a few hours with a few stones and a lot of muscle power, new walls and rooms can magically appear. So my enthusiasm for the construction trade was awakened quite early and to date I think back to my childhood with a small smile on my face, when I smell the specific scent of fresh lime-cement plaster on the walls at the construction site.

During my time at high school, however, my interests developed in a slightly different direction. Not least due to the influence of the teachers and the extremely scarce opportunities to inform oneself about apprenticeships in the school environment, I quickly discarded the idea of a training in masonry. In principle, I was very interested in the field of psychology, but it quickly became clear to me that I was more suited for a job outdoors in the fresh air. So forestry came into closer consideration, especially because in the 11th grade, as part of my seminar paper, which was created in collaboration with a forester, I got a closer look into the designated area. However, the construction site never completely let me go: So at the end of my time at high school, the civil engineering degree was high on my list, as the construction site had always attracted me from an early age and—in my memory—was associated with a special charm.

In the 12th grade, my last year at high school, finally the big turn-around came. Through a YouTube video of a young female master mason, with whose situation I could strongly identify, I decided to take a closer look at the profession of masonry and was able to subsequently do an internship at my current employer in a spontaneous rush.

Fig. 4.9 Bricklaying. *Source* https://campushandwerk-swn.de/ausbildung/umschulung/berufe-von-a-z/umschulung-maurerin.html

Fig. 4.10 Mortaring. *Source* https://www.obi.de/magazin/bauen/baustoffe/ziegelsteine

Fig. 4.11 KS Masonry Connection. *Source* https://www.ks-original.de/verarbeitung/mauern-von-hand

And? I was immediately on fire and passionate about the profession and became more and more confident in my decision throughout the following weeks, to start a masonry apprenticeship and not to study despite my very good grades and in contrast to some teachers´ and family members´opinion. So after my graduation, I did not enroll at the university, to study civil engineering, but followed my dream to learn the profession from scratch—namely as a mason.

4.3.2 Highs and Lows in Everyday Work

Before my first day of work, the excitement (and quite honestly also the fear of the completely new challenges, the physical challenge and the colleagues) was huge…but in retrospect most of the concerns were completely unfounded.

Of course, the initial period was physically not easy at all. After work, I came home the first few days and weeks and just managed to get to the couch, only to fall asleep there shortly afterwards. I didn't get much of my free time at home during the first period, the remaining hours besides work I mostly spent sleeping. But that was only at the beginning of the training. After a little acclimatization, the physical strain was no longer a problem. Although I was especially in the initial period, and still am in certain situations, not quite as strong and efficient as my male colleagues, I can help myself quite well with some tricks and technical aids. And of course, my colleagues support me in every respect when I need their help.

At the beginning, I was viewed a little skeptically by my all-male colleagues, but after I had proven my willingness to work and my ambition, I could and still can count on their support to this day. Especially in the first weeks, I tried to give 120 percent instead of 100 and had trouble keeping up with the men as a woman. But over time, my physical fitness increased enormously and it became increasingly easier for me to survive the entire working day under physical strain.

Of course, the beginning was a struggle. In retrospect, however, every second was worth it for me.

It was always important to me that my colleagues treat me like a completely normal work colleague without special rights, and that's exactly what happened. Yes, I too have to carry mortar bags or scaffold brackets from A to B or throw plaster on the walls. Sure, if it's physically not possible for me, then other solutions are on hand, but otherwise there are no special treatments for women. And that's a good thing.

I am a "normal" part of the team and no exception and am treated just like any other male employee. And so it happens that in the evening an after-work beer is offered, which in the construction industry is probably the ultimate sign of cohesion and made me feel to be part of the team from day one (Fig. 4.12).

Not only I, but also some of my colleagues and my superiors believe that the working climate in this profession benefits from women. Often, as a woman, I have a calming influence in the group and bring different opinions and points to the agenda compared to the "male world".

From the very beginning I was very positively received at the school of my apprenticeship and quickly connected with my male classmates. Here too, I often provide a calming influence and my classmates gladly benefit from my balance and order, which

Fig. 4.12 Author on the construction site. *Source* Renate Rossberger

I bring to the school lessons. However, quite often I was mistaken for a young teacher and not for a bricklayer trainee in the school corridors. Even on the construction site, it is evident that women in craft professions represent a minority: I have often been mistaken by the builders for the girlfriend of a work colleague, who is on the construction site for support, or even perceived as the daughter of my boss. From the outside, I was often seen as a young motivated building owner, who helps diligently on the construction site. Actually reason enough to enlarge the circle of women in craft professions with broader awareness in society and above all set it up as normal.

But there are also downsides of working as a female bricklayer. Although the backup of my classmates and my work colleagues, which probably origins to the relatively young age structure of the company, young men tolerate me as a woman in the profession rather than the "established" apprentices, who cannot imagine women in this profession due to years of habit and experience. This is not the case with the few older bricklayers in the company viewing my situation, but I can remember a very special circumstancee on a construction site that still makes me a little sad today. On this construction site, my boss clearly assigned me the responsibility, but the older builder could not accept me as a woman in the "leading role" and controlled and criticized every step of my work. The critical point arrived and he called my boss full of skepticism for control at almost every major work step to assure himself of the correctness of the work processes. Of course, not everything can always run smoothly and setbacks are part of the game. Fortunately, however, situations like these are absolute exceptions in my daily work and I am usually positively received from the outside and respected and appreciated as a full workforce.

4.3.3 The Bricklaying Profession Itself

The working day as a bricklayer is of course not always easy and every day I face new challenges and problems. In recent years, I had to learn to take responsibility for the work I do and for the performance of colleagues on the same construction site. In addition, the profession requires a little manual skill (which actually develops and evolves over time) and a lot of knowledge about the various materials used such as concrete, wood or brick, to name just a few. Organizational talent is required when it comes to planning the work on the construction sites or ordering or delivering the right material on time and calculating the timing of the many different work processes.

But that's exactly what makes the profession special for me: the enormous variety. No day is like the other and that's for the best match with me. It is one of the best things about my job. The different areas of activity add to the variety of daily work keeping it highly exciting. From raw buildings to concrete work and plastering to paving or installing pipes, everything is included—but still a lot for me to learn.

Not only do the tasks challenge daily, but the external conditions demand a lot: Being outside in wind and weather, rain and snow, means a tremendous strain on the body and requires a little adjustment at first. Of course, in spring and summer, you can't imagine anything better than spending the workday outside in the sun, but hard working in subzero temperatures or in pouring rain is just as much a part of it and must not be overlooked.

The longer you are exposed to the often harsh conditions of Mother Nature, the better you get used to it and in today's time, you can counteract well with appropriate clothing.

The same applies to the strain caused by heavy lifting or physical activities in general. Especially as a woman, it means going to your limits on the construction site at first, actually every day, as the work is often very hard. But it is by far not as exhausting as it might have been 30 years ago. With the help of tools like cranes, excavators or lifting platforms, a lot of physical work is relieved and especially bones and joints are spared. From personal experience, I can also say that over time you gain a lot of muscle mass and strength and the work becomes much easier after a few months compared to the beginning.

But maintaining the female body to withstand the heavy strain as long as possible and gently, the body needs particular support and care. Therefore, additional sports have been part of my daily routine for many months to strengthen my back more and relieve my body. I think that as a woman you have to take more care of your body and keep it fitter than some men working in construction.

4.3.4 Future Outlook—Where Are There Still Construction Sites in the Craft?

In my opinion, women in the craft profession and especially in construction are confronted with many obstacles in their way. Especially in terms of reconciling career and family, there is room for large improvements. Currently, the profession of a bricklayer is still based on the classic model of the last century, where the man's employment ensures the provision of the family while the woman runs the household and takes care of the children. Due to the seasonally-dependant quite long working hours, there is hardly any time left in the summer besides work to take care of the household and -if available- the family. In the summer, workdays of almost 10 h are not uncommon, especially with long travel times to the respective construction sites. In this respect, many frame conditions should change in the next few years to elevate the profession's attractiveness for women. The working hours must become more family-friendly and there must also be more time left so that, for example, the compensatory sports needed for long-term health can be properly exercised.

Furthermore, in the next few years, further major steps must be taken in the construction industry to stop the loss of skilled personnel and counteract the shortage of skilled workers. Especially in recent years, the change has been enormous. In the past, craft professions were much more respected, reflected in the fact that many more locals practiced these professions. Nowadays, not only the number of foreign workers on construction sites increase, but at the same time the quality of work decreases, which of course is not surprising when you consider that many foreign workers on construction sites have little or no training. Only a few still master the real craft ART. Sure, attaching precast concrete parts to a place specified by the capo or site manager using supports, or mixing and applying screed plaster according to bag instructions and rubbing it in with a smoothing trowel, almost anyone can do that. Special spray plasters, stucco decorations or elaborate brick masonry are more the exception today and represent a rarity in daily work onsite. Today, construction is about speed and quantity. The true ART of masonry is unfortunately being lost more and more.

At the same time, however, other job profiles are coming to the foreground: The few skilled workers who are still on construction sites today take on leadership positions and manage, monitor, and document construction processes. I think this development will intensify in the coming years. Precisely here, I allocate more opportunities for women on construction sites and in the trades. Especially because—to a certain extent regrettably—the hard manual labor will decrease in the future and will be increasingly taken over by foreign unskilled workers, which means that work techniques must be precisely adapted to these people and thus also simplified, it offers us, the really well-trained skilled workers, the opportunity to take over more management positions and the organization. This, in turn, offers attractive niches for women due to the reduced physical labor.

The trend in recent years has increasingly been towards rapidity and turnkey construction. The automation of work steps is increasingly under focus in order to work more efficiently and save labor.

While this will probably remain the focus in the near future, fields of sustainability and assuming responsibility are increasingly playing vital role, and that's a good thing. Especially with regard to environmental and climate protection, there is still a lot of room for improvement in terms of sustainability in construction. In particular, cement production is very harmful to the environment and therefore a new rethinking is required. Leaving binder cement behind, moving and researching towards more climate-friendly materials. Many companies, like my current employer, already offer alternatives or strive to work more sustainably. For many years we have been working with clay plaster instead of lime and cement-bound plasters and recently this material has also been in increasing demand from builders. A rethinking is noticeable on behalf of the buyers. But in the future, much more attention and financial resources must be put into research to find more environmentally friendly alternatives to raw materials like cement and thus to operate more sustainably. Alternatives must also be researched with regard to the use of environmentally harmful machines like excavators or other construction site vehicles.

Sustainability in the context of the 17 UN sustainability goals involves society and the well-being of the individual. In my opinion, sustainability includes equality and equal rights for men and women, and there is room for improvement—not limited to the construction industry.

4.3.5 Summary and Conclusion

In summary, a major change has taken place in Western society in recent years with regard to the role of women.

In the construction industry, the groundwork has already been laid for more women to work in construction and trades through modern technology and a rethinking by the general population has been achieved. But why are there still so few? It is probably due to a combination of factors explained before. Overall, however, it is largely due to people's general attitude towards the trades. The idea of the trades as dirty, hard work in a male-dominated industry with a rough tone is still firmly anchored in the mindset. It often prevents young women from finding their role and defining their way into a trade. Perhaps this is where we should start and do more educational work, especially in secondary schools, and above all in high schools. Mandatory internships might be useful training stage to make students aware of trades, apprenticeships, working skills and give them an idea of what is important in this field of work.

Social media, under whose influence most young people spend a large part of their free time, should be used more actively to encourage people to enter the trades. Personally, when I decided on my career, I was influenced by female "influencers" who work in the masonry profession and share their daily work on social media, felt very encouraged. I think that many young women feel the same way. Otherwise, the craft will die out. And it would be terribly sad to ignore many more, often hidden, craft talents of young women, as has been the case in recent decades, due to lack of information and education.

Personally, I would always choose the same path for myself. The many experiences and events that I have already been able to gather in the few years in my profession, and the contacts that I was able to make, I would not want to miss. It is simply something very special to be able to create with your hands every day anew. I would be happy if women could also be as enthusiastic as me about the masonry profession or other craft professions in the future. Furthermore, I am of course looking forward to future changes in terms of craft, modern technology and in terms of sustainability and working conditions.

4.4 Craftsmanship Can Be Presented as It is: Open and Future-Oriented!

Karolin Röhring (Master Metalworker)

"What do you actually want to do after school?"—This question started my journey into craftsmanship, although at that time I had no idea that it would lead me there.

I was just about to complete my high school diploma at the vocational college and increasingly had to think about what should happen next. Because apart from a two-week internship with a tax consultant and a three-week internship in an audit firm, the mantra of the city high school was anyway "If you have a high school diploma, you must also study", just like that of my parents, so the general direction was already predetermined. I didn't see myself in a lecture hall but in joyful anticipation of my first own apartment and a bit of independence, I finally moved from NRW to Hamburg in 2016 to start studying business administration there.

I enjoy business, even today, but with each day that passed from studying, I had less and less desire to go to university. Although I lived directly opposite and would only have had to cross the street, but still, this happened less and less after the first semester. Over 400 people in the lecture hall and an unstructured daily routine were amongst facts, I could not get along with at all. Sometimes a lecture started at 8:00 am, sometimes at 4:00 pm, sometimes the whole day was free. For this construct, you need a tremendous amount of self-organization and self-discipline and not everyone can do that (including me). Without routine, I really find it hard to be productive and that is still the case today. However, I have now understood that it is not bad if you prefer a clearly structured daily routine. At the time of my studies, I didn't know that and it made me increasingly dissatisfied to see how my fellow students were flourishing in their studies and I was not. But what to do? Sitting around was also not an option and to be supported by my parents for years did not feel like the hoped-for "independence" either.

I was already tempted to start an apprenticeship back then, but my parents were not yet convinced. Of course, I could have also just decided on my own, after all, I was of age, but it is of course nicer if you also have corresponding tailwind for your decisions.

But well, something had to happen. From my internships during school, I knew that one professional field was not an option for me: pure office work. Why did I then start studying business administration? I don't know either.

I have always enjoyed crafting and tinkering, so craftsmanship sounded more interesting to me, but I had no idea which professions belonged to it and what opportunities I would have with it. The offered online career choice tests I did not find really helpful. So I made an appointment with the Chamber of Crafts to get some advice. The consultation itself was so mediocre, but I was given a contact for an internship in an electrical company and decided to call them. The staff were very friendly and so I was able to come for an interview just a few days later. Since it was only about an internship, the process was very straightforward. And then it started! The first day of work. Since I wasn't going to university anyway, I was flexible in terms of time and could start right away. In the workshop, I helped to wire control cabinets for the first two days and was guided by a female certified apprentice who had also previously studied and then decided to do an apprenticeship. I thought "Great! She does that as a woman and has the same background as me. Then that's really an option!".

The rest of the internship I was on the construction sites and in customer service. And now there was nothing that could stop me. I told my parents that I had decided to do an apprenticeship and by then, their openness grew. The question now was: What kind of training should it be? The internship with the electrician was fun and I could have started there too, but I was determined to go back to NRW and in the course of that, I wanted to get a taste of other professions. The second internship was then in a refrigeration and air conditioning company. Also a really exciting time, with nice colleagues, but somehow it wasn't right for me.

Papa then put me in contact with a friendly locksmith and that was my last internship. On my first day I was totally nervous, I came in and no one talked to me. Then I thought "great, how are you going to survive two weeks here?" But with each day it got more pleasant. My colleagues noticed that I really wanted to work and that I wasn't stupid, and so I got more and more interesting tasks. I was allowed to go to the construction sites, try grinding and welding and much more. As I got to know the guys better I also got to know them better, it got funnier with each day and I decided that this is the work I want to do—here in this team. Now the only question was: do they want to train me? After all, it was already September, the start of apprenticeship came closer, so did the start of school.

But everything went well and I was able to sign my training contract there sign. The first day of school was another exciting experience. After I had found the room, 30 boys stared at me confused in the hallway and one said: "I think you are wrong in this room!".

First I thought: "You? Seriously? We are the same age!", rolled my eyes annoyed and replied: "No, I'm not."—Silence.

However, I must say that I only spent one day in the lower apprentice school and then switched to the middle level because I was able to shorten the training time through my high school diploma with A-Level.

There I settled in relatively quickly. Although no one except one colleague spoke to me, but there were no stupid comments, so it was okay. After the intermediate examination, the situation changed: I paid attention in class and was also able to practice for the practical exam at work, so I was quite well prepared. In contrast to some of my comrades, however, finally the ice broke. They asked me about the production of components or—during the breaks of the theory exam—certain questions. Then after the exam, my class teacher said, whether the girl had to show the boys first, how to file, I was then fully accepted, and everyone suddenly spoke to me.

After this class I switched directly to the final level, because I could shorten the training once again. Here from the start, the acceptance was no problem. On the one hand, the boys were very open-minded and there was another girl in the class. On the other hand, I had also become much more confident in the meantime and wouldn't have taken any stupid comments.

There, I had the most fun during my apprenticeship school time. The group was really funny and the cohesion in the entire class was very strong. The whole thing was topped off by a final ski trip, which I still remember very well.

At work, however, it was mixed in between. Although I must say, never with colleagues. After the initial acquaintance I quickly became a solid part of the team and was given many responsible tasks. For example, after some time I was allowed to completely independently build window grilles or small railings. From the cutting of the material at the saw, over the aligning and clamping, all the way to filing and welding and grinding. If I didn't know something, or if the railing turned out a bit crooked, I could always ask my elder certified apprentice for advice. That was the most helpful, as he showed tricks that would never be found in any textbook. And when I was allowed to go to the construction site to install the components, everything was perfect. Nothing fills you with more pride than driving through the city and thinking: "Man, I built that window grille! And we installed the handrails in that house!".

For me personally, that was a great feeling. However, this brings us to the mixed part: the customers.

The majority of the reactions were positive. Many were very interested and especially older people often asked how I came to become a female metalworker, whether I was doing well in the profession and that they thought it was great. I never took questions of this kind as an attack, but rather as openness towards the present time and as genuine interest. That's why I always liked to explain how it came about. Because for older people it is not a matter of course that women work in these trades and that's okay. After all, the role model at that time was quite different from today.

But of course, there were also very conservative people among the customers and in my case, it was actually only men who disagreed with my career choice. And not only older ones, but also younger or same-aged ones. I still remember my favorite sayings: "I believe you have something in your head, but that you have manual skills… you don't look like it now."

For one customer, I welded a gymnastics bar for his child's play equipment and he came to pick it up. After he walked past me purposefully to ask a male colleague about the whereabouts of the bar and I informed him that I could certainly tell him, as I am employed here, he was then thrilled: "It looks great and exactly to measure, super!" That made me proud and I replied: "Thank you, I built it!" Then came: "Ah okay, yes here it is also sloppy work."—Thanks, you A-hole!

But "Can I speak to your colleague" is probably the standard phrase that female craftsmen have to listen to. At least that's what the exchange of experiences says.

However, it was nice that my colleagues always stood 100% behind me and my work. That means, when they noticed that I was being treated unfairly or condescendingly, they always intervened. I appreciate that, as many girls have reported different experiences.

All in all, I have very positive memories of my training. After my second shortening, I completed my apprenticeship in about 1.5 years and then switched to my parents' company.

Shortly before that, I started sharing my workday on social media, specifically on Instagram (https://www.instagram.com/metallbauerin_karo). And that helped me enormously to stand more confidently behind my work. Because: The more pictures I shared, the larger the network became. I made contacts with other women in the craftsmanship and exchanged ideas. It motivated me to see that other girls are also pulling through, even though there is sometimes strong headwind. In addition, more and more young people started writing me wanting to know how such a training actually works—and they were not just boys! This motivated me in staying active on Instagram. The first haven of call for young people today is the internet or social media. This is where information is gathered and a first impression can be won. There is a decisive advantage: there is first-hand information. In almost every professional field, you can find representatives who offer insights into their daily work, answer questions and show challenges that await you in the respective profession.

Especially when one does not dispose of a background in craftsmanship, both parents and relatives are active in other areas, the gap to decide for an apprenticeship in the crafts is quite large. Many would not even consider, as without appropriate role models, no impression can be formed of what one does in craft professions and which career opportunities do exist. Added the still prevalent prejudices.

That's why I think it's incredibly important to use the internet to clear up these misconceptions and create an understanding that the craftsmanship is a modern fields and everyone is welcome. I regularly receive messages from older people, for whom women working in "male professions" was not normal in their youth. For the most part, these are very positive, followed by comments like: "I think it's great that women are doing this now," or "more women should do this, keep it up!" So, a rethinking is taking place and that is necessary. Especially in metal construction, many people have the image in mind: dirty man hits glowing steel with a thick hammer on an anvil. This image may still occur occasionally, but it is likely the exception. The rule is modern tools and highly technologized machines. We also use many CNC-controlled machines, e.g. to punch or bend sheets, which are also operated by women. It simply helps to have the opportunity to see testimonials and pictures from reality, and this is easiest when they are shared online. If I have difficult questions myself, or problems that do not obviously fall into my trade, I often find solutions there. Because in the community there is always someone whose

Fig. 4.13 Author at workplace 2023. *Source* Karolin Röhring

specialty is exactly that. The networking among each other, also across industries, is very useful. It leads to business relationships, the own company can be promoted, young talent can be recruited. The craftsmanship can be presented as what it is: open and future-oriented.

For about 5 years I have now been working in the family business (Fig. 4.13). We focus on sheet metal processing and manufacture noise protection solutions and filter systems for the industry. For the once with deeper interest, they may check https://www.jps-metalldesign.de/metallbauerin-karo.

With me we are four women in production and that feels good. Since modern machines and aids reduce physical strain and no one voluntarily endangers their health anymore, differences in strength hardly play a role anymore. My impression is that a mixed team has a very positive influence on the working atmosphere. The interaction with each other is respectful and friendly, no difference is made among colleagues based on gender and it has a positive external effect. Repeatedly, young women have already applied for a school internship, also with interest in an apprenticeship. This is simply easier when there are already other women as role models in the company. I noticed such positive effects in my internship at the electrical company. So, the hardest part is finding the beginning. But I believe it is doable if openness and respect are lived in the company. Of course, you will still have to make an effort in the search for suitable personnel, but you increase the chances of finding them.

For the future, I wish that we stop portraying apprenticeships and craftsmanship training as a "second class" education path. You are not worth less if you do not study, and you also do not have fewer opportunities. There are incredibly good opportunities for advancement in the crafts and to also address this: you do not necessarily earn less than an academic, even though this is often claimed.

Meanwhile, I have obtained the master title and am employed as deputy operations manager, the further training to become a business economist is still ongoing. Even though I initially had a very hard time deciding against studying, I do not regret it for a second. In the end, everyone has to decide for themselves which path they want to take. The important thing is that we respect this decision and understand that "the child will also become something if it does an apprenticeship" and thus also has good future prospects. But I also see this as a socially, societal, human responsibility and task for schools, to offer information, introduce and explain the various educational paths. And yes, with a high school diploma, one can go to university, but it's not mandatory, dear high school.

4.5 Woman in Craftsmanship—Experiences that Have Shaped My Life

Eva Bork (Carpenter)

"Oh! A carpenter—how interesting! That's great!" The reaction of most people is very positive when I am asked about my profession. However, for many people, it is still unusual for a woman to take up a "man's job".

My name is Eva, I live with my family near Bonn on an old farmstead and about 25 years ago I decided to learn the profession of a carpenter. I look back very positively on this instructive and not always easy time, during which I was truly able to learn for life, but definitely had a lot of fun! I was allowed to learn in a great company, a construction and furniture carpentry near Euskirchen. There I had many valuable insights into the fascinating and extensive profession of carpentry! From furniture construction to drywall, doors, windows and stairs, as well as laying floors and restoring old furniture, I was able to get to know the many facets of this profession. For which I am very grateful!

After 3 years of apprenticeship and one year working in the job, I decided to study at the furniture school in Cologne and continued to work as a carpenter during the semester breaks. Today I am no longer active in this physically very challenging profession, but I daily benefit from my knowledge and what I have learned! A woman can do it herself! Whether it's just about driving a dowel into the wall, renewing the floor of a room, pulling up a new wall or renovating one of our holiday apartments. I think it's great to be so independent in this respect and to be able to do things just the way I imagine them. Even though I of course (almost) always appreciate the help of my husband!

Although most people, as already mentioned, reacted very positively to me (back then it was even more unusual to see a woman on a construction site), there was of course the other faction—who had written on their faces: "A woman?!?" I remember (in retrospect I find these little anecdotes rather amusing) two situations in particular.

Once my boss sent me to a neighboring town, where I was supposed to clad a courtyard gate with new boards. After he had instructed me and discussed the tasks on the construction site with me, he went on his way to let me do my job. Now apparently the (somewhat older and quite conservative) homeowner had internalized that from now on a woman -alone!?- was working on his construction site and apparently had the task of cladding his courtyard gate anew. He quickly decided to get out his folding chair, sit down behind me for safety's sake and watch every move I made. You can probably imagine how "relaxed" it is to work under such circumstances. Fortunately, the gentleman had not many objections apart from a few good tips…

Another time I drove to a customer in the Eifel. The lady had called because the door of her living room cabinet would not close properly anymore. Together with my colleague a few weeks earlier for another repair, I had been there. It was very obvious on our first visit that the lady was very pleased about my young colleague, quite in contrast to the presence of his female companion—I was treated like air. Accordingly, the enthusiasm was great when she joyfully opened the door and instead of the expected colleague, I was standing at the door. After a short greeting, I was allowed to come in and started working. Unfortunately, the cabinet was a bit old and although I adjusted the hinges as far as possible, the problem could not be completely solved. New hinges were needed, which in this case could not simply be replaced, but had to be newly inserted. When I tried to explain this to the lady, I looked into a disapproving face and the answer came promptly: "Aha, so now the hinges are to blame!" I offered her that we could gladly exchange the hinges and I would take the door with me. Or, that she should get in touch with my boss if she wanted to reconsider it.

When I arrived back at the company, my boss already came towards me: "What did you do to that woman? She is beside herself!" The lady had already called the company and apparently made it clear that in her opinion, not the cabinet hinges, but the employee was to blame for the irreparable door and that someone else should please come by to take a look at the situation. I explained the situation to my boss and that new hinges would be needed for the cabinet to close again. He believed me, but to calm the customer down, he sent another -male- employee to the Eifel, who took another look at the cabinet door and confirmed my statement. When the lady heard it from him, she was apparently satisfied… And I thought that such prejudices mainly come from men. But the customer's displeasure that instead of the expected young colleague a woman stood at the door was simply too great that day. Well, so much for the anecdotes from everyday professional life. I could tell many more stories about interest, surprise, great encouragement and various experiences with people in connection with my decision to work as a woman in the trades.

13 Years ago, my husband and i bought a roughly 300-year-old half-timbered farm near bonn. the house was love at first sight!

Here we could really benefit from my previous vocational training. Before moving in, we completely renovated the residential building. For us, the ecological and monument-appropriate aspect was in the foreground. Following the motto:

- As much as necessary and as little as possible!—we set to work modernizing the beautiful old house but preserving its originality and naturalness as much as possible or restoring it (Figs. 4.14, 4.15, 4.16 and 4.17).

Fig. 4.14 After the wall heating elements were laid on the reed insulation,everything could be covered with clay and later turned into a smooth wall surfacePlastered. 2012. *Source* Eva Bork)

Fig. 4.15 The clay plaster is applied to the walls in several layers. The half-timbered beams are a practical ladder replacement! 2012. *Source* Eva Bork)

Fig. 4.16 Wall breakthrough to connect 2 rooms. The upper lintel and thestructure of the plaster show that there had been a door there before,which had been bricked up. 2012. *Source* Eva Bork

Fig. 4.17 The old shutters were sanded down and later repainted. 2012. *Source* Eva Bork

Fig. 4.18 One of my favorite tools! The concreted area was removed and later redesigned with cobblestones. 2012. *Source* Eva Bork

We were extremely fortunate to have had the support of excellent and helpful craftsmen and a great civil engineer who assisted us with coordination. The "eco-construction industry" was uncharted territory for us at that time. Over the following months, we learned how to work with clay and clay bricks, different clay plasters and paints, renovated and restored ceilings, walls and floors, laid heating pipes and cables, and installed horizontal barriers (We nominated the term "horizontal barrier" as our personal non-word of the year 2013. Endless hours of work! And apart from a strip of protective film that protects the masonry from moisture, you could see absolutely nothing. No obvious construction progress…We were very happy when this construction phase was over!).

And at all points where we didn't know how to proceed, we were properly instructed and trained so that we could carry out the next work steps independently. That was a great experience! From the "normal" construction industry, we were not used to craftsmen passing on their knowledge and supporting in such a great way!

After about 1.5 years of construction, we moved in and over the following years gradually renovated the associated 4 apartments. There were always smaller and larger construction sites. Such as the redesign of the courtyard (Fig. 4.18), the renewal of the barn and house roof (Fig. 4.19), the demolition and new construction of an old shed in the side building, or the plastering of the facade.

During this time, I was again able to gain experience about the still somewhat "strange phenomenon—woman on the construction site". For example, the roofer who was pleased that "my husband" had already completely uncovered the roof so that he could start his work directly. My statement "not only my husband" clearly confused him and it took a while for him to realize that I had also been on the roof.

I believe most people think it's great when women work in craftsmanship. My personal feeling is that men in particular sometimes feel unsettled by this. Over the years, I

Fig. 4.19 While uncovering the house roof. Autumn 2019. *Source* Eva Bork

have repeatedly found that it is important to clearly communicate your point of view. As in other professions, it is very easy to talk past each other. In such "male professions", as a woman you are almost always appreciated, but the other gender usually prefers to appear a little superior and all-knowing. Because even when it was obvious that "Man" did not know the answer to one or another question -which is completely legitimate and normal- or did not understand my question or statement—I almost always got an answer with a lot of knowledge, no matter whether it fit the question!

I would find it great if in the future even more the professional competence would be in the foreground instead of the gender. Good and bad craftsmanship exists in both males and females!

4.6 Never Give up—Obstacles Make You Strong. Dare to Try New Things with Courage!

Dipl.-Ing. (FH) Sonja Ebling

As a child, I was more interested in manual work and building with Lego bricks, which led to the creation of the first houses and bridges. In the surrounding gardens, we built waterways from rivulets after rainfalls. We also built dams and small bridges from stones and branches along the streams in the area (Fig. 4.20). In school, I would have liked to do crafts, but to everyone's dismay, I had to try my luck at knitting, crocheting, and embroidery, with rather moderate success.

I was encouraged outside of school by my family, my jigsaw was a gift from my grandfather and at my father's workbench, I was able to gain experience with saws, hammers, and clamps, etc. I was also very interested in technical drawings, especially the ink drawings of my great-grandfather (some of them from the 1890s) fascinated me. I created smaller, sometimes elaborate technical drawings early on. After graduating from high school, I didn't know where my further path should lead. So, the apprenticeship was supposed to help me make a decision, the career counselor recommended a training either as a carpenter or a construction draftsman. Now the question was crafts-

Fig. 4.20 Sonja at 2.5 years old, 1975, on the construction site for her parents' house extension. *Source* Sonja Ebling

manship or technical drawing, drawing won the race, that was the first step towards civil engineering. After several discussions—and some significant resistance—I started training as a road construction technician in the public service. The image of authorities was rather poor at this time.

The training included not only the creation of technical drawings but also complex calculations and the surveying of roads and buildings. My trainer proudly showed us his then very innovative calculator, which could calculate a road axis overnight and also graphically display it. These were my first encounters with the digital planning world in road construction. We didn't quite see the great advantages of digital processing, we were to be taught better. We spent the best times in our training and further education center, in addition to theoretical lessons, sections on the training construction site were also planned, here everyone had to participate. Making road edges, setting curbstones, paving work, but also masonry work and the formwork of concrete lintels were on the program, not everything was so easy, but it was fun (Fig. 4.21).

I was able to complete the training after three years as the best graduate in my year. I was then transferred from Frankfurt to Darmstadt. There I initially drew by hand and carried out the first small calculations on the PC. In the further course, I was able to familiarize myself with the CAD program for developing road designs in location, height, and cross-section.

Thanks to the support of my first supervisor, I was able to develop quickly. Unfortunately, he moved to another location, so that I could also switch to a new field of activity after his departure. Due to my quick comprehension and the support and encouragement of a committed engineer, I was able to further expand and deepen my skills. With a lot of patience and persuasion, my two "Mentors" convinced me that there is more to me than just a technician. Unfortunately, in addition to my two advocates, there were many who made it very clear to me that I would not be able to com-

Fig. 4.21 Sonja (right) on the training construction site. *Source* Sonja Ebling

plete my studies and would give up after 1–2 semesters. Nevertheless, at the age of almost 27, I decided to enroll in the "International Civil Engineering" program at the University of Applied Sciences Mainz. I still remember the first day, excited about what awaited me and especially who would be sharing the school bench with me for the next few years. Of course, I was by far the oldest in the course, but what surprised and pleased me was the relatively high proportion of women. At the beginning of my studies, I had to overcome many hurdles. My school days were almost 8 years behind me and learning was difficult for me. Nevertheless, I did not want to give up my studies, just to show my predominantly male and very critical colleagues that I can and will succeed. This was also my motivator to continue my studies even in challenging times. In addition to my studies, I initially created the financial basis for my study section abroad by working in an engineering office, then at the cash register of a gas station due to the tight schedule. During my studies, I met many interesting people. Some of them accompanied me during my studies, some are still a fixed part of my life. In addition to the consistently positive feedback from my professors, we formed a small study group and supported each other, especially in the preparation of exams (especially concrete and steel construction). My ambitious goal of successfully completing all exams in Mainz before my stay abroad in Dundee, Scotland, I was able to achieve with this support.

Initially in Scotland, despite the preparatory English courses during my studies in Mainz, there were language barriers to overcome. Especially the different dialects—from Scottish, Irish, Welsh, Nigerian but also Greek and Spanish influenced accents—at least initially shaped everyday life both privately and in studies. As a result, I was able to successfully complete my studies with a "Bachelor of Civil Engineering", including exams in areas such as geotechnics, static proofs of oil platforms, and wastewater technology.

The diploma thesis was then the conclusion of my study time. In summary, studying in Mainz and Dundee was a very important and correct decision for me.

Even today, a half 5 €—note in my wallet reminds me of one of many beautiful hours during my study time. Together with my friend Bianca, we celebrated our last successful exams in an Italian restaurant. When paying, the waiter simply tore our change, the aforementioned 5 €—note, in half and handed each of us with the comment "You will never forget this evening". After the initial shock, 5 € was a lot of money for us students, we were very pleased with this idea and still like to think back to this evening and the good times. Even while writing my diploma thesis, a colleague drew my attention to a job advertisement for an engineer in Darmstadt. Initially, I had other plans for my professional career after successful studies, but the entry opportunities for civil engineers were low, there were at least 50 to 100 partly experienced applicants for a job advertisement. So I applied and was able to convince my superiors in the interview that I am the most suitable candidate for this engineering position in road planning. In addition to my own planning of cycle paths and conversion of junctions, I processed the award of planning services for a bypass. In addition, I was given special tasks, in addition to qualifying as a safety auditor, I was able to deal intensively with the topic of building law creation, thus further steps for my future were set. Over time, I was able to gain in-depth knowledge in all subject areas, the PC-based programs were now a major part of my daily work. In addition to the CAD program for design processing, the first controlling instruments were also already being used to document and optimize the project process. After just 4 years in professional life as a civil engineer, I was given the leadership of a small team with a focus on creating building law. Not all of my four male team members made it easy for me to start as a young female leader, after all, most of the team was significantly older than me. The beginning was certainly not easy, but as is well known, every beginning is harder. Over time, both the team and the range of tasks has increased. In addition to creating building law, now also route planning was part of the team's tasks.

The range of tasks of planning and creating building law in the course of federal, state, and county roads was expanded with the next restructuring to include topics related to highways in Hesse. All employees now had to take over the complete task portfolio. Due to my extensive professional experience and my career, I was able to challenge and promote each individual employee—no matter in which position. With the last restructuring, I took over the management of a specialist department for planning for the South Hesse region with a total of 32 employees, spread over three departments at two locations. The tasks are still diverse, only the highway planning is no longer part of our task field. The focus of my job is now the approval of planning projects, technical exchange with third parties, project control and financing as well as personnel responsibility, which I share with my three department heads.

In recent decades, the world of work has changed a lot, digitalization is increasingly becoming part of everyday life with all its advantages and certainly also disadvantages.

Planning is becoming increasingly complex and elaborate, the CAD-based programs support especially in duplication, but are unfortunately still not self-explanatory and easier in handling. Time planning with MS-Project represents possibilities for project control, but cannot replace the important togetherness. The next stages of project planning with the BIM methodology is added to the agenda in road construction, there will certainly be advantages and disadvantages to recognize.

The compatibility of work and family is developing in clearly positive directions, but also the opportunities especially for women to hold leadership positions, has sent clear signals.

Currently, I can look back on 32 years in professional life, my career from training as a road construction technician to the leader of a successful and highly motivated team with 32 employees is incredible, but at the same time a good sign that with personal commitment and support from colleagues and superiors a lot is possible. A big "Thank you" for that.

Therefore, I can encourage everyone to have the courage to try new things, to believe in themselves and to accept and offer support.

4.7 Craftsmanship Has a Golden Foundation

Jule Janson (Concrete and Reinforced Concrete Builder, Dipl.-Ing. Civil Engineer) in an interview with Bianca Weber-Lewerenz on February 28, 2024

My Story……is shaped by my passion for craftsmanship, for working outdoors with daily new encounters.

What part of the building did you work on today? *Oh dear, today it wasn't such a big, exciting thing. The construction crisis is tangible and there are many of small construction projects these days. Today I was busy with the fire protection modification of windows and masonry. This is also part of our daily job in concrete and reinforced concrete construction.*

What was your path to construction, your professional stations such as training, jobs up to your current activity and function? *I graduated from high school in 2017. Immediately afterwards, from 2017–2019, I completed my apprenticeship as a concrete and reinforced concrete builder. Despite many hurdles, a lot of headwind against a high school graduate, who decides to do a craft apprenticeship, and the lack of orientation at the gymnasium through career counseling on training paths, I followed my heart. With the successful apprentice degree in my pocket, I continued from 2019–2023 with my studies to become a civil engineer (FH). I chose a university of applied sciences also because it is the most practice-oriented. The period of Covid was "optimal" for me during this phase, as I could pursue my work-study job and various jobs and earn money. All in all,*

Fig. 4.22 Jule Janson—Campaign ambassador of Handwerk.de. *Source* https://www.handwerk.de/infos-zur-ausbildung/berufsinsider/jule-janson

it was clear to me at this point that my workplace must be outdoors and an office job is out of the question for me. Since 2023 I have been working as a concrete and reinforced concrete builder (Fig. 4.22) *in my hometown Mühlacker, Baden-Württemberg. Here I am employed in my mother's company and very proud of the construction company originally founded by my grandfather. For me, the saying goes: "Craftsmanship has a golden foundation". That's why I work with full passion in the daily versatile craft job, no day is like the other, you constantly meet new people, my work depends on the work of so many different trades and is mutually dependent, the weather changes and you have to be flexible on the construction site. But the fulfillment that I feel, I wouldn't trade for anything.*

What does diversity in construction mean to me? *The following thoughts come to mind spontaneously: Every day there is a different task, the activity, the fields of application, the components and the construction sites are always changing. Often you don't know what's coming up in the following week or where you will be ordered to go, which is why a high flexibility and spontaneous vein is required. I also think of the diversity of technical development of machines in daily work. Mostly, devices with batteries are used instead of devices with electric cables. The electronic mason's platform, which can be operated steplessly at the push of a button, is a very big support. The wall to be built "grows" quasi with the platform. In the past, a lifting platform could only be adjusted manually, now much more efficiently and without having to leave the workplace. Another area is that there should be more women in the crafts, but I am not in favor of boosting this through quotas. Because craftsmanship should inspire, and women should be inspired by the task itself. It should not be in the foreground that women work more in*

Fig. 4.23 Jule Janson passionately involved in construction—Campaign ambassador of Handwerk.de. *Source* https://www.handwerk.de/infos-zur-ausbildung/berufsinsider/jule-janson

the crafts to increase the female quota and only with this objective. That would be the wrong approach.

Do I wish for more diversity, if yes, in which areas? *More women, but as I said, not just because of the quota for women, but more women who can and want to do this work and live it with passion* (Fig. 4.23)*!*

How did I get into my profession in construction? *After school, I was a seeking soul, like many others around me, too. I had no idea about career choice. The high school offered no advice on training paths and other educational opportunities away from university such as the diversity of apprenticeships. This is the reason why I strongly suggest a general rethinking because disparaging the trades or not considering trades at all as an option for high school graduates is far from any career advice and also not contemporary. For the subjects that were interesting to me, mathematics/sports were once up for debate, I was strongly interested in aerospace for a while. The latter only fell away because of the physics load. My love for mathematics led me closer and closer to technology. After high school, I turned directly to manual work, which my family business on my mother's side, a construction company, made significantly easier.*

What were the highs, highlights, and lows? *The absolute gigantic highlight so far was the award of the silver medal! I won this at the "WorldSkills"—competition*[1] *that bases ona global world association and organization that announces a competition among 50 skills. The skills go beyond construction trades and include everything that belongs to Industry 4.0. This is pure diversity! Lived craft diversity! Incredibly inspiring are the many trips and meetings with the same age group. There, we all share the same goal: Winning! WorldSkills Germany offers innovation, promotes young talent and career development. The non-profit association promotes talents and education. The professional competition not only sends signals into the industry with the award, but also boosts one's own ego. I am incredibly proud of my silver medal! The highlights of my professional career include my training and everything that offered itself afterwards with competitions, diverse opportunities, especially the diversity of human encounters. As for the lows, I can think of one: at the beginning of the training I reached my limits, doubted, is it the right thing or not. Then everything was so new, the whole environment, the tasks, the expectations and the people on the construction site. Today, looking back, I can say that dealing with stupid remarks, which women often get, becomes easier with increasing professional activity, it bounces off you and I can easily give a counter.*

What are anecdotes that I like to remember? *There are no specific anecdotes. But it was simply incredibly exciting, historic and unforgettable that I won the "WorldSkills" silver medal.*[2] *The numerous competitions after the journeyman's examination, the further promotions after many wins to the German Championship, and then straight into the national team of the German construction industry! What a feeling!*

Which words best describe me? *To put it in a nutshell: I was ambitious as a child, I'm not shy about speaking my mind, I'm quick-witted. Maybe this even intensified through my training. Who knows?*

What qualities have always helped me in construction? *The inner ambition, and the fact that I told myself right from the start: I'm going to finish this, especially during my high school years I wanted to prove it to myself and of course to everyone else around me, who had less confidence or simply didn't think much of the craft. It's important to state this publicly: unfortunately, high schools do not provide information and advice about different training paths, professions in the craft. Construction is still too often, in my experience, bad-mouthed by teachers from the start. Construction is portrayed as a loser.*

---

[1] https://worldskillsgermany.com

[2] https://worldskillsgermany.com/praktisch-unschlagbar-beton-und-stahlbetonbauerin-jule-janson-kennt-ihr-handwerk-aus-der-praxis-und-nicht-nur-in-der-theorie

Were there mentors or special people in my life who motivated me? Yes, there were and still are. My parents were skeptical at first. Then, when it worked, they gained more confidence and my sister encouraged me. My former training boss was an absolute pillar, a motivation machine. He never doubted me and my abilities. "Go go go" was the motto.

How are digital transformation and entrepreneurial culture linked? To answer this question, I would rather refer to larger companies than our construction company. Our family business is too small with 8 people, I think. It is generally a very interesting, exciting topic, and important to drive the construction industry as such innovatively and strengthen its competitiveness. We in our small business have no concept for this so far. For us, people, values, and togetherness come first. We are also continuously looking for new ways to work more efficiently, productively, structured, and without data loss.

What Opportunities Does the Digital Transformation (BIM, AI, Metaverse…) Offer? *Generally speaking, the benefits are obvious, and there are some who have already had good experiences with it and continue to do so. My personal view is that this change takes a lot of time for knowledge, application fields, and the establishment and guarantee of a necessary infrastructure. Digital technologies and AI facilitate agreements between trades, enable a significant improvement in communication. I notice that the digital change and innovations in construction improve the image of the construction industry as a whole. Everything is becoming more digital, construction is highly conservative and is considered a latecomer. By actively shaping and tackling this change, the construction industry also sends an important signal to the outside world, namely: Construction is moving forward!*

How is sustainability being implemented in construction? (Reuse of recycled material, Cradle to Cradle, use of BIM and risk management, to avoid material waste) *I see a general urgent need for action in construction in the area of environmental protection. But, the disposal costs are skyrocketing, how are we supposed to pass these on in the company? Building is becoming increasingly expensive, but more expensive due to the increasing bureaucracy to cope with environmental protection regulations. The effort in offices due to this bureaucracy is so infinitely high that one must distinguish between sensible and nonsensical here.*

How can the shortage of skilled workers be countered? *The main problem lies in the image of construction from the perspective of our society. The external reputation is bad, there is too much opposition. The outflow and the attractiveness towards professions and training in the industry is noticeable. Much more needs to be done to counteract and steer against this in order to urgently win new skilled workers in construction. I am happy to take the image campaign of the ZDH*[3] *as an example, as I am involved*

[3] https://www.zdh.de/ueber-uns/imagekampagne-handwerk/

myself: there is a recognition effect, it visualizes professions in construction, it brings them closer, inspires. The option "construction" comes across as innovative, cool and clever. The construction association has used this image campaign again and again, also the construction guild in the district, so the campaign remains sustainable.

Which networks help those who are still unsure, but are interested in technical professions, e.g. in construction, to find their way there? *There are some female influencers who give insights into their professional daily life using their Instagram profiles. There are often exciting stories about the people themselves, their individual paths to the craftsmanship job, the typical everyday environment, processes, questions from outside. I was also involved here at the very beginning, but I have to say that it is not manageable due to time constraints and rather annoying. There are enough other influencers in the craft.*

Note: Otherwise, for all those who have more and in-depth interest, there are professional portals[4] and association websites like ZDH, which report on the diverse developments in the craft, put women in the craft in focus[5] and with "Small hands, big future" advocate for the next generation from childhood for construction professions[6], associations that strengthen modern craft, blogs, TikTok and Instagram profiles, hands-on ministers at craft fairs[7], and future dialogues on craft. Because: Change can only be shaped together.

4.8 After the Studies Was Before the Apprenticeship—Un Desafío Maravilloso

Antonella Menrath (B.A. Romance Studies & Economics and Plant Mechanic) in an interview with Bianca Weber-Lewerenz on October 31, 2023 and January 3, 2024, closely at the end of her apprenticeship graduation in the field of heating ventilation plumbing.

Who Am I, What is My Background?

Currently, I (Fig. 4.24) *am completing my apprenticeship as a plant mechanic at the company SHK Friedrich Morsch GmbH & Co. KG. Before that, I studied Romance Studies & Economics and earned my Bachelor of Arts degree. Some might ask, why on earth,*

[4] https://www.bauberufe.net/beton-stahlbetonbauer.html

[5] https://www.zdh.de/ueber-uns/fachbereich-soziale-sicherung/frauen-im-handwerk/#c4680

[6] https://www.zdh.de/ueber-uns/aktion-modernes-handwerk/bundesweiter-kita-wettbewerb-startet-in-sein-10-jubilaeumsjahr/

[7] https://worldskillsgermany.com/vizekanzler-und-ministerpraesident-mitteldeutschen-handwerks-messe/

Fig. 4.24 The author Antonella Menrath, 2023. *Source* Agency Malix Mannheim

why a training after studying? For me, it makes complete sense, because the path to construction is only logical for me and suits me as a person. Question in return: does any of the readers speak Spanish as fluently as I do?

« My Story» is one that does not follow a regular red thread, so I am all the more excited about the interview questions and how I can inspire others for the construction industry.

What Does Diversity in Construction Mean to Me?

The first things that come to mind are: human diversity, women in construction, diversity of data, creativity in construction, diversity of project sizes ranging from single-family homes to large to mega construction projects, especially in plant construction, where I work. All have one thing in common: here, diversity is lived. Construction has an incredible diversity of trades, and has more than one sector. I would like to make a critical remark right at the beginning: I wish for much more diversity in construction. This can be achieved by additional motivation for women and by inspiring them for jobs in construction. The diversity in the training paths is enormous. However, more PR and advertising for the second educational path is needed. Here, role models, influencers for example with appearances on social networks and video portals like Instagram, TikTok can help.

How Did I Get into the Construction Trade?

The "big bang" comes from my father, who himself learned the profession of plant mechanic. He then took over a company and as a child, I grew into it. The time of

Corona and home office work were the reason for me to realize: "I want to do something with my hands and above all get out of the office and out to the customer". The big advantage was that I could get a taste of it in my father's company.

What Were the Highs and Lows?

For me, the highlights include every single project success, every completion and satisfied customers.

I can briefly outline the lows in the job: every beginning is hard, school and cramming theory in indoor spaces. It is therefore all the more difficult to have to go back out to the construction site afterwards—in wind and weather. Of course, everyone knows, every beginning is hard, you don't know your way around, everything is new.

Anecdotes that I like to Remember?

I can't think of a specific one. But when a customer opens the door, she or he is often surprised: "Oh, a woman is coming, I've never had that before." My conclusion is, they are all consistently positively surprised and happy when there is a craftswoman in the house who successfully completes the repair. On Instagram, for example, I read feedback like "Hey, I also started in construction." This quickly leads to a motivating multiplier effect and I feel once again that the success is there and the work is worth it (Fig. 4.25).

Fig. 4.25 Antonella Menrath in her everyday construction site life, 2023. *Source* Agency Malix Mannheim

Which Words Describe Me Best?
In short: reliable, helpful, curious, i like to question, love to learn new things, like to prove myself.

What Qualities Have Always Helped Me in Construction?
Being open, flexible, actively contributing ideas, being active and delivering concrete suggestions.

Were There Mentors or Special People in My Life Who Motivated Me?
Yes, indeed there is a role model, namely Sandra Hunke, who is very well known in Germany and has made a strong impact as an influencer with her sharp Instagram profile and commitment. She has been in the profession for a long time, more than 10 years, and is simply an all-around cool personality, I think.

What does it take to balance woman—construction—starting a family—further career? *In my view, stereotypes in the profession should generally be cleared up. Construction sites and work in construction have changed a lot, it's not just about working in the dirt and only doing heavy physical work. No, new technologies bring innovation, simplification in human work, individual processes are much more efficient and productive, for example, systems can be read out with a laptop. Working in the dirt is partly true, yes, but the profession has gained so much diversity that it offers many advantages to women. These are exactly the areas of responsibility where women have a greater strength when it comes to using an eye for aesthetics, designing, looking at what families and female needs are and how these can be implemented in living and working spaces. Instagram helps to report transparently about this, to create public access, to stimulate communicative exchange. New fields of work, away from the construction site, are emerging: product training and training, designing with the digital twin. There are so many opportunities to discover a job, a field of application for yourself as a woman, to tackle MINT professions and to develop professionally. A topic that can be endlessly discussed and concrete approaches can be found.*

What Opportunities Does the Digital Transformation (BIM, AI, Metaverse..) Offer?
For my field of activity, i see many opportunities: efficiency, direct start, faster connections with those involved in construction.

There are endless opportunities and potentials for the construction industry in general: in addition to the flow of information and data in exchange with colleagues, these technologies offer a lot of relief in the job. They enable even more visualization and transparency, visual inspections, online measurements, the direct provision of all Data in the office for further processing (commercial processing, incorporation of changes and submission of supplements, etc.*). All in all, this allows for a quick direct start, work optimally aligned with the timeline, and holistic efficiency.*

What conditions are needed in construction that have so far been overlooked? *Great question, it immediately and spontaneously comes to mind that toilets are missing on construction sites. Especially on small construction sites, because men urinate standing somewhere and it is assumed in the construction site preparation that 99% of the people on site are men! But no one expects female staff, the craftswoman and technician! In addition, I see great potential in the digital processing of customer orders. Small craft businesses are still fearful, but I see the trend: the younger the employees, the higher the curiosity about new technologies in the company.*

Why is Construction Lagging Behind in Artificial Intelligence (AI)?
The construction industry is an industry with many employees still "of the old school". They are top fitters who deliver great performance, but from their inner attitude they are rather conservative, old-fashioned. There is a lack of new audience that demands innovation in craftsmanship and new technologies, methods, more efficient and productive ways of working. The shortage of skilled workers speaks for itself, no wonder.

On a Scale of 1–10, How Willing Do You Think Construction Companies Are to Integrate Digitization?
I would guess 6.

How High—on a Scale of 1–10—is the Need for Clarification About Digitization and AI?
I estimate 8, because many still don't know what is possible with digitization and ai. they are rather rejecting and not familiar with the advantages and the added value.

What limits and risks does a company encounter when implementing digitization or what opportunities arise? (Technology, people and skilled personnel, corporate structure, ethics, mission statement, law, politics, etc.):
Typical limits are not only in the head and the general thinking, but simply and straightforwardly limits in the practical implementation during the working day. It requires a high effort in terms of time, the employees have to familiarize themselves in addition to their regular work. In my company, for example, an external company offered invoice program training. And we have specifically one person for digital tasks in the company.

How can sustainability in construction (building, resources, climate, environment, society, human, well-being, diversity, education etc.) be achieved (keyword "17 UN sustainability goals")?
Sustainability on construction sites is becoming more and more of a topic. This involves questions about reducing the immensely high waste and construction waste production, waste separation is increasingly and more strongly observed, in plant construction spe-

cifically it is about collecting the old dismantled pumps for recycling. How does sustainability manifest itself in my daily work? Here is another approach, to specifically adjust heat pumps so that they run ideally and reduce consumption. Special experts come to the construction site, who check the installed system after completion. The control works via *Bluetooth.*

When Talking About Trends, Tendencies in the Construction Industry in the Digital Age, Where Would a New Thinking and Learning Culture Help?
This hits the nerve of the times, as there is a lot of catching up to do. The recruitment of employees and skilled workers in the construction industry is a huge topic, the existing shortage of skilled workers is the biggest obstacle, as well as the generation and training of new knowledge. LinkedIn, social media should be used much more, these are great quick access to knowledge for everyone, from anywhere and at any time.

How Can the Shortage of Skilled Workers Be Countered?
Perhaps by promoting apprenticeships and trainings via *social media. But even there, the company needs experts who exclusively deal with it and know it well. From my point of view, this is the only way to pave the way and gain access to young people and to more enthusiasm.*

If You Look Around in Your Environment (Family, Friends, Colleagues, Construction in General), How is the Enthusiasm for STEM Jobs (Mathematical, Scientific, Technical)?
I live in a craftsman family environment, which brings a high appreciation for craftsman professions. Even those outside the family, so in the circle of friends and acquaintances, are very enthusiastic and have few prejudices.

Do You Use Networks, Initiatives, Blogs, TikTok, Instagram?
Yes, I am a very specific Instagram user, have my own profile there, to inspire for the craft profession, to provide access to everyday professional life and to dispel prejudices (Fig. 4.26).

Fig. 4.26 Antonella Menrath passionate about recruitment, STEM and women in construction, 2023. *Source* Agency Malix Mannheim

4.9 Inside the Office, Outside on the Construction Site: These Are Worlds Apart!

Leonie Häuser (Master in Road Construction Craft) in an interview with Bianca Weber-Lewerenz on December 22, 2023

My Story…. How Did I Get into the Construction Trade?

Let's go, when it comes to looking behind the scenes of this wonderful job, or inspiring women for this profession, i am on fire.

My profession led me (Fig. 4.27 and 4.28) *through various stations, such as college entrance qualification, completed training as a retail saleswoman and active in this field from 2008 to 2011. My professional development quickly brought me more responsibility and leadership as a department manager in the food sector with 10 people in the period 2010–11. From 2011 to 2018 I was a commercial employee in heating, plumbing, ventilation and air conditioning technology. From 2018 I worked in the commercial project processing at the company KTS Bauunternehmung GmbH. Subsequently, I got the opportunity to enter construction management. In 2023, I achieved the "icing on the cake" with my master's degree in road construction craftsmanship, where in the same*

Fig. 4.27 Leonie Häuser, 2024. First female forewoman in Baden-Württemberg in 2022. *Source* Leonie Häuser

Fig. 4.28 Leonie Häuser, 2024. *Source* Leonie Häuser

breath the construction and project management followed. Currently, I am interested in real estate and project investments.

What Does Diversity in Construction Mean to Me?

From my point of view, an incredibly complex field: the diversity of various tasks, challenges, the involved experts and participants, every day there is something new, no day, no weather conditions are the same. The problems that need to be solved are diverse, the many contacts with the most diverse people and authorities, the fields of activity. I also immediately think of the internally so diverse and complex interfaces, the different trades. The diversity of materials is endless.

Do I Wish for Even More Diversity, if Yes, in Which Areas?

Oh yes, what a question! In terms of digitization, we are quite backward. But sometimes less is more. It's like a mission to complete a project with high customer satisfaction, quality, technically correct, within budget, economically good with resources handled, entrepreneurially productive, well calculated and successful. It requires people who are motivated to implement this technically and in good personal cooperation.

Here an example immediately comes to mind: the laying of fiber optic cables is subsidized by the federal government, i.e. all requirements must be met. To ensure this, good cooperation among the participants, smooth and good communication is needed, leading to a successful conclusion, where the overall project implementation is economically fit.

What Were the Highs and Lows?

The highs and highlights I see in my work are in the self-initiative and successful co-design and software improvement, as well as automated processes at KTS. For me, good internal communication and smooth working also play a big role. Thus, successful completed construction measures result in benefits across all department boundaries over and the shared pride. A personal highlight was my master's degree.

For me, the highs also include lows; I see these in the general crises that always arise when something is not right in interpersonal relationships, or when deviating from the contractual planning—with the goal of achieving even higher economic efficiency, so just for the sake of profit.

Anecdotes that I like to Remember?

There is no specific one. I just find it crazy, at first there were many women working in the office, and now increasingly outside on the construction site. These are simply different worlds!

Which Words Best Describe Me?

Determined, sociable, always solution-oriented, always keep going and accept setbacks, resilient, always thinking holistically, considering the big picture, not afraid of challenges.

What Qualities Have Always Helped Me in Construction?

As already mentioned above, and also an open interaction with each other, or tackling and communicating a topic.

Were There Sponsors or Special People in My Life Who Motivated Me?
Yes, and a very important one in my personal professional development: the managing director of KTS Bauunternehmung GmbH, Ivica Katic. He saw my will and ambition from the beginning, he was regarding my own advancement T H E contact point—as a sponsor, mentor, great entrepreneur and outstanding person.

What does it take to balance woman—construction—starting a family—further career? *The first thing that comes to mind is that as an employer, childcare must be arranged flexibly, that successful coordination with the partner contributes. For me it implies the question itself: everyone has to find their own way.*

What Opportunities Does the Digital Transformation (BIM, AI, Metaverse.) Offer?
For my field of activity, this means much more efficient work and significant time and cost savings.

In relation to the construction industry in general: new impulses, if everything works and the added value and advantages are recognized (e.g. cloud, internet reception). The opportunities also lie in the team, where everyone can participate but also needs to be "picked up", that means curiosity and technical affinity need to be strengthened. It is important to strengthen the will and readiness—depending on the affiliation to the respective generation, age. This in turn depends on the inner attitude. If management communicates the advantages specifically, it can strongly motivate teams.

What Conditions Are Needed in Construction that Have so Far Been Neglected?
Every change process takes time. The next question immediately arises: Is the investment worth it? Is it demanded of me as a company? Because effort and benefit must be right. The company KTS, I worked for, wants to drive forward, be innovative, go new ways. It requires a look at the time factor, dealing with all interfaces.

Why is Construction Lagging Behind in Artificial Intelligence (AI)?
Partly the reason lies in the rejection attitude, or it is a generational problem. Above all: change takes time. Changes in legislation and economic conditions also change the need. There are always pioneers, but laws and guidelines such as funding guidelines, file formats dictate the way.

How High—on a Scale of 1–10—Do You Estimate the Willingness of Construction Companies to Integrate Digitization?
In large companies, i estimate about 8–10, in smes the willingness can also be very high, but the normal daily business has priority there.

How high—on a scale of 1–10—is the need for clarification on digitization and AI? *Trade fairs and events offer widespread clarification, access to new and more* *knowledge. I estimate 6–7. It depends on the user, there are plenty of offers, it just needs to be used.*

What limits and risks does a company encounter in the implementation of digitization or what opportunities arise? (Technology, people and skilled personnel, corporate structure, ethics, mission statement, law, politics, etc.)
Here I clearly see the language barrier between the different nationalities in construction, which goes hand in hand with understanding in reading and documenting. Unfortunately, there is a high disinterest and a refusal to engage. There is also a lack of a legal framework with clear guidelines.

How can sustainability in construction (building, resources, climate, environment, society, human, well-being, diversity, education, etc.) be achieved (keyword "17 UN sustainability goals")?
Here i think of the control of the disposal of material, the clean separation of the excavation and a generally necessary and practiced cycle thinking in construction.

How is Sustainability Implemented in Road Construction? (Reuse of Recycling Material, Cradle to Cradle, Use of BIM and Risk Management, so no Material Waste)
If the Legislation Clearly Stipulates that ....% of Recycling Material Must Be Reused.

How Can the Shortage of Skilled Workers Be Countered?
One of the most important points at the moment: advertising and orientation towards apprenticeships in the trades should start in schools. It is important that the advantages of construction jobs are highlighted, that information about salary, performance-related benefits and flexible working time offers are clearly and transparently communicated.

If You Look Around in Your Environment (Family, Friends, Colleagues, Construction in General), How is the Enthusiasm for STEM Professions (Mathematical, Scientific, Technical)?
Here everything revolves around the fun of construction. And that depends on the style of upbringing, are you outside a lot, a lot in the forest? Do you have a strong connection to nature? Does the weather not matter to you, because you know, that the right clothing is the most important thing? If storm, snow and rain don't matter, nothing stands in the way of later outdoor work. Personal memories of my grandpa come to mind with these thoughts: I was out with him on the sled. My positive childhood memories seamlessly transitioned to considerations like, how do we deal with the world, how do we want to world....

And i like to redesign, even my own path (Fig. 4.29) *self-realization, networking, and building additional portfolios are very important to me.*

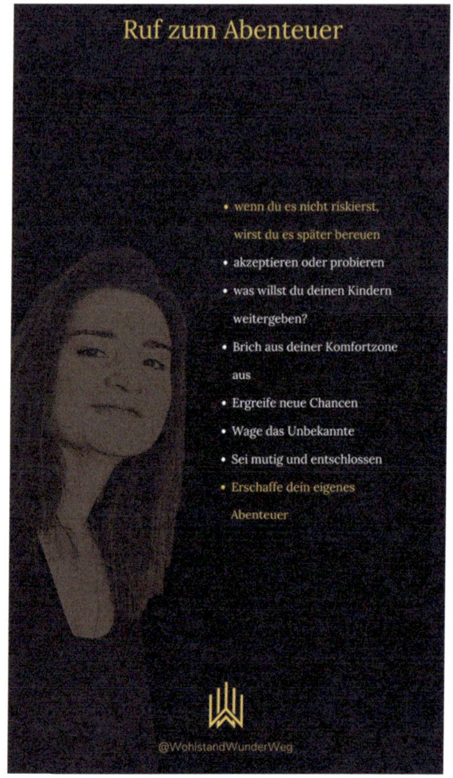

Fig. 4.29 New Instagram profile "WealthWonderWay", March 2024. *Source* Leonie Häuser

4.10 Nothing is Impossible—with Drive in Construction

Bianca Weber-Lewerenz (Dr.-Ing. Dipl.-Ing. (FH) Civil Engineer and Bricklayer)

The cut came with the high school graduation: "Finally, real life begins", free choice of profession, free choice of location for training and study, going out, dancing, new friendships. Self-confident, full of passion, empathetic, respectful, tolerant, creative, I faced the agony of choice. I immediately tackled the pre-internships on the construction site and in the architectural office, which were a prerequisite for my 1st choice, studying architecture. I quickly realized that working outdoors, a high practical relevance, building walls, concreting ceilings, supervising a construction site met my wishes more than architectural ideas, drawing plans, and CAD work. The decision for studying civil engineering was therefore quickly made, specifically for the highly practically oriented type of University of Applied Sciences. There was one hurdle: in 1997, university admission was still subject to a numerus clausus, which is why I fully utilized the waiting time with an apprenticeship in masonry, as a bricklayer. As a high school graduate, I had the advantage of completing this in a shortened 2 years and adding the parallel training as a "business assistant in the craftsmanship" during this time. The next hurdle did not wait long: which

Fig. 4.30 Apprentice Bianca with foreman Ulli, construction site for new primary school, Ballrechten-Dottingen, 1996. *Source* Bianca Weber-Lewerenz

company will take me on for the apprenticeship? Women in construction were still very rare at that time, a woman as a bricklayer hardly imaginable. Construction companies were biased, as regulations were not clear enough. A phone call with the Federal Ministry of Labor and Social Affairs—still in Bonn in 1995—was necessary to confirm that part of the workplace ordinance for construction site facilities, namely §4, "Toilets and changing rooms must be set up separately for men and women", did not apply for the duration of my training by mutual agreement with the contractual craft business. I thought, what do I need my own toilet for a few seconds of use spread over the day? And changing clothes before or after the men is also feasible. How did I find the craft business that contractually accepted me as a woman for this previously purely male-dominated training profession? I remember the big, joyful eyes of the foreman (Fig. 4.30), who already supported me for the pre-internship in the company Hopp and Hofmann, Freiburg. To be on the safe side, I got additional motivation from him, which I then carried into the meeting with the management in preparation for the apprenticeship contract. Standing in the rain, on a construction site, he said: "You have my full support, when do you start?".

His words confirmed my passion for construction once more: The visible results at the end of the day, construction was the right career choice. And when I sat with the team of the apprenticeship school academy in our lightweight concrete canoes and we mastered the regattas 1996 in Dresden under the Blue Wonder[8] and in Holland, I could no longer imagine a life without construction.

The enthusiasm for nature, being in the great outdoors and coming into contact with a variety of materials, is already in the early stages. As a child and in my youth, I was

[8] www.beton.org/events/concrete-canoe-regatta/history-winners/archive/dresden-1996/

Fig. 4.31 Apprentice Bianca with the bricked round arch journeyman's final piece, Practical Academy Freiburg, 1997. *Source* Bianca Weber-Lewerenz

creative, liked to tinker, with carpenter-grandpa I sanded wood, clamped logs between clamps or in a vise, smelled glue, adhesives, primers, paints, oils, marveled at the variety of tools in the carpenter's bench and on the wall board, sawed, planed, played with the curled shavings, felt the wood grain with the hand—despite splinters—and loved to sweep the workshop for my life, because it smelled so good of wood and I liked to see the dust swirls in the cross light and sunshine. By the way, this complete state with all sensory impressions of hearing-smelling-seeing-feeling is ad hoc retrievable and visible to my inner eye.

With baker-grandpa on my father's side—yes, my father also carried the baking and confisserie passion into the family—in the bake shop, hands in the dough, licking cream from the bowl, dusty with flour up to the neck, or watching in front of the oven how the color of the cookies darkened. From an early age I played a lot in nature, was wind- and weatherproof, came back with blue frozen hands (despite gloves) from hours of sledding and romping in the snow, had a "kitchen landscape" in the garden, created my own garden beds, learned how to press bush bean seeds into the ground, how deep and at what angle to set the asparagus knife and maintain the asparagus mound. It was quite normal to build and fly kites made of wood and parchment paper with my father, to build first natural stone dry walls and to be happy about the lizards, salamanders and slow worms. The early musical education, i.e. the child-friendly introduction to various musical instruments and learning rhythm and beat, crafting and making music for preschool children made everything round. Growing up in the midst of a parental home with many international contacts, friends from other countries, with other skin colors, speaking foreign languages came and went at our place. This is certainly due to my homeland, where France, Switzerland and Germany meet at the tri-border region in South Germany. At home, philosophical, societal, and political questions were and are still eagerly discussed, debated, laughed, critically questioned, visions developed, society co-shaped. Later I was then immersed in my own design ideas and drawings for interior design and drew my own floor plan designs of campers with imaginative equipment.

Fig. 4.32 Site manager Bianca during her site tour with representatives of the upper management of VINCI Construction, Annual Health & Safety event VINCI Construction, construction site US Panzer Barracks, Böblingen, 2004 *Source* Bianca Weber-Lewerenz

A person standing next to a stack of bricks Description automatically generated A person standing next to a brick arch Description automatically generated (Fig. 4.31).

In 1997 I celebrated my apprentice's graduation as the 1st female bricklayer in Baden-Württemberg. From 1997 to 2002 my studies in civil engineering took me from Constance, Mainz, Bielefeld to South Africa. Followed by professional stations and career paths in construction, including at VINCI Construction (Figs. 4.32 and 4.33).

At the end of my studies, it was already clear that I wanted to carry out a research project in the form of a dissertation at some point, my then planned doctoral topic was simply incredibly exciting and endlessly researchable in construction: "Risk Management". However, earning money after studying was the first priority.

From 2007, a new opportunity arose and the absolute turning point in our life flow, my husband and I relocated our life and work center to Beijing, China until 2014 (Figs. 4.34 and 4.35).

Completely crazy and far from the daily construction working routine, I founded my own mobile coffee bar "CoffeeArtCafe" with its own website, offered my service in schools, embassies and at events (Fig. 4.36), provided the coffee and tea stand at every Christmas party of the German Embassy, designed and sold my own postcard, chocolate (with the pastry chef of the Kempinski Hotel Beijing, I created delicious varieties) and Rooibos tea lines, wrote my first book, completed management studies with a distance learning university program and eventually set up my own construction consultancy. In addition, for many years, I had a great intercultural job employed with a local company and working as the first contact point for new expats, i.e. foreigners, who want to set-

Fig. 4.33 Site manager Bianca with customer representative, construction site US Consulate, Brussels, Belgium, 2006. *Source* Bianca Weber-Lewerenz

Fig. 4.34 Author in Beijing, China, 2014. *Source* Bianca Weber-Lewerenz

Fig. 4.35 Olympic Summer Games Beijing, China, 2008 and construction site New Wangjing SOHO, Beijing, China, 2012. *Source* Bianca Weber-Lewerenz

Fig. 4.36 CoffeeArtCafe Stand, International School Event, Beijing, China, 2010. *Source* Bianca Weber-Lewerenz

tle in Beijing and start work, life, house or apartment search, kindergartens and schools. Here I assisted in the search for a new home, apartment or house, kindergarten or school selection, various registrations, and visits to international hospitals and the shopping infrastructure. Since I am passionately rooted in construction, from 2011 I focused on my construction consultancy activity: the construction accompanying consultation for the building extension and new construction of the German Embassy School Beijing, worked as a senior consultant for the Real Life Project, advising on the 'gutting, renovation and reconstruction' of more than 300 old Chinese traditional courtyards (4 sides with courtyard), was commissioned by three international corporations for feasibility studies on facility management in China, helped the German Centre Beijing (a branch of LBBW Bank that supports German companies in their business start-up and establishment in China) in their search for new rental spaces. My own consultancy brought me to the on-site contact and local consulting support for Bilfinger for their German-Chinese joint venture foundation with the two partners Siemens and the Chinese company BITCC in Beijing.

Further training in construction and further education with distance learning were preparation for the next challenges (Fig. 4.37).

From 2014, with the return to my German homeland, I was self-employed with "Bianca Weber-Lewerenz Engineering" in high-rise, office and industrial construction—here specifically for turn-arounds, contract management, project start-ups (Fig. 4.38). Passing on construction experience to the younger generation, occasionally immersing myself in their perspectives, learning new methods from their study experience was an incredible enrichment. Working through construction contracts with all details up and down in comparison with the service to be provided, actually delivered and additionally commissioned often led to contract amendment successes in the million range.

Against the background of digital transformation, the constant focus on values and ethics in cooperation on construction sites, the way professional and personal construction project processes are handled, and the rapid development of Artificial Intelligence (AI) experienced first-hand in China, I started my dissertation research in 2019. The triad of construction, ethics and innovative technologies such as AI formed my topic, and thus

Fig. 4.37 Highlight: Meeting with star architect Frank O. Gehry at the MIPIM fair Cannes, France, 2009. *Source* Bianca Weber-Lewerenz

Fig. 4.38 Celebrating successes on construction site, project start-up construction site US Embassy, Berlin, 2016. *Source* Bianca Weber-Lewerenz

a new scientific niche, which I have since shared at conferences, in discussion rounds and as a speaker and webinars (Figs. 4.39, 4.40 and 4.41).

One thing that has emerged from close exchange with like-minded people and those with similar professional backgrounds is how civil engineering departments deal with interdisciplinary topics that extend beyond their own subject boundaries, and the vehe-

Fig. 4.39 31. RESER Conference, lecture and encounter with the President of the RESER Community, Education Campus Heilbronn, 2021. *Source* Bianca Weber-Lewerenz

Fig. 4.40 Hybrid lecture for the buildingsmart theme day 2021. *Source* Bianca Weber-Lewerenz

Fig. 4.41 Discussion round, Engineer Chamber Day Saxony for their 30th anniversary celebration, Oschatz, 2023. *Source* Bianca Weber-Lewerenz

mence with which many places oppose innovative topics and cross-disciplinary cooperation, and which is still stuck in people's heads, provides prosperous grounds for separate book literature. Unfortunately, this does not benefit the young generation of civil engineers who are striving for the job market, where innovative employers are full of expectation for new knowledge and skills in the digital age. Often society, as shown by the interest and participation in topic-related dialogues and shaping of contemporary conditions, is further ahead and braver than academic education. Don't we need specifically these new paths to reach new innovative solutions?

Digitalization and AI have a strong influence on people and society, they cause incredibly strong incisions on us. Many aspects we already see, many are coming towards us, many are becoming visible little by little or are still hidden, and we should understand this as an opportunity for a constructive co-design of human-technology interaction. The challenges can only can be overcome together. Compared to other industries, the construction sector could step up a notch in this field. My scientifically applied research work offers a new approach and invites discussion. As the first representative for ethics in AI in the construction industry and the exploration of "Corporate Digital Responsibility in Construction—Ethical Principles in Dealing with Digitalization and AI", I gave the child a name and founded the "Excellence Initiative for Human-Led, Sustainable AI in Construction" in 2020[9]. Since then, I have been vigorously driving the sustainability landscape in construction[10].

Interdisciplinary dialogue is needed, it is long overdue in construction. A constructive possibility for bridge building technology-philosophy is the interplay of ethics—digitalization/AI—construction. Where it is about technically feasible, questions arise about ethics and responsibility. Such knowledge creates trust, strengthens awareness, reduces fears and enlightens. Therefore, I am rather perceived as a "paradise bird", because it is extraordinary that a civil engineer questions the data-driven technologies and innovations philosophically, goes to the ethical moral bottom, and researches in the circle of ethicists, philosophers, theologians, IT experts. It is just like raising a wall or in the concreting process: no standstill. In 2023, I completed my doctorate oral exam in my interdisciplinary interface research and completed my doctoral thesis in 2024 at the Rheinisch-Westfälische Technische Hochschule (RWTH) Aachen University, at the Institute for Sustainability in Civil Engineering (INaB) (Fig. 4.42).

What were the influencing factors and inspirations that led me that led to construction and have held me ever since? As a young adult, I built a natural stone wall in the entrance and terrace area of my parents' house, laid floor tiles in the wine cellar. My enthusiasm for languages taught me English, French, and Chinese. After Art and French A-level courses,

[9] www.bwl-engineering.com/iethics

[10] www.sustainabilitylandscape.rwth-aachen.de/activities/promotion-project-innovation-promotion-in-construction-4-0-through-the-corporate-digital-responsibility-cdr-approach-a-new-scientific-research-field-for-the-digital-breakthrough

Fig. 4.42 Dissertation defense and oral examination on December 13, 2023, RWTH Aachen University, Aachen. *Source* Bianca Weber-Lewerenz

I spent a lot of time in France, where parts of my family live. If I look a few steps further back, I was passionate about art education, psychology, and art history, up to the restoration of works of art. My gut feeling tells me that this is where the big bang for my commitment to enthusiasm for STEM subjects lies.

I enjoy an international life and cuisine, an international circle of friends around the globe. My love for languages, crazy new approaches, and technologies allows me access and enthusiasm for many cultures, life, world, and time topics (such as technologies, innovations, values, ethics, world religions, rememberance culture and contemporary witness work, education, entrepreneurial responsibility, 17 sustainability goals). It is important to me to make topics that demand more awarcness and orientation publicly accessible, to push perception, to strengthen awareness. I fulfill this as a book author (Weber-Lewerenz 2022a, b, 2024a, b), book co-author (Weber-Lewerenz and Vasiliu-Feltus 2022; Weber-Lewerenz 2022a, b, c 2023), panel discussion participant and with scientific publications.

The refugee crisis in 2015 brought the admission of 200 male refugees to my place of residence. That means 200 individual escape routes, 200 traumas, 200 unique backgrounds. From 2015, I committed myself as an interpreter and internet/mobile service consultant. Since then, I have been providing support as a freelance integration guide in daily life/professional questions and maintain contacts with former refugees and returnees in their homeland, such as Iraq and Afghanistan. When I returned from China to Germany at the end of 2014, I knew about my own hurdles starting life in China and not understanding the language. When coming back to Germany, I felt like a guest who had to settle in again, so it was an inner call for me to help refugees in 2015 with their start.

Since our return, we have been consciously celebrating the years in China with friends for the Chinese New Year, living traditions in everyday German life, and want to continue to be ambassadors for German-Chinese friendship.

What is the beauty of it all? Change is the eternal constant in life. Over 25 years of professional experience in civil engineering, 7 years of living and working in China. What grounds me? Learning about construction from the ground up and the pushing the Construction branch forward. And, so happy, that one of the valuable contacts could be maintained since the beginning with my very first foreman Ulli, who took me under his wing.

References

Weber-Lewerenz, B. (2022): Value accents in Construction 4.0 - Ethical Observations in Dealing with Digitization and AI, Springer Nature Publishing, Wiesbaden. https://doi.org/10.1007/978-3-658-38238-4.

Weber-Lewerenz, B. (2022): Accents of Added Value in Construction 4.0—Ethical Observations in Dealing with Digitization and AI, Springer Nature Publishing, Wiesbaden. https://doi.org/10.1007/978-3-658-39407-3.

Weber-Lewerenz, B. (2024): Innovation Empowerment in Construction 4.0 by the CDR-Approach. Dissertation. Springer Nature Publishing, Wiesbaden.

Weber-Lewerenz, B. (2024): Diversity in Construction 4.0 - The Gamechangers. Strengthening Innovation through Diversity, Springer Nature Publishing, Wiesbaden.

Weber-Lewerenz, B. (2022a): Creating the Foundation for Diversity, Inclusion and Sustainable Digital Innovations by Education and Training. In: Impact of Women Empowerment on SDGs in the Digital Era. Special Edition by Ingrid Vasilliu-Feltes, IGI Global. https://doi.org/10.4018/978-1-6684-3637-0.ch012.

Weber-Lewerenz, B. (2022b): Perspective Chapter: The sovereign way: How Diversity in Construction is Key of Success in the Digital Age. Book Chapter. In: Digital Transformation - Towards New Frontiers and Business Opportunities by Antonella Petrillo, Fabio De Felice, Monica Violeta Achim and Nawazish Mirza. IntechOpen Book Series. pp. 49 ff. Published 05 October 2022. https://doi.org/10.5772/intechopen.102662.

Weber-Lewerenz, B. (2022c): Education and Training in the Digital Era - A Compass for "Building Forward Better", Book Chapter in: Digital Transformation - Towards New Frontiers and Business Opportunities by Antonella Petrillo, Fabio De Felice, Monica Violeta Achim and Nawazish Mirza. Pp. 33 ff. Published 05 October 2022. IntechOpen Book Series. https://doi.org/10.5772/intechopen.102678

Weber-Lewerenz, B. and Vasiliu-Feltus, I. (2022): Empowering Digital Innovation by Diverse Leadership in ICT, In: Impact of Women Empowerment on SDGs in the Digital Era. Special Edition by Ingrid Vasilliu-Feltes, IGI Global. https://doi.org/10.4018/978-1-6684-3637-0, ISBN13: 9781668436370. https://www.igi-global.com/book/impact-women-empowerment-sdgs-digital/285517.

Weber-Lewerenz, B. (2023): Thinking Differently: Integrating Existing Buildings and Monument Protection in Smart Cities -Experience Shared from User Practice, In: Impact of Digital Twins in Smart Cities Development. In: Special Edition by Ingrid Vasilliu-Feltes, IGI Global Publishing. https://doi.org/10.4018/978-1-6684-3833-6.

Part II

SKILLS

5 Intercultural Bridge Building: Ukraine—China—Japan … Have Their Say

5.1 Strategies to Scale-Up the Construction Innovation

Kateryna Pylypchuk, Strategist in Modulina Straw Panels

In the wake of rising environmental concerns, the construction industry stands at a pivotal crossroads, seeking innovative solutions to reduce its carbon footprint and embrace sustainability.

Modulina Straw Panels, a manufacturing company based in Klaipeda, Lithuania, founded by Zilvinas Bitinas, represents a significant leap toward eco-friendly building practices. Utilizing agricultural byproducts, such as straw, for insulation within prefabricated timber frames, Modulina has redefined traditional construction methodologies (Fig. 5.1), emphasizing decarbonization and efficiency.

To emphasize the methods through which Modulina Straw Panels has effectively catalyzed selected markets, I will delve into several strategies that have underscored the profound impact of this unique bio-based solution compared to traditional building materials. It is our conviction that the adoption of these strategies by any emergent solution within the construction sector can significantly expedite the journey towards achieving a broader impact.

5.1.1 Strategy Number 1 Product First: Foundational Beginnings

Modulina's journey commenced in 2015 with the development of proprietary production machinery and facilities in Lithuania, specializing in the creation of prefabricated load-bearing straw panels (SPs), primarily aimed at single-family homes. Years of research

B. C. Weber-Lewerenz (ed.), *Diversity in Construction—The Gamechangers*,
https://doi.org/10.1007/978-3-658-45308-4_5

Fig. 5.1 Modulina Straw Panels implemented within the private project. (Modulina)

and development led by the company's founder and a team of machinery engineers culminated in a special technology for the panels' prefabrication. This technology utilized fewer wooden components compared to other market players, achieved perfect straw compression, and attained optimal heat and sound insulation metrics. The company's strategy was firmly rooted in technological development. Unlike its competitors, Modulina initially refrained from expanding into other markets. The focus was on perfecting product qualities to meet market demands, ensuring that the straw panel—envisioned as a building material for a future that combats global warming—would be recognized for its superior quality and value. Furthermore, it aimed to prove that straw panels (Fig. 5.2) could last for decades without issues, providing significant added value to occupants of the constructed spaces.

5.1.2 Strategy Number 2 Find the Beachhead: Feasible Market Exploration

After several years of constructing family houses in Lithuania (Fig. 5.3, 5.4 and 5.5), the founder and the team recognized the potential to extend their product's impact beyond national borders. Denmark was identified as a "ready to start" market for Modulina. This strategic move was informed by Denmark's significant legislative shifts, incorporating ESG regulations for development and construction companies. This realization led to comprehensive market analysis, including numerous interviews with local market players to identify a beachhead market. As a result, the Tiny House project was developed to be launched for demonstration at the Circular Lab during the Rockslide Festival in Danmark and to conduct health and allergy measurements. Potential partners were invited to join this social initiative, showcasing bio-based building materials to inter-

Fig. 5.2 Modulina Straw Panels implemented within the private project. (Modulina)

Fig. 5.3 Modulina Straw Panels implemented within the private project. (Modulina)

ested customers and aligning strategies with local preferences and regulations. This has become a key focus to attract all the community around the project which would be feasible for all the local players as well as to the Modulina Straw Panels in Danmark.

Fig. 5.4 Modulina Straw Panels implemented within the private project. (Modulina)

Fig. 5.5 Two Modulina Straw Panels uploading from the track to the construction site. (Modulina)

Another strategic market beachhead was explored by Modulina Straw Panels in Germany. Specifically, an inquiry from a Berlin-based architect about the cost of building a straw panel house evolved into a bilateral cooperation. In this arrangement, Modulina

Straw Panels became the material provider for constructing a showroom building, and the architect agreed to use his own house for future demonstrations and research on the building's features and its impact on human health—a classic win–win scenario.

Modulina is strategically expanding its market presence by cultivating in-house teams stationed overseas. These teams are tasked with leveraging their local networks to introduce Modulina Straw Panels as innovative solutions in their respective regions. By collaborating with partners eager to establish production lines, Modulina aims to transfer sales and partnership management to these entities, facilitating market penetration in Italy, France, Sweden, and Ukraine. This approach fosters a dynamic, diverse team across Europe, united in steering the straw-based building narrative toward new directions.

5.1.3 Strategy Number 3 in Straw We Trust: Advocating for Sustainable Building Practices

At the core of Modulina's ethos is the championing of sustainable building practices, centered around the principle that straw, an abundantly available agricultural byproduct, holds unparalleled potential for eco-friendly construction. Straw's dual capabilities as a carbon sink—preserving CO_2 when properly utilized, and as a potential source of carbon emissions when burned—underscore its significance. By incorporating straw into our building solutions, we effectively sequester tons of CO_2, contributing to atmospheric health for decades.

This understanding forms the foundation of Modulina's potent message: "In Straw We Trust." Echoing the familiar adage "In God We Trust," this slogan shifts the focus to a reliable building material, emphasizing the urgent need to transition from traditional construction methods to bio-based alternatives. Straw, readily available and sustainable, stands as a prime candidate for this shift, promising a future of green construction. Our message aims not just to advocate for straw's use in construction, regardless of the builder, but to highlight its accessibility and long-term benefits, encouraging a departure from conventional materials in favor of eco-conscious, straw-based solutions.

5.1.4 Strategy Number 4 Green Rebuilding Aid: Helping Ukraine to Unveil Global Circular Project Opportunities

Through initiatives like the REBUILDING UKRAINIAN CITIES GREEN, Modulina demonstrates the potential of straw panel technology for post-war reconstruction, offering sustainable solutions to communities in need (Fig. 5.6). Additionally, our participation in workshops hosted by the European Straw Bale Association (ESBA) and FIABCI international, under the UN charter "Building Better Lives," underscores Modulina's commitment to promoting sustainable construction, especially in Ukraine's rebuilding

Fig. 5.6 Kateryna Pylypchuk at the Modulina Straw Panels booth at Architectural Practice Event in Kyiv, Ukraine, promoting the Rebuilding Ukrainian Cities Green Initiative. (Modulina)

efforts. These activities underline our dedication not only to advancing green building materials but also to educating and advocating for eco-friendly construction worldwide. It's vital to note that by leading redevelopment efforts, Modulina Straw Panels become an ambassador for the European natural construction industry. We provide a platform through supported urban hackathons for pilot project development, engage veterans in learning new business opportunities with Modulina, and foster collaboration with local farmers. The establishment of the Modulina Straw Panels Mobile Aid delivers optimal solutions for rural communities and assists veterans in reintegrating into civilian life after challenging times. The Ukrainian project has paved the way for building relationships with people, companies, and experts eager to contribute to aid and rebuilding projects. This has enabled us to evolve our mission-driven model with stakeholders from European countries, the USA, and the Asia–Pacific region.

5.1.5 Strategy Number 5: From Competition to Recognition: Fostering Innovation for Being Acknowledged

In the pursuit of sustainability, it's essential that innovation not only be nurtured but celebrated. Modulina's progressive stance on sustainable construction has not only set a new standard within the industry but also captured the attention of global platforms dedicated to recognizing groundbreaking achievements in sustainability and environmental stewardship. Being nominated for the Global ESG Awards (Fig. 5.8) and making the shortlist

Fig. 5.7 Single Family House in Birstonas, Lithuania built of the Modulina Straw Panels has won the Overall Principal Active House Award 2024. (Modulina)

for the Built By Nature prize among 275 bio-based resource companies, as well as being featured as Overall Principal Active House Awards 2024 for the "Single Family House in Birstonas, Lithuania," project (Fig. 5.7), are significant milestones that showcase Modulina's impactful contributions to sustainable building practices.

Participating in these competitions serves a dual purpose for Modulina. Firstly, it provides an invaluable opportunity to benchmark our innovations against the highest global standards. This competitive analysis not only reaffirms our leadership in the domain of sustainable construction but also illuminates areas for future growth and development. Engaging in these arenas challenges us to continuously evolve and adapt, ensuring that our solutions remain at the forefront of technological and ecological advancements.

The recognition garnered from these accolades goes beyond mere prestige (Fig. 5.9). It plays a crucial role in amplifying our message of sustainability and the viability of straw panel technology as a leading solution for green construction. Such recognition lends credibility, encouraging stakeholders, from clients to collaborators, to embrace more eco-friendly building practices. It underscores Modulina's commitment to excellence and its significant role in driving the construction sector towards more sustainable practices. Additionally, these accolades facilitate networking opportunities, opening doors to collaborations with like-minded innovators and industry leaders, further propelling the adoption of sustainable construction methods.

Fig. 5.8 Modulina Straw Panels has been nominated as a nominee at the ESG Recourse Category during the first Global ESG Awards. (Modulina)

Fig. 5.9 Modulina Straw Panels has become the New European Bauhaus Friend. (Modulina)

5.1.6 Strategy Number 6 What's Old is New Again: Educational Initiatives and Collaborations

Education and collaboration are foundational to Modulina's mission to advance sustainability within the construction industry. By emphasizing the enduring relevance of traditional materials like straw and clay—used by our ancestors for millennia to construct walls and houses—we instill confidence in these time-tested methods, underscoring that while the technology may appear old, it requires new knowledge and learning to be effectively applied today. Through engaging educational workshops and partnerships with organizations such as the European Straw Bale Building Association (ESBA), for FIABCI international members (Fig. 5.10), Global Event attendees (Fig. 5.11), and promoting educational opportunities within the New European Bauhaus community,

Fig. 5.10 Modulina Straw Panels has been represented by Kateryna Pylypchuk at FIABCI Global Leadership Summit in Riyadh, Saudi Arabia, in December 2023. *Source* FIABCI-Saudi Arabia Linkedin Page

Fig. 5.11 Modulina Straw Panels has been represented by Alissa Bankovska at European Straw Bale Building Association Gathering in Brunderup, Denmark. (Modulina)

Modulina not only disseminates its specialized knowledge but also cultivates a practice community centered on sustainable building techniques. These efforts are vital for increasing awareness, transferring knowledge, and motivating a shift towards more environmentally friendly construction practices.

Modulina's dedication to education and collaboration is a testament to its broader vision of a sustainable future energized by cutting-edge construction solutions. A notable project in our educational endeavor was the creation of a specialized book on straw panel construction in collaboration with a Lithuanian builder renowned for his extensive experience with Modulina straw panels. Additionally, Modulina is innovating with the

New Fairytales Series, embedding core values and principles of sustainability for the next generations. Titles like "What's Wrong with This Three Pigs Fairytale" offer parents new perspectives on sustainability and straw construction, while "The New Three Pigs Fairytale" aims to nurture eco-conscious mindsets in children.

Another significant initiative is the "Voices of Innovation: Shaping the Future of Construction" blog series. This platform allows Modulina Straw Panels to spotlight innovative technologies and solutions in the construction industry, featuring insights from some of the brightest minds globally. By posing thought-provoking questions about the interplay and benefits of high-tech and low-tech sectors, Modulina positions itself not merely as a product provider but as a catalyst for educational advancement and meaningful discourse within the industry, advocating for its development in harmony with the natural environment.

In conclusion, the strategies outlined in this chapter—ranging from leveraging historical methods in modern contexts to fostering innovation through competitions, educational initiatives, and strategic collaborations—illustrate Modulina's multifaceted approach to not just participating in but leading the sustainable construction movement. These strategies highlight the importance of understanding the past to innovate for the future, the value of recognition in elevating sustainable practices, the critical role of education in changing perceptions, and the power of collaboration in amplifying impact.Modulina's journey, as depicted through these strategies, reflects a deep commitment to environmental stewardship, community engagement, and the promotion of green building materials and practices. By embracing the mantra "What's old is new again," Modulina showcases the timeless relevance of natural materials like straw in contemporary construction, thereby redefining the narrative around sustainable building.The company's recognition on prestigious platforms and its active participation in global discussions about sustainability underscore its role as a thought leader in the industry. Furthermore, Modulina's educational efforts, from publishing insightful literature to creating engaging content for all ages, demonstrate its dedication to fostering a broader understanding and appreciation of sustainable building methods. As we close this chapter, it's evident that Modulina's strategies are not just pathways to achieving business success but are also blueprints for making a positive environmental impact. The company stands as a beacon for the construction industry, illustrating that sustainable practices can lead to innovative solutions that benefit both people and the planet. In navigating the challenges and opportunities of green construction, Modulina continues to build not just eco-friendly structures but also a legacy of resilience, innovation, and advocacy for a more sustainable future.

5.2 Stay Flexible and Strengthen Multi-WIN!

Christine Lin

In 2003, I started working as a tenant representative in the residential department of CBRE (Fig. 5.12). My main task was to assist foreign employees in renting an apartment or villa in Beijing. This is a seemingly simple task, but in reality, it is a great challenge,

Fig. 5.12 Meeting with Christine in the client's office. *Source* Christine Lin

as the employees abroad came from all over the world and the clients I served came from India, Malaysia, Singapore, Hong Kong, Macau, Taiwan, Norway, Holland, Germany, France, and Italy, Denmark, Australia, the United Kingdom, the United States, and other countries. Due to different cultures, customs, aesthetics, etc., I had to quickly learn and understand the various cultural customs and understand the very individual requirements of each family. It was necessary to create the most suitable rental apartment plan according to the client's rental budget, aligned with family size, proximity to workplace, school, hospital, etc. On the other hand, it was necessary to negotiate all rental conditions with the local owners to meet the needs of the tenants. The landlords were divided into companies and private owners. Usually, the corporate landlord provided standard services such as standard furniture, standard decoration, and standard rental prices, and there was not much room for negotiation. However, they offered a very attentive after-sales service, such as house maintenance, housekeeping, a nice gym with a pool, security system, etc. Private owners are flexible, the rental price was cheaper, and private owners could make personalized decorations and furniture. The supply was according to the wishes of the tenants, however, the after-sales service was not strongly guaranteed and there was a risk of contract breach by the owner. In accordance with this situation, as a broker, I also had to carry out due diligence and risk control in advance, meet the expectations of both parties, and also participate in the renovation, repair, and acceptance inspection of apartments or villas to ensure that the rented living space is in good condition for delivery to the future tenant. This is a job with great patience, good communication skills, thoughtful agreements, and professional negotiation skills. I dedicated myself to this work with great enthusiasm and strived to get the best accommodation and the best deal within the client's budget, and became a "great negotiator" that the clients liked and respected. I also liked to share a lot of information about popular restaurants, beautiful shopping centers, and local free markets as well as tourist attractions in Beijing with the clients, which made the life of the newly arrived foreigners easier, more entertaining, and more interesting. To build good relationships with foreign clients from different nations and to help them quickly settle in, I practically became a bridge between Chinese and foreign cultures and the overall life exchange. During this professional experience, I used the "golden time" of real estate development in Beijing and bought self-used properties and also properties held as financial investments.

In 2006, I switched to the tenant representative department of the office building department and worked as an office leasing broker for the 500 largest companies in the world. I acquired knowledge about office buildings in a very short time, including the technical indicators. and the construction status, background, and operational status of the developers, I often went to the construction site to conduct inspections, checked the building equipment and construction progress, and met with the property management team. Through my years of brokerage experience, I understood very well that in addition to providing correct basic professional information to customers (rentable area, rental price, building quality, quality of management service, etc.) it is even more important to integrate resources, be flexible, and always have a win–win or even multi-win negotiation idea in order to achieve the greatest benefit for the customers. Through my high learning ability and professional spirit, I quickly qualified for these new business areas. In the fourth month after my entry into the office department, I independently pursued the demand of Cathy Pacific Airline for the rental of 2500 square meters of office space. After 6 months, I completed the entire process from building search, negotiation of business terms to the final signing of the lease. The biggest highlight of this case was that the basement of the selected building was connected to the subway station. As it was a newly constructed building, I negotiated an extremely favorable billboard in the connecting corridor between the building and the subway. The builder was very satisfied because it was an unexpectedly pleasant term, and the building owner was also very satisfied, as the good brand and image of the tenant also had a very positive advertising effect for the later rental business of the owner.

During the financial crisis of 2008, foreign-funded companies gradually withdrew from Beijing. Even the Toyota leasing case, which the entire team had been working on for 10 months, also had to be terminated. Overall, sales fell by 60% that year. In this bleak market, it became clear to me that I had to shift the customer base from foreign to domestic companies. I actively participated in the tender for the leasing contract of Alibaba Company, the largest local e-commerce company, which needed about 4000 square meters of office space in the Chaoyang Business District (CBD). The completion was extremely difficult, my strategy was to conclude an exclusive agency contract with Alibaba for a term of only three months, which meant that I should present the final negotiation conditions to the landlord to show whether I could meet Alibaba's satisfaction. If Alibaba would accept the offer I negotiated, I would have won. Otherwise, Alibaba could work with other agencies three months later. After three months of hard work, I won the case with excellent rental conditions and excellent expansion conditions that exceeded the customer's expectations. Because of this transaction, I received top sales awards (Fig. 5.13) and high praise from the same professional groups and colleagues.

After several years in the office market, in addition to the traditional office rental business, I also entered the office purchase business. In 2016, I completed the procurement process for the Ctrip headquarters in Beijing in the Electronic IT Industrial Park worth 180 million RMB (approximately 23 million euros). This transaction had the following two

Fig. 5.13 Christine receiving Top Sales Reward by CBRE. *Source* Christine Lin

highlights: 1. Electronic IT Industrial Park shared the five-year tax reduction policy of the government, which saved Ctrip a large amount of money. 2. In addition, I integrated the facilities of the Indigo Shopping Center, a high-quality commercial major project, which is 3.5 km away from the industrial park. I made an agreement with the Indigo owner to provide free shuttle buses between the shopping center and the shopping center. to provide. The visit to the industrial park during the lunch break for Ctrip employees was a very clever way to offer employees more choices for eating and shopping. Meanwhile, thousands of Ctrip employees also ensured a steady flow of consumption on the Indigo side.

As a Christian, I always approach my work and my customers with the attitude of gratitude, diligence, continuous exploration, and the win–win mindset, satisfying the needs of customers even more than expected with the best solutions and achieving self-growth and self-development—sufficiency at the same time!

5.3 Diversity Using the Example of Japan: Women in Engineering

Burghilde Wieneke-Toutaoui Prof. Dr.-Ing.

Both in Japan and in Germany, the proportion of female students in engineering and of female engineers in industry and research is low. As both countries face demographic challenges and a shortage of highly qualified workers in the fields of engineering and IT, there is an interest in improving this situation, particularly through policy and industry (Fig. 5.14).

Fig. 5.14 Discussion round with female engineers in Yokohama. *Source* Burghilde Wieneke-Toutaoui

However, progress in this area is slow, which has motivated the author to examine the factors in Japan that hinder it, such as societal norms, corporate cultures, or labor market structures, and thus also find more suggestions for Germany. An overview of some of the results of her field research is given below.

5.3.1 Situation of Working Women in Japan

Two keywords keep coming up when it comes to the situation of working women in Japan: the Global Gender Gap Index, in which Japan ranks an inglorious 125th out of 146 countries (World Economic Forum 2023) and the idea of *ryōsai kenbo* 良妻賢母 (good wife, wise mother), which developed in the Meiji era (1868–1912).

The Global Gender Gap Index "measures annually the current state and the development of gender parity in four key dimensions (Economic participation, Education, Health, and Political participation)" (World Economic Forum 2023). While Japan scores well in the areas of education and health, the low economic participation and political participation of women lead to an overall poor ranking of the country.

The concept of *ryōsai kenbo* emphasizes the tasks of a mother to educate her children and to be a good (house)wife and seems to be deeply ingrained in the thinking of women and men (Koyama 2013). It does not originate from Confucianism, as often claimed, but was propagated during the Meiji era and had and has many consequences for the education of girls and the employment programs for women, comparable to the "Raven mother" narrative in Germany.

In 2019, Yamaguchi (Yamaguchi 2019) has intensively dealt with the available data on the Japanese labor market and analyzed various hypotheses about the causes of unequal opportunities for women in the Japanese labor market:

- Japanese companies still follow the model of lifetime employment and thus the ideal of unconditional loyalty to the employer, associated with a salary that increases depending on the duration of employment. For this purpose, graduates are only hired directly after graduation. A second chance, for example for women who temporarily leave the workforce due to family obligations, does not exist. Instead, employees who have left a company—for whatever reason—are considered "disloyal persons". Although the model of lifetime employment is no longer a reality in most Japanese companies is, the hiring strategies are still based on it.
- Japanese employers often claim that there are no women suitable for promotion. Yamaguchi refutes this with his data.
- Employers fear that the effort for further training, which is relatively high in Japan due to the less practical education system and the requirement that employees should be able to perform many tasks in the company, is not worth it for women, because they will quit anyway after getting married or giving birth to children.

The gender pay gap is particularly large in Japan, so women may not feel accepted in the workplace (Earnings and Wages 2021). Female job starters are also systematically referred to administrative positions (ippanshoku -般職) (fixed working hours, few opportunities for advancement), as opposed to career positions (sōgōshoku総合職) (long working hours, changing workplaces, opportunities for advancement). Administrative positions are often not even offered to male job starters. This also leads to women feeling their tasks are not commensurate with their qualifications, finding the work itself not challenging and feeling underpaid, so they are more motivated to leave the company.

Another reason is the workload at home, which, unlike in other industrialized countries, is hardly divided between the spouses. In addition, having a domestic helper in average families in Japan has not been common since the 1960s, so the burden of household chores cannot be distributed elsewhere. The tax system in Japan, which, like in Germany, is based on the concept of the breadwinner, does little to motivate women to stay in their jobs as full-time employees after marriage.

In addition to these reasons for the poor situation of women in the Japanese labor market, there are the reasons that are also common in Germany: lack of and/or poor childcare facilities, inflexible working hours, mothers' responsibility for their children's school careers.

Fig. 5.15 Female engineering students at Nara Women's University in the Production Technology lecture. *Source* Burghilde Wieneke-Toutaoui

5.3.2 Female Engineers

In 1913, female university students first appeared in Japan. It wasn't until 1959 that the first woman received a PhD in engineering. In 1964, the Olympic Games took place in Tokyo, at a time when Japan was experiencing rapid economic growth. This is one of the reasons why there were now more female students at technical colleges and doctoral students in engineering (Kodate and Kashiko 2016) (Fig. 5.15).

Until the beginning of the last century, most women who wanted to continue learning after graduating from school studied at two-year colleges, often women's colleges, whose graduates also had better chances of getting a job (as so-called office flowers) (Kodate and Kashiko 2022).

It can be assumed that the partially gender-segregated education system contributes to the fact that there are only a few women in STEM subjects (Fig. 5.16 and 5.17). Kodate et al. (Kodate and Kashiko 2016) attribute this to the fact that the production of scientific knowledge was historically the task of men:

> "There are two Japanese terms, *rikei* and *bunkei* . *Rikei* can be directly translated with natural science subjects, while *bunkei* refers to refers to humanities subjects. The division into *rikei* and *bunkei* reflects the dichotomy between masculinity and femininity" (Kitada and Harada 2019).

According to Ikkatai et al., (Ikkatai et al. 2020) the subjects required for university entrance exams in Japan vary depending on the field of study. At most high schools in Japan, students choose either natural science or humanities subjects in their first or

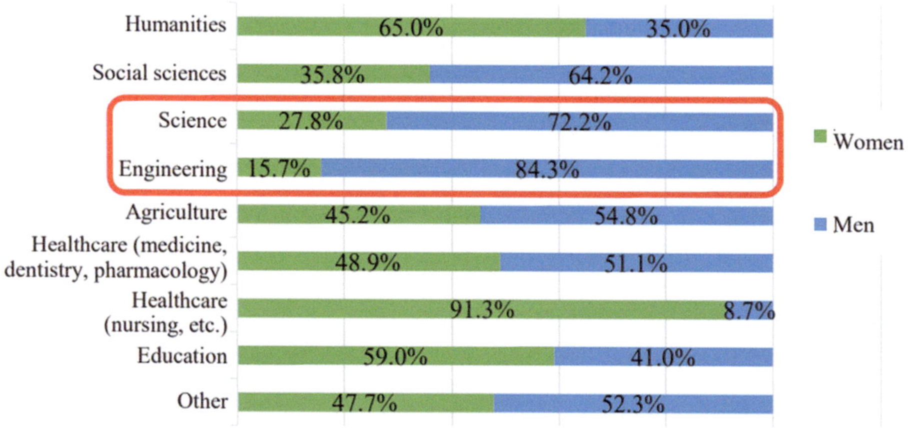

Source: FY2021 School Basic Survey, MEXT

Fig. 5.16 Proportion of female students at universities. *Source* Gender Equality Bureau Cabinet Office 2022

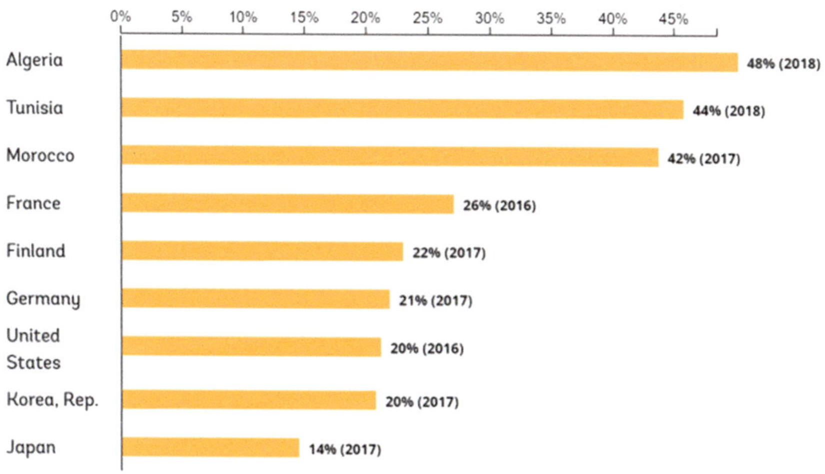

Source: UNESCO Institute for Statistics (http://uis.unesco.org/). Data as of March 2020. Data Retrieved from World Bank Gender Data Portal.
License Type: CC BY-4.0.
https://genderdata.worldbank.org/indicators/se-ter-grad-fe-zs/?fieldOfStudy=Engineering%2C%20manufacturing%20and%20construction&geos=JPN_KOR_CHN_FIN_FRA_DEU_DZA_TUN_MAR_USA&view=b

Fig. 5.17 Proportion of female graduates in engineering sciences. *Source* World Bank

Fig. 5.18 First-semester female students in the engineering program at Nara Women's University present results of the introductory course. *Source* Burghilde Wieneke-Toutaoui

second year (when they are 15 or 16 years old) (Fig. 5.18). Estimates suggest that only about four percent of female students took advanced physics courses in high school from 2005 to 2011, even though most physics and engineering faculties require physics for their entrance exams. Therefore, high school students must make decisions about their future field of study early on. As a result, the general perception that Japanese students have of academic subjects may be more important than for teenagers in other countries, where they can choose their subjects of study when they enter university (Ikkatai et al. 2020) . Since humanities are generally considered "easier" in school, students may choose the seemingly less stressful career path early on in the competitive school system.

There are few female role models for students to find in engineering and natural science faculties, as the proportion of women in academic staff at all levels is low.

Another aspect is the public perception of women depending on their fields of study (Ikkatai et al. 2020). In this study, more than a thousand people were surveyed, representative of the Japanese population. Figure 5.19 shows how the public assesses the suitability of courses of study for women and men, with some courses of study (e.g., mechanical engineering) showing a large difference between the genders.

5.3.3 Common Aspects in Japan and Germany

There are numerous scientific studies on the reasons why the number of women in science and particularly in engineering is low in Germany and Japan. Many of them focus

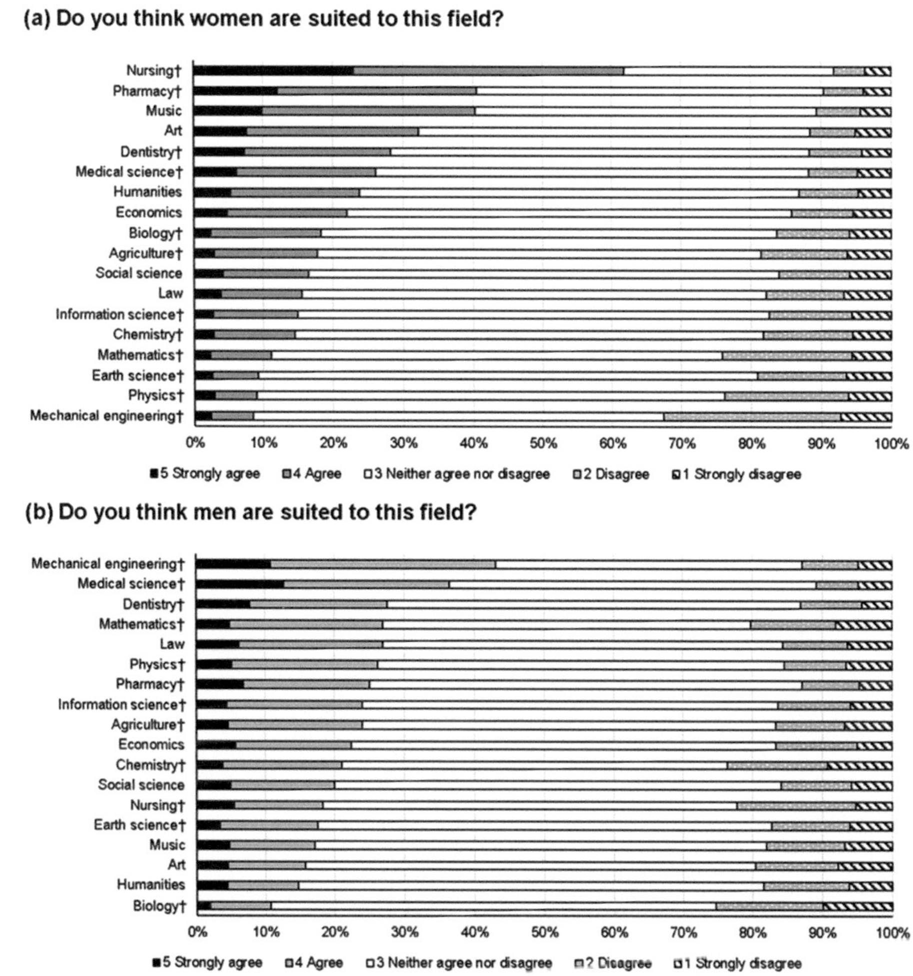

Fig. 5.19 Gender-specific public perception of 18 academic disciplines, expressed in terms of suitability for women (a) and men (b). The subjects are in the order of the arranged with the highest percentage approval at the top, n = 1.086. † names STEM subjects. *Source* Ikkatai et al. 2020

on the segregation of male and female areas in society. The following are some points that should also be discussed.

Both countries have an industrial tradition based on the narrative of male inventors who have built large companies (Siemens, Benz, Porsche etc. in Germany, Toyoda Kiichirō, Tokuji Hayakawa (Sharp), Torakusu Yamaha, Michio Suzuki etc. in Japan). These men are not role models for young women to identify with.

In both countries, the discussion about how to increase the number of women in engineering is almost always linked to the topic of women in leadership positions. This could be counterproductive, as the target group may be more interested in the problems to be solved, e.g. improving the environment, than in a career. "Career" is often associated with power over people, which is not a motivation for young women (and also for most engineers, whether female or male).

In both countries, male engineers have a relatively high status in society and earn well. It is often observed that as the proportion of women in professional fields increases, the image of these professions and the average salary decreases. Why should men in these professions therefore promote improved access for women (Hausmann et al. 2015)?

In professional fields where the proportion of women is relatively high, the conditions for reconciling work and family life seem to be better. It could therefore make sense to encourage more women to choose these areas.

In both countries, there is a general belief that girls are less talented in mathematics and that mathematics is an indispensable prerequisite for engineering, the so-called math barrier, which seems to be a strong exclusion mechanism for women in engineering.

In both countries, the responsibility for raising children is still seen as lying with the mothers. Childcare facilities are often not adequate to support full-time work. For this reason, both countries have similar part-time rates among working mothers. In Japan, part-time work is precarious, so why should a young woman invest in the relatively demanding and expensive engineering studies if she expects to have to work under these conditions after the birth of a child?

5.3.4 Measures

Some interesting measures have been taken in Japan to improve the situation:

The Gender Equality Bureau in the Cabinet Office supports two important activities to promote the interest of young girls in STEM subjects: **Natsugaku** (summer schools) and the **Riko Challenge** (Gender Equality Bureau Cabinet Office 2023).

The natsugaku was launched in 2005 and is supported by companies and institutions. The aim of the 3-day annual summer schools is (Japan Association for Women's Education 2023):

> "Our goal is to bring girls closer to the fascinating culture of science and technology in a wide range of research areas and to encourage and support them through communication to learn and develop their future career paths in the fields of science and technology."

The *Riko-Challenge* (*riko challe*) is roughly equivalent to Comparable to Germany's Girl's Day, albeit smaller and not concentrated on one day of the year but spread throughout the year (Gender Equality Bureau Cabinet Office 2022): As part of this campaign, events are conducted, including visits to STEM companies, practical workshops at universities, and symposiums for primary and secondary school students. It also provides STEM information on the internet, e.g., testimonials from women in the STEM industry, and in-person events where girls are introduced to STEM-related professions.

The Japanese publisher Kodansha Ltd. has a website named **Rikejo** ("What Is "Rikejo"?" 2022, rikejo.jp), which provides a range of information about STEM for girls, e.g., a method for studying science subjects, messages from STEM professionals, a science column, and STEM events.

Nara Women's University, a public university, opened the Faculty of Engineering in April 2022, the first such department at a women's university in Japan. The study program has proven interesting to young women from all over Japan and had more applicants than the 48 offered study places. This also applies to the third academic year, which will soon begin. The curriculum provides a basic engineering education in the first years, and students only have to decide on a specialization later in their studies when they feel better prepared for this decision. So far, not a single student from the first class of this engineering course has dropped out, which speaks highly of the encouraging and empowering environment that Nara University offers (all information from interviews).

A final note: Japan has so far sent two female astronauts, including one engineer, on a total of three missions into space (Casper 2022) and another woman has just been selected for astronaut training (Park Si-soo 2023) whereas neither part of Germany has so far managed to send even a single woman into space (Anderl 2022).

References

Anderl, S. (2022). Eine verpasste Chance. https://www.faz.net/aktuell/wissen/weltraum/keine-deutsche-astronautin-im-esa-kernteam-18483036.html (Accessed 10. March 2023).

Casper, E. (2022). Aim High: Female Astronauts in Japan. https://www.deine-korrespondentin.de/aim-high/ (Accessed 10. March 2023).

Earnings and Wages (2021). OECD. "What Is "Rikejo"? On International Day of Women and Girls in Science." 2022. https://www2.deloitte.com/jp/en/blog/group-diversity/2022/d-nnovation-blog-di202202-women-in-tech.html. (Accessed 21 June 2023).

Gender Equality Bureau Cabinet Office. "Riko Challenge (Rikochare)." https://www.gender.go.jp/c-challenge/about_rikochalle/index.html (Accessed 14. June 2023).

Gender Equality Bureau Cabinet Office (2022). Current Status and Challenges of Gender Equality in Japan. https://www.gender.go.jp/english_contents/pr_act/pub/status_challenges/index.html (Accessed 28. May 2023).

Hausmann et al. (2015). Entwertung Von Frauenberufen Oder Entwertung Von Frauen Im Beruf? Köln Z Soziol 67 (2): 217–42. https://doi.org/10.1007/s11577-015-0304-y.

Ikkatai, Y. et al. (2020). Gender-Biased Public Perception of STEM Fields, Focusing on the Influence of Egalitarian Attitudes Toward Gender Roles. JCOM 19 (01): A08. https://doi.org/10.22323/2.19010208.

Japan Association for Women's Education. 2023. "NPO STEM Career Path Project for Girls." https://gstem-cpp.or.jp/ (Accessed 14. June 2023)

Kitada, M. and Harada, J. (2019). Progress or regress on gender equality: The case study of selected transport STEM careers and their vocational education and training in Japan. Transportation Research Interdisciplinary Perspectives, Elseviewer, Vol.1. https://doi.org/10.1016/j.trip.2019.100009.

Koyama, Shizuko (2013). Ryōsai Kenbo: The Educational Ideal of "Good Wife, Wise Mother" in Modern Japan. The intimate and the public in Asian and global perspectives v. 1. Leiden, Boston: Brill. http://erf.sbb.spk-berlin.de/han/BrillAsiaEBooks/doi.org/10.1163/9789004244351 (Accessed 21. June 2023).

Kodate, N. and Kashiko, K. (2016). Japanese Women in Science and Engineering: History and Policy Change. Routledge contemporary Japan series 60. London: Routledge.

Kodate, N. and Kashiko, K. (2022). Career Paths Dependent and Supported: The Role of Women's Universities in Ensuring Access to STEM Education and Research Careers in Japan. In The Palgrave Handbook of Women and Science Since 1660, edited by Claire G. Jones, Alison E. Martin, and Alexis Wolf, 195–222. Cham: Palgrave Macmillan.

Kodate, N. and Kashiko, K. (2022). Career Paths Dependent and Supported: The Role of Women's Universities in Ensuring Access to STEM Education and Research Careers in Japan. In The Palgrave Handbook of Women and Science Since 1660, edited by Claire G. Jones, Alison E. Martin, and Alexis Wolf, 195–222. Cham: Palgrave Macmillan.

Park Si-soo (2023). Japan Selects First New Astronauts in 14 Years to Support Artemis Program. https://spacenews.com/jaxa-astronauts/ (Accessed 10. March 2023).

World Economic Forum (2023). Global Gender Gap Report 2023. https://www.weforum.org/reports/global-gender-gap-report-2023/ (Accessed 21. June 2023).

Yamaguchi, Kazuo (2019). Gender Inequalities in the Japanese Workplace and Employment: Theories and Empirical Evidence. Singapore, SINGAPORE: Springer Singapore Pte. Limited. http://ebookcentral.proquest.com/lib/staatsbibliothek-berlin/detail.action?docID=5789729. (Accessed 21. June 2023).

6 Start in the Early Stages—Do What Your Heart Beats for

Dr.-Ing. Dipl.-Ing. (FH) Bianca Weber-Lewerenz

Do what your heart beats for.

Doesn't everyone ask themselves this when choosing a school and deciding on an education, and also later in the middle of life, in the middle of their career? Where does the enthusiasm for a craft profession come from? For working with hands, crafting and creating work with your hands?

Natural interests, abilities and talents emerge already in early childhood. Children understand with their hands. "*Accompanying child development means to me wanting to understand children*", says Christel van Dieken, educational consultant, advisor for kindergartens and primary schools and founder of "Waterkant Workshop Pedagogy".[1] She knows, " *Creativity, curiosity and craftsmanship: They are important for early childhood development .....Kindergartens should be places of research and experimentation .... but the reality in kindergartens and schools is different. Because here, the acquisition of knowledge is central.*"[2] Children "grasp" their environment literally with their hands; this offers a healthy and happy development—and in the long run also the motivation to learn a craft profession. Also, what is heard reinforces: You can be anything! For example, on the Reading Day in cooperation with kindergardens in Hamm, where the president of the Chamber of Crafts with the metal construction master Karolin Röhring

[1] https://www.waterkant-academy.com/ (Accessed 15. January 2024).

[2] Dipl.-Päd. Christel van Dieken in an interview with Anna Planken, June 2022, for the craft podcast "#3 Children understand with their hands" (www.handwerk.de/podcast).

B. C. Weber-Lewerenz (ed.), *Diversity in Construction—The Gamechangers*,
https://doi.org/10.1007/978-3-658-45308-4_6

Fig. 6.1 Kindergarten children playfully approach crafts. *Source* Sarah Rauch/HWK Dortmund. www.hwk-do.de/mini-craftsmen

(see Chap. My Story) read to children and introduced them to crafts in a child-friendly and playful way.[3] At the same time, it encourages reading, as a fundamental part of learning, (Fig. 6.1). Similarly, project days and initiatives like "Crafts make School" in primary schools tackle it. Here, sawing, filing, drilling is done to the fullest[4,5] and former craftsmen share their decades of experience with the little ones. Double win–win. Those looking for educational material for their teaching will find it: Crafts make School.[6]

Looking back on 2023, it becomes clear how strongly crafts, digitalization, sustainability, new approaches to renewable energies are anchored in teaching content: the Pythagorean theorem, electric motor and material-supported writing are among the most popular teaching units.[7]

Already the family environment one is born into provides signposts. In the cradle, enthusiasm for a certain activity, for touching and understanding with hands, for example led by a grandparent and parent working in a craftsmanship. Here, the foundations for later professional interest are laid: by working in the studio, at the workbench, in the

---

[3] Chamber of Crafts Dortmund, HWK President reads Presented to kindergarten children. Press report 16.11.2023 by Lina Stoll (www.hwk-do.de/mini-handwerker).

[4] https://schule-kroonhorst.de/kooperationen/377-handwerk-macht-schule.html (Accessed 15. January 2024).

[5] https://hermann-ehlers-grundschule.de/04-2023-handwerker-projektwoche/ (Accessed 15. January 2024).

[6] https://www.handwerk-macht-schule.de/ (Accessed 15. January 2024).

[7] https://www.lehrer-online.de/aktuelles/aktuelle-nachrichten/news/na/die-top-10-der-unterrichtsmaterialien-2023-handwerk-macht-schule-1/ (Accessed 15. January 2024).

family business, experiencing the processes, communication, experiencing by using all senses.

An affinity for professions in STEM subjects is often laid in the cradle, can be promoted by trying out internships and stages. To promote female talent, there are, for example, initiatives and commitments such as Girl's Day and VDI STEM Role Models,[8] which introduce professions in STEM subjects, provide first insights and professional orientation by presenting attractive STEM job profiles.

For Prof. Dr. Gerald Hüther, neurobiologist, education activist and author, early childhood STEM education stands for "*Enthusiasm is a fertilizer for the brain*". The involvement of grandparents, parents and all those who influence the child's career choice and support it in this is groundbreaking, strengthens self-confidence and provides fertile ground for manual work. It is a flowing process in which values, knowledge acquisition, creativity, pleasure in trying out, enthusiasm and inspiration for working with hands—craftsmanship—and appreciation come together.

Hüther sees himself as a "bridge builder" between scientific knowledge and societal life practice. Promoting children, but correctly. "*There is no child who does not want to achieve something … and every child chooses what suits him best.*"[9] The joy of learning—in the German education system with its rigid curricula and evaluations, it is all too often lost. Already in kindergarten and school, children are educated to meet the expectations of others. Furthermore, they are confronted with the principle of sorting out, along the duration of training, apprenticeship and working life. Hüther therefore calls for more freedom for children and adolescents to be able to discover their own interests and inclinations.

Initiatives such as Girl's Day and STEM in schools and companies, craftsmanship makes school[10] enable an open exchange, answering open questions and having orientation as a student. Especially at grammar schools, there is a lack of professional advice on training and professions in the trades. On the contrary, there is automatically assumed that the training path leads directly to colleges and universities. This experience is confirmed by co-authors and the editor of this book. It would be desirable, and serves equally at this point as a recommendation for action: the crafts, training and job profiles should be specifically presented and not only exclusively academic education and courses of study. This means lived diversity!

[8] https://www.komm-mach-mint.de/komm-mach-mint/bmbf-gefoerderte-projekte/bmbf-gefoerderte-projekte/mint-role-models-und-mintalente (Accessed 15. January 2024).

[9] Prof. Dr. Gerald Hüther in an interview with Anna Planken, June 2022, on the craft podcast "#2 Promoting children, but correctly" (www.handwerk.de/podcast).

[10] www.handwerk-macht-schule.de (Accessed 15. January 2024).

"New Work" in Craftsmanship: Success (and Happiness) Factor Family. When Core Competencies and Fun Come Together

7

7.1 New Work—Creating Environments Where People in Engineering Can Flourish. My Journey to New Work

Dr.-Ing. Katharina Ritzer

It is an exciting time for engineering, as we are at the beginning of an era where new work methods and environments are changing our professions daily. I learned early on in my parents' hotel-restaurant what it means to be passionate about what you do. Especially in a family-run business, it usually means "we give everything or nothing". To endure this in the long run, it requires an environment where you feel comfortable, trust each other and can constantly develop. This very intense time has significantly shaped my interest in contributing to environments where people can thrive. However, my path then led me a bit away from the hospitality industry, because I wanted to become an engineer. I completed my bachelor's and master's degrees in mechanical engineering at the Karlsruhe Institute of Technology. In the end, I decided to do a PhD at the IPEK—Institute for Product Development of the KIT. An important decision for me, because there I was able to gain my first leadership experience as group leader of development methodology and management and successfully test my visions for an environment where engineers can thrive. In addition, during my doctoral studies, I gained insight into many different companies. Because for me it was important to apply and test the EDiT method (Enabling Distributed Teams) I developed for improving the collaboration of geographically distributed development teams in practice. Now, as a doctorate engineer at the ISEM—Institute for Smart Engineering and Machine Elements of the Technical University of Hamburg (TUHH), I not only have the privilege of participating

B. C. Weber-Lewerenz (ed.), *Diversity in Construction—The Gamechangers*, https://doi.org/10.1007/978-3-658-45308-4_7

in groundbreaking projects, but also the responsibility of contributing to the design of a forward-looking work environment at a new institute, where the WE is in the foreground.

In my personal life, a new dimension has recently opened up—motherhood. As a young engineer who has recently become a mother, I have found that the challenges of parenthood have further developed my skills for effective and efficient work. The responsibility for a new life has changed my view of my professional tasks and strengthened my ability to set priorities. The limited time I now have for my professional obligations has motivated me to be more productive and to focus on the essential tasks. In an environment characterized by constant innovation and change, it is crucial to set clear priorities and focus on the tasks that create the greatest value. This ability, which I have developed as a young mother, has proven to be invaluable in successfully managing the challenges of the demanding engineering routine. These new experiences have changed my perspective on the world of work and reinforced my belief that "New Work" is not only a necessary development, but also an opportunity for engineering.

7.1.1 Success Factors of New Work Methods and Environments

The importance of diversity and inclusion in Construction 4.0 is becoming increasingly obvious, as innovative ideas and solutions emerge from diverse perspectives. To promote this diversity, work environments must be created that enable the development of each individual. From my experience, I would like to discuss some success factors of new work methods and environments that allow this:

The first success factor is the **Flexibility of working hours**. As a young mother, I know all too well how important it is to have flexible working hours to meet the challenges of career and family. Companies must understand that trust in their employees and their abilities often leads to higher productivity and creativity. A flexible working schedule allows engineers to efficiently complete their tasks, without jeopardizing the balance between work and personal life. The flexibility in my working hours allows me to optimize my energy use. I have learned to strategically plan my working hours to align periods of high concentration with the needs of my child or to take the necessary breastfeeding breaks. This not only increases my efficiency but also promotes my creativity, as I can gather my thoughts in quiet moments. Companies should recognize that this flexibility is not only beneficial for young parents, but can be a meaningful enrichment for all employees.

A **flexible workplace** plays a crucial role in the development of engineers. The opportunity to work from various locations not only promotes the compatibility of work and private life, but also increases individual productivity. The option of working from home, not being tied to a fixed location, gives me the freedom to adapt my workplace to the needs of my family and thus the opportunity to continue my career as a mother. The flexibility of the workplace contributes to engineers being able to optimize their creative processes. A work environment that matches personal preferences, whether at home, in a co-working space or in a traditional office, creates the ideal atmosphere for focused work. This flexible approach not only opens up the possibility to adapt one's own workplace to individual needs, but also promotes the integration of work and living spaces. Overall, a flexible workplace creates an environment in which engineers can balance their professional and personal requirements. This adaptability not only unleashes individual potential but also strengthens diversity and innovation in the engineering profession. In addition to flexibility, the presence of necessary equipment such as a suitable rest room is incredibly important. More than once I have spent my breastfeeding break in a toilet. It is obvious at first glance that this environment does not lead to relaxation, let alone stimulate my creativity. Companies that promote this flexibility and consider these needs are not only investing in the well-being of their employees, but also in a forward-looking and dynamic engineering culture.

In this context, **digital tools and technologies** have helped me optimizing communication and collaboration with my team. Virtual meetings and collaboration platforms allow me to work effectively with colleagues in different time zones and locations without physical presency. The integration of such technologies into everyday work not only promotes efficiency, but also opens up new horizons for collaboration beyond traditional boundaries. When it comes to digital tools and technologies, however, it is not about integrating as many as possible into everyday business, but about making a small number of profitable tools accessible across the board and ensuring that the added value and use of these tools are understandable for everyone. Tools that I myself would not want to do without are: Microsoft Teams—for daily exchange with my team, Todoist—for planning my to-dos on a daily basis, Miro—as support for remote creativity workshops and Calendly—for automating appointment scheduling. I have tested many other tools, but always end up with these few but useful standard tools. Furthermore, I never travel without my external camera. Because a good image and thus a good transmission of facial expressions and gestures in remote meetings leads to improved visual quality, promotes clearer communication and creates a more professional atmosphere, which optimizes the overall experience for all participants.

The conscious **engagement with the shared "Why"** is of essential importance to develop clear goals and visions within a team or an organization. By taking time for this reflection, we create a foundation on which our shared values and motivations become clear. The shared "Why" acts as a connecting element that inspires team members to identify with a higher purpose. This collective identification with the "Why" forms the basis for a shared understanding of goals and allows us to clearly define the direction of our efforts. These clear goals in turn serve as a guiding star that navigates us through challenges and drives us on the path to a shared vision. The shared reflection on the "Why" thus acts as a powerful tool to understand not only individual motivations but also to create a strong foundation for shared success and continuous development. The discussion should not only take place once, but continuously. A "Why" can change, but then it must be talked about and not only when the request is made by the supervisor. Especially the initially uncomfortable appearing situations, in which a "Why" does not seem to be the same for everyone, are the beginning of a fruitful discussion.

Based on defined goals and visions, the dynamics in engineering require continuous **alignment of priorities**, not only on an individual level, but also in close collaboration with management. Joint prioritization allows for a clear focus on strategic goals and promotes transparency in the decision-making process. As an engineer who is developing in various areas of life, I have recognized the importance of this collaboration. An open dialogue between employees and management not only provides clarity about expectations, but also promotes a shared understanding of the individual challenges of each team member. The alignment should take place on various levels. A daily meeting supports the continuous compliance and especially the readjustment of priorities. A daily meeting provides an effective platform for the team to set priorities together. In these short, regular meetings, current challenges can be discussed, progress shared, and the focus for the day clearly coordinated, which leads to an efficient and well-coordinated way of working. However, it is also very important to deal with the longer-term goals together. Because these are necessary to prioritize the daily tasks at all. If the goals on the horizon are not clear, every team will reach the point of feeling lost. And to get out of this situation, it takes much more effort than to exchange about the long-term goals from the beginning and to work on these together.

In addition, **time for reflection** is of crucial importance to promote sustainable developments in engineering. Engineers often face complex challenges that require creative solutions. The opportunity to regularly pause and rethink what has been achieved not only allows for an evaluation of one's own work, but also promotes the exchange of best practices within the team. At the team level, shared reflection helps to identify efficient work processes, correct mistakes, and promote innovations. The exchange about project successes and challenges not only promotes team cohesion, but also enables learning from each other. By sharing different perspectives and experiences, innovative approaches can emerge and proven practices can be further developed. The reflective team thus not only creates a positive working atmosphere, but also lays the foundation for long-term professional success. The importance of reflection extends not only at the team level, but should also be intensified at the personal level. On a personal level, self-reflection allows engineers to identify their strengths and areas for development. This individual consideration not only promotes personal growth based on a deeper understanding of one's own strengths, weaknesses, values, and motivations, but it also has a strong positive influence on dealing with stress and decision-making behavior in difficult situations. Usually, it doesn't take much for reflection. Three simple questions can have a big impact on a personal level but also on the team level: 1. What amazing things have I experienced today/this week? 2. What could I have done differently today/this week? 3. What can I do differently tomorrow/next week so that this doesn't happen again?

On my own professional path, I have found that providing time for joint goal setting, prioritizing, and reflecting not only improves work quality but also strengthens the sense of belonging in the team. The management or the leader plays a key role here by promoting an open communication culture and creating space for exchange. This shared time is not a temporal "luxury good", but an investment in a sustainable and effective engineering culture, which is supported by shared values and clear goals.

Guest Word: Univ.-Prof. Dr.-Ing. Nikola Bursac—Head of the ISEM
As the Head of an Engineering Institute, I see the long-term benefits of "New Work" for organizations as a central aspect of good cooperation. It is not only a corporate responsibility to create an environment in which people can develop, but also a societal obligation of leaders.

Creating a supportive framework for parents in our society is of essential importance. Children are indispensable for a stable society, and organizations play a key role by enabling the necessary freedom for parents. The flexibilization of working hours and locations is not only a means of increasing productivity and creativity, but also a crucial factor for long-term employee motivation and retention. In a world of constant change and innovation, these are decisive elements for the sustainable success of an organization.

However, the associated increased flexibility leads to an increase in the complexity of leadership tasks. To cope with it, new leadership tools are needed. Examples include: a clear definition of the "why", emphasizing shared values, an effective communication culture, and conscious reflection on one's own actions.

On my personal leadership path, I have recognized that "New Work" brings along increased complexity of leadership tasks in the short term, but in the long term it releases synergies that would not be conceivable in a traditional approach.

A final crucial factor for a supportive work environment in engineering is the **promotion of further education and lifelong learning**. In our fast-paced world, it is essential to constantly develop and learn new skills. Here, companies can actively support by providing access to training, workshops, and conferences. I encourage everyone to embark on a journey to themselves at least once in their life. Be it a meditation or silence retreat or just a week without access to smartphone, laptop, and other media. Such an intensive time with oneself can increase mental clarity, reduce stress, and improve the ability to focus, which can promote productivity and emotional well-being in the long term. Especially at ISEM, we have recognized that the promotion of diversity is not only limited to gender or ethnic affiliation, but also includes the integration of parents and their specific needs. By creating child-friendly work environments and flexible childcare solutions, parents can also pursue their professional ambitions without sacrificing their family responsibilities.

In conclusion, I would like to emphasize that creating an environment where people can flourish in engineering does not only represent a corporate obligation but has a positive impacts on innovation and competitiveness. By understanding "New Work" as an opportunity to integrate the various facets of our identity—whether as an engineer, parent, or member of another group—we shape a world of work that is not only modern and efficient, but more humane. In our research at ISEM (Fig. 7.1), we actively advocate for such design potentials and are convinced that an open and diverse engineering culture lays the foundation for groundbreaking developments. Let's together master these challenges and shape a world of work that is inspiring and fulfilling for everyone.

Fig. 7.1 The ISEM team at the strategy seminar after the joint WHY workshop. *Source* https://www.tuhh.de/isem/willkommen

7.2 My Story: A Professional Life Between Family, Sales, and Engineering

Dipl.-Ing. Industrial Engineer Heike Kling

Currently, I am working as Head of Engineering West and authorized signatory at Hilti Group, actively shaping the future of digital planning and construction with construction robotics and Building Information Modeling (BIM).

I am a graduate industrial engineer and, in my role, I am responsible for more than 90 engineers in internal and external service at Hilti Group.

The challenges and opportunities of the digital construction site are diverse and that is exactly the excitement of working in the construction industry.

7.2.1 A Job with Many Exciting Topics and Challenges

My work as Head of Engineering at Hilti is extremely multifaceted and fascinating. On the one hand, the job poses great challenges, on the other hand, it brings me a lot of joy. I carry the responsibility for the personnel development of our engineers.

This includes employee interviews, the organization of applicant appointments, and the accompaniment of the recruiting process for new employees. In addition, we have cooperations with various universities to make prospective engineers aware of Hilti at an early stage.

Another important aspect of my daily work is the customer and planner contacts as well as the associated acquisition of projects. I also often give lectures at trade fairs, meetings, and conferences on engineering topics. Digital planning and construction are always at the center, as well as the positioning of our engineering services in the market. At the end of the day, I lead the entire Engineering West and I continuously work on its further development (Fig. 7.2).

Hilti as a manufacturer in fastening technology places a strong focus on hardware, which includes devices for construction work as well as fastening and fire protection products. In addition, we offer software and IT services as well as engineering services. Especially in the field of engineering, it is of great importance to make the work of engineers visible accordingly.

For this vision, we pursue different approaches. Some of my employees are working in the field and look after planner and engineering offices. In addition, we have engineers in the field who deliver direct customer service, especially the technical departments of our customers. This ensures that our products and solutions are optimally tailored to the respective target group.

Fig. 7.2 BIM World Munich, November 2023, with presentation of the semi-autonomous drilling robot "Jaibot" by Hilti. *Source* HILTI

7.2.2 From Planning to Practice: An Exciting Journey

An example from practice: In collaboration with planners, it is our goal to introduce our products as early as possible into the planning process. We exchange on the specifications of our products, hoping that these will later be included in the tenders. After the tender, our product ideally comes into the application phase. Here, the sales engineer is in direct contact and on site with the customer and discusses both tender and products contained therein.

Additionally, we offer engineering services for customers and planners, such as dimensioning, drawing, and BIM Design Services. We charge for these services on an hourly basis—all under the motto "our work is valuable".

This is how the process from planning to execution unfolds—an exciting journey in which we actively shape the future of construction.

The feedbacks from our expert and customer networks are consistently positive. We actively shape digital planning and construction in the field of fastening technology.

In Germany, generally we are still relatively at the beginning when it comes to BIM projects. Nevertheless, we are leading in the field of fastening technology, thanks to our BIM use cases and services that we offer and place with the customer.

Why do our customers gladly accept this service?

On the one hand, due to the shortage of skilled workers, and on the other hand, for time reasons. We take care of the use of digital data on the construction site at the same time. One of our most productive achievements is the semi-autonomous drilling robot[1] (Fig. 7.3 and 7.4). This allows us to convert the data from the digital model into real drilling points on the construction site without skilled personnel, highly productive and efficient.

7.2.3 Diverse Opportunities for My Career at Hilti

I look back on a 25-year career in the company, that I have been responsible for about eight years in the field of engineering in the North and West region. This period seems both long and short, as my professional life is characterized by constant change and continuous development. I particularly appreciate the team-oriented cooperation and the focus on success.

To actively shape the success of the company with great commitment is fantastic.

---

[1] The semi-autonomous drilling robot "Jaibot" receives digital plan data from the construction project and implements it quickly and accurately. It drills and marks the holes for the various trades according to the data model, maintains the depth and diameter precisely and sucks up the dust. It synchronizes via the mobile data network, so that the installation progress can be tracked in real time. In this way, construction site processes can be accelerated and improved (Source: https://www.bauenimbestand24.de/jaibot-goes-bim-07122023).

Fig. 7.3 Construction site in Hamburg with demonstration of the semi-autonomous drilling robot "Jaibot" by Hilti, 2021. *Source* HILTI

In addition, Hilti offers diverse opportunities for development.

I was one of the first sales consultants in sales at Hilti and later became the first sales manager of the company in Germany.

My professional journey led me further as an internal sales manager in Kaufering and finally to the position of regional manager and authorized signatory.

Indeed, it is a classic sales career that provides me with the right tools for success in my current role as Head of Engineering thanks to my training.

My studies in industrial engineering with a focus on mechanical engineering have provided me the necessary technical know-how. Since Hilti, as a direct sales company, also aligns engineering in a sales-oriented manner, my sales experience in combination with my engineering studies is a great advantage.

7.2.4 Chance Led to the Dream Job

The decision for a technical/sales career was rather random for me. After my high school graduation at a Catholic girls' high school with a social science focus in Bavaria, I came across the Munich University of Applied Sciences, which actively advertised the "new" course of study "Industrial Engineering" at various high schools. That sounded exciting at the time and I was also attracted by the possibility of being a pioneer and doing something completely new.

Thus, I switched from a purely female environment to university into a course of study that was attended by 99% male students. Initially a shock, but it turned out to be the right "school" to work on my self-confidence, standing, and assertiveness.

Fig. 7.4 Heike Kling as a speaker at the Construction Summit, Hamburg 2023. *Source* Heike Kling

My professional journey began in a medium-sized company in drive technology, where I worked as a sales engineer for France. After nine years, I took over the sales management in a company for woodworking machines, before I joined Hilti.

In my current role, I am not only busy with my actual work, but also on various stages. I lecture at events such as the Construction Summit (Fig. 7.4), the BIM World and many other trade fairs and conferences on trending topics such as digitization and innovation in the construction industry. My focus is particularly on value engineering.

What exactly is value engineering at Hilti?

Value engineering is a structured method for planning and developing products. The "value" refers to the benefit for our customers in relation to the resources used.

In short: We aim to be involved in the planning process early on to fully exploit the optimization potential in terms of material and time.

As a result, we are significantly ahead of others in this area and I will ensure that we continue to stay one step ahead in this field.

7.2.5 My Family as a Source of Energy

My family environment represents the source of drawing energy and support for the successful implementation of all these topics.

For many years, as a single mother, I had to be perfectly organized to meet and implement my professional demands and requirements.

This usually only works to a limited extent as a single parent. That's why I have relied on the support of au-pair girls and boys from the very beginning. And when I moved to Berlin/Brandenburg, I was able to rely on a perfect care system through daycare, kindergarden, after-school care, and a nanny.

Today, shared activities with my now adult children (18/27 years old) and my dog are an important part of my free time.

The exchange and activities with friends are very important to me. In addition, I love to be on the ski slopes or enjoy relaxing hours in my garden.

My greatest passion is to travel through Europe with my motorhome whenever possible. This allows me to relax and at the same time get to know many new places. Sometimes I also take a look at construction sites outside of Germany to bring inspiring ideas back home to my everyday work.

Part III

INNOVATION & TECHNOLOGY

Human Transformation—Human and Technology

8

8.1 The Human Side of Construction Innovation: Paths towards the Integration of Technology and Human

René Breyer

8.1.1 Introduction

In a world where constant innovation is part of daily business, the construction industry is experiencing a remarkable upheaval. As a traditional trade that is now turning to modern technologies at an increasing pace, it is being swept up by the wave of digital change. The change, which shakes the foundations of many an old school, brings not only new tools and processes, but also demands a new way of thinking—a mindset that values human intuition and experience as much as the efficiency and accuracy of modern machines.

A new era of construction has dawned, marked by disruptive innovations such as Building Information Modeling (BIM), Artificial Intelligence (AI), autonomous construction machinery, and comprehensive digital networking. Introducing these technologies in all their complexity and potential seems almost superhuman, and yet it is precisely the human being who is at the center of this upheaval; as a designer, user, and developer.

"The human side of construction innovation: Paths towards the integration of technology and human"—the following contribution aims to explore such diverse and dynamic sphere of influence. How can these two fundamental forces—the human mind and digital innovation—be harmonized to make an industry not only more efficient and profitable, but also more human and sustainable?

We face the challenge of not only allowing a paradigm shift, but actively shaping it. The key lies in integration; an integration that aims to understand technology as an

B. C. Weber-Lewerenz (ed.), *Diversity in Construction—The Gamechangers*,
https://doi.org/10.1007/978-3-658-45308-4_8

extension of man, to enhance human ability through technical performance and thus to establish a new age of construction that meets both our deeply rooted values and the requirements of a fast-paced, globalized world.

This article does not offer easy answers, but it outlines a path, a mindset, that enables the bridge between man and machine to be built and strengthened. It is a call for dialogue and action, a dialogue that is based on mutual respect and recognition of each other's strengths, and an action that has the courage to rethink tried and tested structures and to see man not as a tool, but as a partner of technology.

8.1.2 Technology-Driven Changes

Continuous change is a characteristic of our time, in which progress no longer takes place linearly, but exponentially. In the midst of this transformation is also the construction industry, which has relied on the craftsmanship of its skilled workers for decades, but now lingers on the eve of a new technological reality. Innovative technologies, which were previously only found in demanding engineering and architectural offices, have now found their way onto construction sites and into offices around the globe, fundamentally changing the way of construction.

The technological waves, driven by BIM, AI, Robotics, the Internet of Things (IoT) for example, overturn existing procedures and open up space for new approaches. But to carry out the technical transformation and collect its fruits, the construction industry must rethink its processes and reshape human work.

Everything starts with understanding! So what characterizes the technologies that are so influential and significant for the future course of the industry that they obviously become? They extend beyond mere automation and touch on the core aspects of human creativity, planning, collaboration, and analysis. It's not just about what these technologies can do themselves, but how they can transform the entire industry in the hands of engineers, architects, designers, and construction managers.

BIM is a outstanding example of this shift. With a centralized digital platform, BIM offers a way to not only design better buildings, but moreover to track and control the construction process. It enables a seamless connection between design, execution, and maintenance, unifying the knowledge and insight of all project participants. BIM ensures that all those involved in the construction process speak a common language, reducing misunderstandings and strengthening the coherence of the entire project (Bryde et al. 2013).

AI and Machine Learning (ML) are slowly but steadily taking on a role in the construction industry, serving as decision support systems that enable professionals to make informed predictions and optimize planning processes. By recognizing recurring patterns and performing complex data analyses, they facilitate the identification of risks and offer opportunities for higher efficiency in all construction phases (Meßner 2021).

Robotics and automation have the potential to take over routine tasks and extend human performance. By performing tasks that are dangerous, tiring, or cumbersome, they contribute to improving work quality and increasing safety. But they additionally enable skilled workers and engineers to focus on more complex, strategic aspects of their profession (Bauen aktuell 2021).

IoT offers an unprecedented level of information transparency and diversity. It connects machines, tools, and materials into an intelligent network that communicates in real time, enabling construction managers to control and monitor processes. Thus, IoT lays the foundation for predictive maintenance and significantly increases the responsiveness of a construction site (Mittelstand-Digital Zentrum Bau 2021).

3D printing offers significant benefits to the construction industry by increasing speed and flexibility in the production of components. This technology contributes to material and energy efficiency that sets new standards. It revolutionizes architecture by transforming design processes: designs can now be quickly, flexibly, and individually realized, redefining the creation and production of building structures and forms and providing fertile ground for innovation (BauNetz 2024).

Geographic Information Systems (GIS) revolutionize the planning and implementation of construction projects by intelligently utilizing and making available multi-layered data. This technology improves the visibility and understanding of location-based data, leading to more informed and, thus, more responsible decisions (Jarosch 2023).

The above-mentioned technologies—which do not claim to be exhaustive—initiate a change whose impact will affect the construction industry in all its facets. It's about seeing people not just as those who operate these technologies, but as the creative minds behind their development and application. Each technology is designed to support and enhance human work—they provide resources with which professionals can expand their skills and reinvent the construction process.

On the way to a future-oriented construction industry, these technological driving forces cannot be considered in isolation; their integration into ongoing work processes and their acceptance by professionals are crucial for success. By complementing human abilities with digital means, a synergy is created, that allows us to literally build the future anew, despite tradition.

8.1.3 Human Transformation

We live in a time when technological progress is rapidly changing the construction industry, bringing new challenges but also new opportunities for employees. Digital tools and procedures revolutionize the way we build, creating new requirements and opportunities for professional development.

The scope of challenge herein consists in finding and implementing methods enabling the workforce to use the new technologies to their advantage. This requires not only investments in technical infrastructure, but especially the promotion of human capital. It's about creating a work environment where innovation and collaboration can thrive and where every professional is ready and able to actively participate in shaping the future.

The following sections provide more details on how such a work environment can be created through personnel development, leadership strength, diversity, inclusion and thoughtful change management. Particular attention is paid to the introduction of practical training programs, the development of a strong feedback and communication culture, and the implementation of support systems that enable employees to keep up with change.

Considering humans at the center of digital transformation, we affirm the belief that the strength of an industry is not solely defined by its technological achievements, but above all by the innovative power and commitment of its employees.

8.1.3.1 Education and Lifelong Learning

The ability to understand and use new technologies in their full facets requires a solid educational foundation upon which further learning can be built on. In the course of technological change in the construction industry, educational programs must therefore not only convey current technical knowledge, but also lay the foundation for continuous learning.

Adaptation of training curricula: Apprenticeship schools, universities and training companies face the challenge of rethinking their curricula. They must ensure that future professionals are familiarized with the digital tools and methods used in modern construction. It is important that traditional skills and new competencies are promoted equally.

Professional further education and retraining: For those already in professional life, offers for further education and retraining need to be expanded. These should be available both in person and online to make access as flexible and comprehensive as possible. Such educational system ensures that all participants can continue to develop and grow with the changes in the industry.

Learning at the workplace: In addition to formal education offers, learning at the workplace is an essential component of professional development. Employees should be encouraged to participate in internal training, but also to expand their knowledge on their own initiative, for example through online courses or mentorship programs. Supervisors and colleagues play an important role as knowledge transmitters and supporters.

Continuous education as a company value: Companies must anchor continuous education in their corporate culture and maintain as key value. This requires a culture that promotes and rewards learning—for example through appropriate incentive systems and career paths that recognize and reward professional development. Investing in employees, their skills and their careers represents an investment in the future of the company.

Overall, education and lifelong learning form the decisive foundation upon which the sustainable integration of new technologies in the construction industry can be built. The strengthening of individual competencies and the creation of a culture that values further development are crucial to achieve such educational milestones.

8.1.3.2 Psychological Support in Change

Adapting to technological innovations and changing established work processes can be a source of stress and uncertainty for employees (Ries 2018). It is essential that companies invest not only in technical training, but in the well-being and mental health of their employees.

Stress management programs: Employers should offer stress management training that teaches techniques for coping with work-related stress. Workshops or seminars can provide practical methods for dealing with pressure and change and promoting a healthy work-life balance.

Mindfulness and resilience training: Programs that build mindfulness practices and resilience help employees to respond more flexibly to changes and to see challenges as opportunities for growth. The ability to deal constructively with change and uncertainties is improved by strengthening inner resilience.

Psychological counseling: Access to psychological counseling, whether through internal employee counseling programs or through external service providers, gives employees the opportunity to seek support in difficult phases of change. It signals that the company has an open ear for the concerns and emotional well-being of its staff.

Building a supportive corporate culture: A culture that acknowledges the psychological aspects of change and actively addresses them is fundamental. Through regular discussions about emotional and mental health, by building peer support groups or by providing coaching offers employees can be encouraged to speak openly about their experiences.

Train-the-trainer approaches for multipliers: The training of multipliers within the company, who serve as contacts for collegial support in change processes, can help to build a supportive network. These multipliers act as links between the workforce and management and can significantly contribute to the spread of a positive attitude towards change.

Psychological support is an integral part of any strategy that accompanies technological change. It is increasingly about equipping employees with the necessary tools for coping with the increased level of complexity and the need for agility and flexibility. While in the past the avoidance of physical overload was in the foreground, today's demands, due to an increased density of information and the requirement for constant availability, require targeted measures for mental health promotion. Only when employees can mentally adjust to the new ways of working and possess the necessary tools to manage the resulting stress, are they able to be part of the transformation process and actively shape it (Von Hattburg and Troxler 2020).

8.1.3.3 Employee Engagement and Team Dynamics

Employee engagement and a healthy team dynamic are essential to fully exploit the strengths of human transformation. In a rapidly changing technological landscape, the way team members work together and align on common goals becomes a key success factor.

Promotion of teamwork: To create a high-performance environment, companies must establish conditions that favor effective and efficient collaboration. These include regular team meetings, joint brainstorming sessions, and the Promotion of open discussions about existing challenges. The use of collaboration tools can facilitate cooperation across different locations.

Establishment of Communities of Practice: The formation of topic-specific communities, known as Communities of Practice, allows employees to share knowledge and experiences and learn from each other. They serve as platforms for spontaneous and structured knowledge exchange and promote independent further education.

Involvement in change processes: The active participation of employees in shaping changes increases their motivation and commitment. By participating in innovation projects or transformation initiatives, the feeling of playing a valuable role in the company is reinforced.

Culture of trust and psychological safety: Creating an atmosphere is crucial letting all team members feel safe and supported. A culture of trust allows employees to bring in new ideas without fear of failure and promotes a climate in which mistakes are seen as learning opportunities.

Feedback systems and employee recognition: By building systematic feedback loops and introducing recognition programs, achievements can be appreciated and suggestions for improvement can be obtained in a timely manner. This contributes to an environment in which the value of each individual and their contribution to the company's success is recognized.

By investing in these aspects, companies lay the groundwork for a high-performing and satisfied team that not only accepts the challenges of technological change but actively shapes it.

8.1.3.4 Leadership Development and Role Modeling

The role of leaders in the process of human transformation can hardly be overstated. They set the direction, shape the culture, and guide the employees through the technological change. Their role is continuously evolving and requires new competencies beyond traditional management skills (Wagner 2018).

Leadership in times of digital transformation: Leaders must not only understand digital tools but demonstrate its application and encourage their teams. They need a deep understanding of how these technologies change work methods and what value they offer.

Development of leadership qualities: To meet the dynamic demands of the market, leaders need to develop a range of soft skills, such as emotional intelligence, adaptability, and an inspiring vision. Training programs and leadership workshops can help promote these qualities.

Strengthening communication skills: Clear and convincing communication is a key feature of effective leadership. Leaders should therefore be trained to present technology strategies understandably, convey the meaning behind changes, and in this way generate commitment and trust.

Role modeling and integrity: In a rapidly changing environment, leaders are more in the spotlight of their teams than ever before. They must serve as lighthouses that, despite the stormy change, provide a clear direction. By confidently dealing with innovations and demonstrating agile action and high adaptability, they convey consistency and generate trust. Their task is to guide employees not only through clear goal setting but also through personal development readiness and to inspire learning ability. Thus, integrity becomes a dynamic size, which manifests itself in continuous further development and in commitment to the welfare of the company (Lorenz 2018).

Empowerment and Delegation: Modern leadership techniques value empowering teams so that they can work independently and creatively on solutions. The delegation of responsibility, which trusts employees and encourages them to find their own ways and make decisions, reveals as an important aspect.

Innovative leaders represent the so-called "engine" of transformation, which not only drives technological implementation, but creates an atmosphere motivating employees to face the challenges of this new era.

8.1.3.5 Transparent Communication and Feedback Culture

Transparent communication and an open feedback culture are key elements of a successful transformation. They help to build trust, reduce fears and take employees on a common journey, the goal of which is to strengthen the construction industry through people and technologies.

Building communication channels: Transparency starts with access to information. Companies should establish effective channels to spread news—from internal social media platforms to newsletters to interactive team apps. The goal must be to ensure a steady flow of information that is accessible and relevant to all levels of the organization.

Regular information events: To meet the need for information and clarification within the workforce, regular events such as town hall meetings or information webinars are important. Such events provide opportunities for dialogue and allow leaders to openly present vision, progress and challenges regarding the transformation.

Establishing feedback systems: The creation of structures that promote regular feedback from employees, allows organizations to continuously improve their processes and cultivate a culture of appreciation. Feedback should not only be top-down but must also provide opportunities for bottom-up feedback, giving employees a voice.

Dealing with employee concerns: Open doors and the provision of clearly defined processes for raising concerns allow employees to speak without fear of retaliation. It is important to take these concerns seriously and actively address them in order to establish a culture in which change is understood as a team-oriented task.

Highlighting success stories: Successes in the context of the transformation process should be highlighted and celebrated, to underline the value of change and signal that every contribution to change has meaning. Such stories serve as inspiration and encourage imitation.

In a world where change is the only constant, transparent communication and an active feedback culture becomes one of the most important tools for maintaining employee engagement and trust.

8.1.3.6 Technology-Supported Competence Development

The use of technology goes beyond the digitization of construction processes and also includes training and development of competencies. Advanced educational technologies offer innovative ways to equip the workforce for the challenges of tomorrow.

Use of VR/AR for training programs: Virtual and augmented reality (VR/AR) are transforming vocational education, by enabling work environments and complex situations to simulate realistically. These technologies can be used for a variety of training purposes, from simulated operation of heavy machinery to navigation through digital construction plans.

Online learning platforms and e-learning: Online training programs offer flexible access to a variety of courses and are thus ideal for the continuous further education of employees. Interactive learning modules and self-directed educational paths allow individuals to develop at their own pace and according to their personal educational goals.

Gamification to increase learning motivation: The principle of gamification—the application of game elements in a non-gaming context—can increase engagement and motivation in learning. Rewards and incentives promote a positive competitive spirit that encourages employees to work on their development while having fun.

Collaborative learning tools: Tools that enable collaboration and exchange between learners promote a communal educational environment. Discussion forums, group projects and peer-review systems help to share knowledge and grow together.

Data analysis for individual promotion: Learning management systems that capture performance data allow tracking the progress of employees and identifying individual learning needs. Based on this, personalized educational offers can be provided that are specifically tailored to the strengths and areas for improvement of the learners.

The targeted use of technology to develop skills create an environment that not only expands the skills of employees, but also actively involves them in the transformation process.

8.1.3.7 Change Management and Organizational Development

Amidst technological radical change, effective change management is crucial for organizational adjustments to proceed smoothly and with sustainable results. It is a strategic process that aims to prepare both the operational structure and the employees for the upcoming changes.

Planning the change process: Every change starts with well-thought-out planning. This includes defining clear objectives, involving the relevant stakeholders and developing a realistic timetable for implementation. A mature change process ensures that both the business and human aspects of the transformation are considered.

Communication strategies for change management: Open and ongoing communication throughout the change process ensures transparency and acceptance. By sharing information about the reasons for the change and the expected benefits, leaders can foster the commitment of the workforce.

Training and support for staff: By integrating employees early into training and workshops, fears can be reduced and understanding of the new technologies and processes can be improved. It also helps if the workforce can actively contribute to shaping the change, to strengthen a sense of ownership.

Measurement and adjustment of the change management process: Evaluating progress is an important part of change management. By establishing KPIs (Key Performance Indicators) and regular reviews, leaders can assess the success of the measures and make adjustments as needed.

Promoting an adaptive organizational culture: A culture that is adaptable to change enables it an organization to quickly respond to market changes and proactively work on new solutions. Values such as flexibility, openness, and lifelong learning play a central role.

The integration and active management of changes are not only crucial for the seamless functioning of modern technologies within the construction industry, but for creating a work environment in which the talents and abilities of the workforce can optimally unfold. Change management and organizational development create the structural and cultural conditions necessary for a successful transformation.

8.1.4 Conclusion and Outlook

This article highlights the importance of strengthening human capabilities and well-being in a rapidly changing work environment. This transformation is a collective challenge that affects all levels of the company, starting from the trainee to the management level. Important aspects in this process consist in the promotion of education and continuous learning, psychological support, improving team dynamics and employee motivation, excellent leadership skills, and an open communication and feedback culture. These are the indispensable elements to shape and achieve a successful transformation in the human sense.

Although this article does not present comprehensive solutions, it outlines a path for how humans and technology can interact and complement each other more effectively. It is an invitation to conversation and action, based on mutual respect and the recognition of each other's strengths, and to initiatives that have the courage to question conventional structures and understand humans not as mere tools, but as active partners of technology.

It should be emphasized that the methods and tools discussed in this article are primarily to be understood as impulses. They provide a first overview and should be examined in more detail upon deeper interest. The presented approaches and technologies have been deliberately outlined only in basic terms to provide initial orientation. A profound implementation requires a detailed analysis and adaptation to the respective company context.

It is essential to recognize that the application of these methods is feasible for smaller companies. Even if the depth of implementation may vary, the basic mechanisms remain. Smaller businesses can benefit from these developments by applying them according to their possibilities and needs, thus contributing their part to progress and innovation in the construction industry.

As we expand the horizon of possibilities and continue to move towards an integrated and digitally supported construction method, we should always keep in mind the well-being of those who use these technologies. So let's shape a future in which machines and humans do not act as opponents, but as partners to bring sustainable, efficient, and inspiring buildings to life.

8.2 Together We Can Achieve More

Dr.-Ing. Kathrin Rüschenschmidt

"Have you ever thought about studying physics?", with these words my physics teacher brought me to the idea of thinking about it. And so physics and psychology competed for the choice of study subject. When the NC (numerus clausus) for psychology was raised in the year I graduated from high school, so that I was 0.1 grade points too bad to start studying directly and in this field, it was these two events that helped me decide to study physics. After my diploma, I did my doctorate and since 2006 I have been working in industry. At this point, the question may arise as to what all this has to do with the construction industry. In my case, not much at first, although physics as such is of course also fundamental to the construction industry. But the topics DEI (Diversity, Equity and Inclusion), Psychological Safety, AI (Artificial Intelligence) and Innovation are cross-disciplinary and cross-industry and thus an interface where we can learn from each other and develop together.

Networks are of invaluable value in finding like-minded people and generating "out of the box" ideas due to different perspectives and experiences. Because it starts with diversity so that successful innovations emerge in the end. Diversity can be diverse: gender, education and life paths, geographical background, age, intelligence, ways of thinking, religion, skin color, sexual orientation, hierarchy and interests are just some that already lead to very different experiences, approaches and idea generation and thus open up enormous opportunities for us to develop further and successfully implement innovative ideas.

When I did my doctorate in physics, the proportion of women was about 10%. I personally did my doctorate because I enjoyed scientific work during my diploma thesis, I had been offered three doctoral positions and I did not yet know what and where I wanted to work. Basically, I chose the easy way for me. I decided on materials physics and started dealing with the exciting semiconductor material silicon carbide. It is not dissimilar to diamond in structure and is therefore also often used as a gemstone. In my dissertation, I dealt with the doping possibilities of silicon carbide, an important manufacturing step for the production of power semiconductors. The power density in silicon carbide semiconductor components is higher than in silicon semiconductor components due to the physical properties of silicon carbide, which makes the material so interesting for use. However, processing the material is more difficult due to its structure. These difficulties repeatedly set me back and challenged me in my experiments during my doctorate. Eventually, I was able to successfully eliminate my main problem after finding the idea in a lecture at an international conference that ultimately opened up the solution path for me. Curiosity, perseverance, and scientific exchange helped me to successful work. In addition, I was able to present myself at annual conferences in the German-speaking area on my topic and was approached towards the end of my doctorate by my future colleagues asking if I would be interested in working for/with them. And so I started in 2006 as a doctorate physicist in the semiconductor industry, where I took on both the role of technology developer and project manager. I learned a lot. Next, I moved to a large German conglomerate, where I continued to develop power semiconductors and led projects, although not in the direct dual role here. In terms of content, my work was similar, but both companies had a very different corporate culture. At my first job, in addition to my substantive work, I learned a lot about project management, problem solving, I learned moderation and communication in an environment that had established an innovative mindset in the company culture. Just trying something out to move forward with the results was the order of the day. The area in which I worked in the conglomerate was organizationally different. There was a research and a pre-development department. As a result, the actual product development, in which I worked, was no longer so technologically innovative, but was more responsible for the execution of product ideas into series production. Systematic problem-solving approaches also gained more and more importance here and I personally experienced a lot of appreciation. I was promoted, among other things, to be prepared for my first leadership role. Nevertheless, I was advised there to consider whether this conglomerate would be the right company for me in the long term. At a further training course, where casual clothing was expected and I showed up in jeans and a hoodie, I was accused of having a pubescent style of dress. And in an exercise during this further training, I encouraged my fictitious team regarding a fictitious, unpleasant management decision not to simply accept it, but to counter it with good arguments. This initiative was linked to the clear announcement that we would make this one attempt and then implement the decision if we were not successful. I am generally a critical thinker and question processes and procedures when I have the impression that they could be better and more efficient through further development and adjustments. This kind of "resistance" was not so well received in my area at the time, although my then department head strongly supported me, as he saw in me an example of a modern leadership style. A first change was noticeable. Since I wanted to develop further, it was clear to me that my professional journey would continue. What would be a new experience for me was unimaginable for some of my then colleagues, as the conglomerate was a safe, established and successful employer, which some could not imagine leaving. I then switched to AVENTICS, a company division sold by Bosch Rexroth to Triton, which develops pneumatic products. In short, things are moved with air in pneumatics. Even though the applications are diverse, the first and last time I had consciously used pneumatics (apart from pumping up my bicycle tires) was the pneumatic control of my Lego excavator, which I received for my 9th birthday (so I probably also had a certain affinity for the construction industry) (Fig. 8.1).

At AVENTICS, I also did not start in development, but used my knowledge in the global quality department with the aim of supporting product development. What I had learned and deepened at the university and in the industry, I could now use profitably here, including my passion for processes, to drive topics effectively and efficiently. Whether development processes, risk management or problem solving: I still find it incredibly helpful and exciting to be able to control complex topics in a targeted manner with the help of processes. It is important to see the process as a guide and control instrument, it should never be the goal of the respective project or project. So I worked in the quality department as a methods expert before I switched back to development just under two years later, where I took on my first leadership position.

For me, leadership combines all aspects from diversity to innovation and puts people at the center. During my diploma thesis and doctorate, I worked on topics very deeply and was able to acquire special expert knowledge. Leadership, on the other hand expands more broadly, with the aim of providing experts with a framework and an environment in which they can give their best and achieve the best results in collaboration. In 6.5 years, I built various teams that were diverse from the start. I started with a methods expert, our then only dedicated project manager and a product developer. In the end, I led a group of two teams, one consisting of product developers and one of project managers, plus the already mentioned methods expert. With the team, I established a team culture in which I also realigned the role of the leader. In the industry, the best expert is often made

Fig. 8.1 LEGO Pneumatic Lego Excavator. *Source* Kathrin Rüschenschmidt

the leader. And so part of my team expected me to make all technical decisions. I was never a pneumatics expert and I am also convinced that the best developer is not necessarily the best boss. Leaders have many tasks, but not those of the expert. Accordingly, I handed over the responsibility due to my experts and lived my role as a leader. With my passion for processes and their potential to facilitate work, we also tried out new methods in the team, e.g. Scrum, an agile project management method that was developed for software development and was very rarely used in the development of mechanical products at that time. We have together curiously tried out a lot, learned a lot and integrated some of it firmly into our everyday life. At the end of this journey, I was Director Project Portfolio Management for our new product developments. During this time, AVENTICS was bought by the American corporation Emerson. There, in my role, I built the above-mentioned team of project managers and introduced a transparent and successful project portfolio management for the overall control of our new product development projects. Working in a large corporation has advantages and disadvantages for each individual, like almost everything in life. One advantage is the wealth of opportunities. To broaden my horizon and to be able to develop further in breadth, I got the chance to take over the management of our site in Laatzen. I held this role for about a year, during which I was also part of the management of Aventics GmbH, the legal business unit within the entire corporation. During this time, I have learned a lot of completely new things and was able to expand my network with great new colleagues who came from completely different areas than the development-related ones. It was a very exciting time, which again culminated in a new opportunity. Emerson has been driving a transformation and reorientation for some time: "*Emerson combines advanced technologies, industry-leading expertise, and an insatiable curiosity about the world to create sustainable solutions for our customers.*" (https://www.emerson.com/de-de).

The business area Discrete Automation, in which I work, also underwent organizational reorientation in 2023. In this context, I was offered a position that seemed to me, my passion and my competencies like a fitting puzzle piece: My current role as Director Innovation Process and Project Management Office in the Strategic Business Units (SBUs) area. Topics such as project management, processes, leadership, methods, talent development, trust, collaboration and goal-oriented value creation have accompanied me all the way and are now the heart of my daily work. They are on the one hand specifically tailored to our company and our business area but on the other hand also of universal character, as it always involves the collaboration of people and human drivers and needs form the basis, whether in the industrial or the construction sector. In collaboration, letting go of your own views and approaches, allowing other perspectives and finding a common path, besides all opportunities also has a not to be neglected potential for conflict. Diversity is therefore not the goal, but the potential and the starting point on the path to innovation. Only when diverse people feel safe and valued, do they dare to authentically and passionately go new ways together. So what does it look like, the path from the "D" in DEI to the company culture of innovation?

Innovation begins where the comfort zone ends. Therefore, it is all the more important that people wear a kind of "protective suit" of psychological safety when they embark on this adventure. The concept in my opinion, the continuation of DEI (Diversity, Equity, and Inclusion) initiatives, which many companies have increasingly focused on in recent years, is crucial to psychological safety. This topic also plays a significant role at Emerson. Since 2013, there have been so-called Employee Resource Groups (ERGs). "*An Employee Resource Group is a voluntary, employee-led group that promotes a diverse and inclusive workplace while working towards a common goal. The group's mission is to promote the professional development of its members, educate other employees about the Employee Resource Group, and participate in charitable activities.*" There are now 8 different ERGs:

- Asian & Pacific Islander Alliance: "*The Asian & Pacific Islander Alliance was founded not only to share the rich and diverse API cultures, but to further support Emerson's commitment to building and creating a more diverse and inclusive workforce.*"
- Black Employee Alliance: "*Promoting a diverse and inclusive environment through partnership with Emerson to attract, retain, and promote black employees and allies, and strengthen the communities we support.*"
- Diverse Abilities: "*To celebrate diversity and differences within the Emerson community by promoting social awareness of perceived impairments and/or disabilities and highlighting the impacts of discrimination and social exclusion on individuals. To foster mutual understanding and go beyond simple tolerance to embrace, celebrate, and utilize the rich dimensions of diversity that Emerson offers.*"
- Pride: "*We are seeking Emerson colleagues to help promote a diverse and inclusive corporate culture and strengthen our commitment to innovation by promoting the recruitment, promotion, and retention of diverse talents. This group aims to bring together all people of different backgrounds and lifestyles to exchange ideas, network, and ensure comfort for all.*"
- Mosaic: "*Our vision is to promote a community that is passionate about various cultures, a community that connects common ideals, promotes cultural diversity, and creates an inclusive work environment for people working outside their home location.*"
- Somos: "*Emerson aims to be recognized as a preferred employer, thereby increasing the leadership representation of Latin Americans and Hispanics at Emerson and in the communities where we operate.*"
- Veterans Resource Group: "*Building an unprecedented community where veterans and veteran champions can support and inspire each other to give their best.*"
- Women's Impact Network: "*Emerson's Women's Impact Network attracts, promotes, and retains women at Emerson by increasing their visibility and inspiring them to achieve their career goals.*"

Furthermore, there are networks that consider the topics of diversity, justice, and belonging in a comprehensive manner and holistically. The sum of the initiatives and activities is diverse. From panel discussions, mentorship programs, and Emerson-internal networking activities, to webinars with external experts on topics such as Psychological Safety or Mental Health, to collaborations with universities and schools, to name just a few.

When Lal Karsainbai became CEO (Chief Executive Officer) of Emerson in 2021, the topic of cultural change got the key topic on his priority list, knowing that such change requires time, work, and attention, and with the goal of generating long-term value for employees, our customers, and Emerson (like Satisfaction, climate protection measures, and profitability, to name just a few). In one of his first information events for employees, he introduced his leadership style as a key carrier of this culture, which greatly excited me: He relies on trust and empowerment of employees. Trusting teams and empowering individuals to take responsibility for their work, in his experience, yields the best results. Similar to the topic of "safety", the topic of "DEI" is anchored in our company values.

What does DEI mean?

D = Diversity. This is the basis for collaboration, where many experiences and perspectives lead to a wide range of creative ideas and thus represent the innovative potential.

E = Equity ≠ Equality. People are not equal and therefore equality does not help them. Figure 8.2 illustrates this in a simple way:

I = Inclusion = Every employee feels heard, respected, and as an integral part of the company. The word "feels" is important here. It's not just about integrating people, but about them feeling this way. And it's important that we can be authentic and don't have

Fig. 8.2 Equality and Justice. *Source* https://www.internationalwomensday.com/Missions/18707/Equality-versus-Equity-What-s-the-difference-as-we-EmbraceEquity-for-IWD-2023-and-beyond

Fig. 8.3 The 4 Stages of Psychological Safety. *Source* Thimothy R. Clark

to pretend to belong. We are good as we are and we feel safe. This is the first stage of Psychological Safety "Inclusion Safety" according to Thimothy R. Clark.

Psychological Safety is an important building block for Mental Health and in the current time with all the difficult conditions such as disruptions in the supply chains, the Corona pandemic, the war in Ukraine, and climate change, to name just a few, it is indispensable. Thimothy R. Clark described the following model in his book "The 4 Stages of Psychological Safety—Defining the Path to Inclusion and Innovation" (Fig. 8.3):

After the first stage described above comes the second stage (Learner Safety). This stage examines whether one feels safe enough to ask questions and make mistakes. The culture of error in companies is extremely important and unfortunately often not addressed in a differentiated enough manner. For example, Bosch and historically also AVENTICS once had a zero-error culture. This meant that the products should reach the customer without errors, which is a good goal. The result, however, was a culture in which employees felt they were not allowed to make mistakes. But this is simply not in human nature, we all make a lot of mistakes. The question is how we deal with them and what we learn from them. I remember a situation where a colleague came to me and admitted a mistake. As a result, we lost some time in a long-term test, but were able to quickly eliminate the error and restart the test. Without his courage, we would have lost much more time. Therefore, I thanked him for proactively admitting his mistake. He was extremely irritated that he was essentially praised for a mistake. This showed me how difficult we find it in our performance society to establish a good error culture.

The next step (Contributor Safety) is now about whether we dare to make a contribution to create values. Do we feel are we confident enough about our competencies? And do we stand up for our views even against resistance, without ceasing to listen and possibly adapt our approach and suggestions in a meaningful way?

The final stage, leading to holistic psychological safety (Challenger Safety), is about the courage to question the status quo and shape changes. According to Thimothy R. Clark, this stage is "the license to innovate". That is, in order to truly become innovative and to completely leave the comfort zone in terms of content, the highest level of psychological safety is necessary as a protective coat. Dear reader, do you feel so safe in your environment that you can question the "status quo" at any time, openly, with all your own vulnerability?

I myself have environments in my company where I feel sufficiently safe, but this does not apply to all areas and/or situations. And in my role as Director of Innovation Process & Project Management Office, it is also up to me to work on the company culture, including psychological safety. So, innovation starts with culture, even though many people might first think of new technologies. Artificial Intelligence (AI) is one such technology. At the beginning was the development of artificial intelligence as a very innovative invention. And the development continues steadily. Nowadays, people can work with people, people with machines, and machines with machines. And both people and machines are constantly learning with each other, among each other, and from each other. Similar to how a person builds their knowledge based on their early childhood experiences, AI also learns based on its starting parameters. While there are good and bad people, the machine is neither good nor evil. But it reflects the people who initiated it and on whose data basis the machine continues to learn. Therefore, it is incredibly important that many and diverse people are involved in this process. A machine can learn discrimination just like a human. By providing a broad, diverse, and as prejudice-free as possible data basis initially, we can use the performance capacity and speed of AI, which is far superior to ours, in a meaningful way to support and improve our lives. It is an innovation that can now help us accelerate innovation processes. AI can promote creativity by quickly and extensively testing ideas. It uses many existing data points and learns from the past to make predictions. But it does not invent in the way humans can. Human and machine are the combination that can bring out the best (or worst) with combined strengths. AI offers humans security when everyone feels included, but it can also lead to people feeling more insecure because it discriminates and/or excludes, which blocks the power of innovation. But regardless of the opportunities and risks, artificial intelligence is here and whether we see it as a blessing or a curse, it is here to stay. Therefore, we should all make it our task to minimize AI-learned prejudices initially and continuously reduce them.

In summary, from diversity to a sense of belonging to an innovative environment where we question the status quo, a lot has happened in recent years and decades and the journey is not over yet. To improve our lives in many areas, we need to generate and implement new ideas for technologies, business models, emotional and social "jobs to be done" with curiosity. It is an exciting journey, during which I personally have learned a lot and continue to learn every day. Networks play a special role, where exchange takes place beyond daily collaboration. Networks can be internal to the company, external to the company, and also cross-industry. They offer the opportunity to look beyond the horizon with other interested parties, to learn from each other and to support each other. A wonderful networking opportunity in the industrial context is offered for women and interested men by the "*FEMWORX (formerly WomenPower)—Career Congress for Women in Industry. In addition to the program of lectures, panel discussions and the accompanying exhibition offers workshops, the opportunity to learn about career prospects, studies, coaching and mentoring offers, as well as other labor policy topics and trends.*[1]" FEMWORX is a congress within the framework of the annual HANNOVER FAIR, a globally important industrial fair. "*The HANNOVER FAIR is the internationally most important platform for all technologies around industrial transformation—with excellent innovations or unusual products.*[2]" The HANNOVER FAIR and FEMWORX are valuable platforms for Emerson, to show us, our products, our philosophy and strategy, and our company culture. In 2023, as part of the then still WomenPower named congress, we organized a panel discussion on the topic "*Embracing Diversity: Exploring Cultural Differences to Drive Innovation*". Our 5 participants came from 4 different countries and from 4 different departments. Moreover, we had this year the only male speaker in the program. For him it was an unusual experience, in an event with industrial focus to be a man in the minority. The resulting insecurity for him he clearly articulated at the beginning of the discussion and he reflected himself that the open handling of his vulnerability gave him a sense of security for the course of the event. Politically, WomenPower 2023 also took another step in the right direction. Julia Willie Hamburg, Minister of Education and Deputy Prime Minister of Lower Saxony spoke at the opening of WomenPower at the annual award ceremony of the Engineering Power Woman Awards. And Chancellor Olaf Scholz took part in a photo session with the WomenPower Advisory Board/Team before the opening of the HANNOVER FAIR, to demonstrate the importance of WomenPower (now: FEMWORX) (Fig. 8.4).

Innovation is my passion, both functional, emotional, social, business, as well as mindset and behavior culture. Mindset and behavior culture are the foundation and the protective suit, with which people dare together into unknown worlds and are not slowed down by fears in their authentic creativity. As a leader, I see my task primarily in providing this protective suit and to give a strategic direction. For me personally, the exchange with many different people outside my comfort zone, but also so enriching, that I dare out every day again and take on the adventure of collaboration. And every shared success fills me with pride, meaningfulness, motivation and joy. And so I am particularly pleased to be able to write this cross-industry guest contribution, because together we can achieve more!

---

[1] www.femworx-congress.com (Accessed 15 December 2024).

[2] https://www.hannovermesse.de/en (Accessed 15 December 2024).

Fig. 8.4 Author at her award ceremony with the Engineering Power Woman Award 2023. *Source* FEMWORX

8.3 Current Opportunities for Diversity in the Context of Building Information Modeling (BIM), Artificial Intelligence (AI), Metaverse, and Augmented Reality (AR)

Dr. Manuel Mühlbauer

Working with people who have very different professional backgrounds and relationships is never easy. The technologies BIM, AI, Metaverse, and AR/Virtual Reality (VR) assist in the effective communication between the disciplines in the industries of creative, construction, and real estate. The application of digital technologies facilitates the coordination and alignment of designs and constructions with digital twins of buildings and details while simultaneously appreciating the reality of other professional contributions.

In construction, BIM allows the creation of virtual 3D models of buildings from different sub-models of architecture, structural planning, and technical building equipment (TGA). On the other hand, BIM can facilitate efficient planning of maintenance work and data provision in the real estate sector. Moreover, these digital twins of buildings enable the determination of quantities and simulation of sustainability aspects (BIM2SIM[3]). AR can be used to project virtual models into the real environment, for example, to provide construction sequences and instructions in real time on the construc-

[3] https://www.usgbc.org/education/sessions/bim-sim-10401906 (Accessed 01. March 2024).

tion site.[4] The digitization of the construction industry is perceived in Germany as an option to overcome the current hurdles in productivity development and process automation and to consolidate the current market situation.[5]

When it comes to data, many immediately think of data science and data engineering—the efficient and automated collection, cleaning, and visualization of data for decision-making purposes. Different properties and features can be read from digital twins or geographic information systems. AI can, for example, predict rental prices and property values or assist in optimizing design decisions. These analytical approaches are called discriminative AI, which distinguishes itself from generative AI that can generate images from different inputs (for example with Midjourney and Dall-E 3[6]), videos, texts (think of Chat GPT[7]) or even 3D objects.

These technologies have countless applications and are constantly evolving. Therefore, it is important to understand digital transformation as a gradual process and not to look for the upheaval from a division of labor in the construction industry to fully automated processes. After all, it is about creating a culturally vibrant building culture that profitably complements analog and digital processes.

BIM, AI, Metaverse, and AR play a crucial role in revolutionizing the thinking and learning culture in construction. By using these technologies, construction projects can be virtually planned, coordinated, and which leads to a more efficient way of working. Construction workers have the opportunity to learn in a digital environment and continuously expand their knowledge. Furthermore, AR applications provide an immersive learning experience by overlaying real-time information on the actual construction project. Overall, BIM, AI, Metaverse, and AR contribute significantly to improving the thinking and learning culture in construction.

8.3.1 Clear Advantages Through Advanced Digital Technologies

There have been enormous advances in design, planning, and construction execution in recent years. Particularly through the use of BIM as a bridging technology, which allows architects and construction companies to create and manage virtual 3D models of buildings. But how can BIM help optimize design decisions and make construction projects more efficient? And what role does AR play in visualizing planning and construction processes?

---

[4] https://www.handwerkundbau.at/bauen/ki-am-bau-anwendungsgebiete-und-herausforderungen-17266 (Accessed 01. March 2024).

[5] https://blog.digitall.com/de/wie-digital-ist-die-baubranche-im-dach-markt (Accessed 01. March 2024).

[6] https://www.blogmojo.de/ki-bildgeneratoren/ (Accessed 01. March 2024).

[7] Https://chat.openai.com/auth/login (Accessed 01. March 2024).

BIM revolutionizes the way buildings are planned, designed, and built. It allows architects and engineers to store all relevant information of a project in a central 3D model. This model contains not only geometric data but also information about materials, costs, schedules, and other qualities. This allows all project participants to access a common database and coordinate their work.

Through improved collaboration and communication between the various actors in the construction industry and construction tech startups,[8] BIM enables a more efficient execution of construction projects. One of the greatest strengths of BIM lies in its ability to cover the entire lifecycle of the building and contribute to the clarification of interfaces in planning, execution, and management of buildings through its diversity in handling. By creating digital models, architects, engineers, contractors, and other actors can seamlessly integrate their work results. This leads to more efficient coordination of all project phases—from the construction of the BIM model, through coordination to digital defect management and data transfer into facility management.

Another advantage is the reduction of risks and costs through early error detection and correction in the virtual environment. It can minimize or even avoid costly changes during the construction phase. It not only saves time but also conserves the budget. In addition, BIM offers better traceability of changes, costs, and plan updates. The positive and model-related communication, increased planning transparency, and clear role distribution reduce friction losses and potential for conflict. It is therefore no surprise that more and more companies worldwide are relying on this innovative technology. This is where AI comes into play.[9]

8.3.2 Analyzing, Evaluating, and Predicting Data with Artificial Intelligence

With AI, architects and planners analyze large amounts of data from past projects, trends, patterns, and features e.g. of different building typologies. Based on this information, digital tools can provide recommendations for the design of buildings, for example, to improve energy efficiency or user-friendliness in the Green Building segment.[10] In addition, AI can help in parametric design by automatically generating building designs by varying different parameters[11] (Fig. 8.5). Different cultural expressions of construction projects can be combined with BIM and AI with enormous efficiency gains. Through this type of architecture optimization, construction companies can map the entire con-

[8] https://www.bauindustrie-bayern.de/bautech-start-ups.

[9] https://bit.ly/3HVmkGr (Accessed 01. March 2024).

[10] https://www.realestate.bnpparibas.de/marktberichte/investmentmarkt/deutschland-market-focus (Accessed 01. March 2024).

[11] https://constructionblog.autodesk.com/ai-construction/ (Accessed 01.March 2024).

Fig. 8.5 Generative design of a station represented with AI. *Source* Manuel Mühlbauer

struction process step by step and optimize it[12]—up to the networked construction site (Fig. 8.6).

In addition to AI, AR also offers exciting possibilities for architecture and the construction industry.[13] AR allows virtual 3D models to be projected into the physical environment. This opens up entirely new perspectives in the visualization of planning and construction processes. Architects can present their designs directly on site and offer customers an immersive experience.

Construction workers can use AR applications to utilize virtual models to better understand instructions during the construction process. In addition, AR can also play an important role in checking the quality of buildings. With the help of AR glasses, inspectors can more easily detect errors or defects in a building by overlaying virtual information on real structures (Fig. 8.7). It is expected that these technologies will also play a larger role in the real estate industry in the future.

[12] https://practice-research.com/presentations/prs-australia-june-2018/typogenetic-design (Accessed 01. March 2024).

[13] https://bit.ly/3SxdaF3 (Accessed 01. March 2024).

Fig. 8.6 Networked construction site of the future. *Source* Manuel Mühlbauer

8.3.3 With Digital Processes, Real Estate Developers and Builders Work More Efficiently

The real estate industry has made much progress in recent years to benefit from the latest technologies. In particular, housing associations, real estate developers, and builders use BIM in facility management and for managing the plant data from the technical building equipment and automation (TGA). These technologies allow real estate developers and managers to optimize their work processes and avoid recurring tasks in regular tenders.

Areas, costs, and maintenance requirements can be networked in digital space books and quickly queried. This also allows planning processes and construction interventions to be efficiently coordinated and escape and rescue plans to be adjusted. BIM also offers numerous advantages for the long-term management of real estate. Since all relevant information is digitally recorded, maintenance and repair work can be better planned. This leads to higher efficiency in the implementation of maintenance measures and to the reduction of downtime.

The buildingSMART specialist group Open-BIM in housing construction has developed a common understanding of the BIM method and describes the most important use

Fig. 8.7 Construction site visit with AR. *Source* Manuel Mühlbauer

cases in the field of housing. A corresponding white paper forms the basis for Identification, Description, and Publication of Added Value for the Housing Industry.[14]

Furthermore, BIM facilitates collaboration between various professionals such as engineers, electricians, and installers without structural privileges. All participants have access to the same information and can thus collaborate better.

8.3.4 Site Assessment and Feasibility Studies

By using BIM, it is possible to calculate construction projects accurately at an early stage. This allows the requirements from the site due diligence to be clearly and detailedly implemented in data-rich 3D models. The simulation of various development options, construction progress, and construction costs allows for an effective assessment of future construction projects. By working with BIM early on, building law can be calculated accurately and real estate projects can be efficiently adjusted to a budget.[15] In addition, BIM allows for a more accurate budgeting, as the costs for maintenance and

[14] https://bit.ly/bSD-Whitepaper-Wohnungswirtschaft (Accessed 01. March 2024).

[15] https://earlybuild.ai/ (Accessed 01. March 2024).

repair can already be determined in advance. Another promising approach is the use of artificial intelligence to predict rental prices and property values.

By analyzing large amounts of site information, historical market data, and sociodemographic factors, AI provides accurate forecasts about market profitability. It allows real estate investors to make informed decisions about buying or selling real estate based on individual location exposés.[16] In addition, landlords can optimize their rental prices based on current market trends. The awareness of different perspectives and market niches allows to critically question existing assumptions and make evidence-based decisions (Onsite, Earlybuild).

Finally, AR offers an innovative way to virtually visit properties. Potential buyers or tenants can use an AR glasses or a smartphone to display realistic 3D models of buildings. They can walk through rooms, change furnishings, and simulate various scenarios. This makes property viewing more flexible and convenient, especially for interested parties who are far away or have little time. It is therefore advisable for real estate developers and managers to integrate these technologies into their business process. Those who do this will be able to stand out from their competitors and meet the growing demands of the real estate market and better adapt their offerings to the market conditions.

8.3.5 Metaverse and AR

In the world of real estate sales and marketing, a lot has changed in recent years. AR is revolutionizing the shopping experience in the real estate sector with walk-throughs (Fig. 8.8), fly-overs, and guided tours of digital twins as well as interactive and personalized shopping experiences.

Metaverse offers groundbreaking technology that allows people to immerse themselves in virtual worlds and thus experience expanded networking and communication possibilities. This innovative platform not only revolutionizes the social aspect of our society, but also creates new opportunities for companies and educational institutions, e.g. for the training craftsmen.

There, AR offers great added value for training and training methods, through their flexible and individual process especially during business hours, participants involved benefit from it. By guiding employees through virtual support systems, even complex processes can be learned more easily. An example of this are technical repairs: Instead of relying on extensive manuals, employees can be guided step by step through the process using AR applications. This not only leads to a more efficient training, but also to fewer errors and a higher quality standard.

One of the most impressive applications of the metaverse is the collaborative work in a virtual environment. Whether colleagues are at different locations or even on different continents work—thanks to the metaverse and New Work they can easily interact with

[16] https://deepimmo.com/ (Accessed 01. March 2024).

Fig. 8.8 Interior of a foyer for representation in the Metaverse. *Source* Manuel Mühlbauer

each other and work together on projects. This means less travel, more flexibility and ultimately more efficient results.

But not only AR plays an important role in modern sales and marketing—AI also has a big influence. AI can be used to personalize customer offers and predict buying behavior predict. By analyzing customer data such as purchase history, preferences and demographic information, companies can create tailored offers and present their customers with more targeted advertising and personalized customer offers (recommendation system like Amazon).

Finally, AR enables more precise planning and implementation of projects. Architects or interior designers can, for example, create virtual models of their designs and project them at a 1:1 scale into the real space. This not only gives them a more accurate idea of what their project will look like, but can also identify and correct potential problems early on.

8.3.6 Further Thoughts on Technology Use for Building Planning, Construction and Implementation

The German location faces the great challenge of quickly and comprehensively providing affordable housing in the metropolitan areas to provide. To effectively meet this

Fig. 8.9 Robot-assisted prefabrication. *Source* Manuel Mühlbauer

challenge, we must use innovative solutions. A promising possibility is to use BIM in standardized housing construction. Through the use of the latest technologies and processes, we will be able to create housing more quickly and cost-effectively that meets the needs of our society.[17] But not only the real estate industry needs to change—politics and society must also contribute to bring about sustainable change.

AI has made enormous progress in recent years and has become an indispensable tool for companies, e.g. in innovations.[18] Through the ability, to analyze large amounts of data and recognize patterns, AI enables data-driven decision making. A major advantage of AI lies in the automation of repetitive tasks and processes. This not only saves time, but also costs for companies. Routine tasks can be efficiently performed by intelligent machines be done, so that employees can focus on more demanding tasks can concentrate.

Overall, there are many convincing arguments that AI represents a great added value for companies. From data-driven Decision-making through automated processes (Fig. 8.9) to personalized customer service and optimized production processes[19]—AI offers numerous opportunities to make companies more successful and efficient (Fig. 8.10). It is therefore essential for every company to engage with

[17] https://bauingenieur-studium.net/kuenstliche-intelligenz-im-bauwesen/ (Accessed 01. March 2024).

[18] https://boardofinnovation.com/tools/ (Accessed 01. March2024).

[19] https://constructible.trimble.com/construction-industry/the-benefits-of-ai-in-construction (Accessed 01. March 2024).

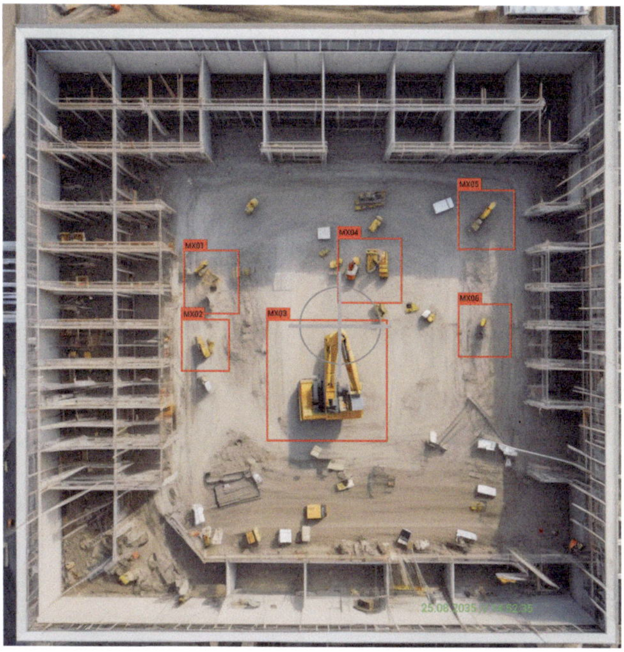

Fig. 8.10 Construction site monitoring with drone. *Source* Manuel Mühlbauer

the potentials of artificial intelligence and to incorporate this technology into their business strategy.[20]

In geographic information systems, AI identifies risks in the infrastructure and AR provides planners and decision-makers with additional information for complex spatial planning projects (XPlanning). The metaverse enables interactive learning in virtual environments in the education sector (e.g. Project MyScore at RWTH Aachen University Germany), while AR improves learning through virtual objects and description of real objects. In sales and marketing, AR provides an interactive understanding of the planned real estate projects and AI individual project information, which equips the buying experience with improved information. Smart Access as a key trend in the real estate industry[21] and Human-Centered Building Automation (HCBA)[22] round off the overall package in the proptech market.[23]

---

[20] https://buildingsmart-verlag.de/produkt/bim-und-ki-wissenschaft-und-unternehmenspraxis/ (Accessed 01. March 2024).

[21] https://swiftlane.com/newsletter-archive/property-technology-trends-2023/ (Accessed 01. March 2024).

[22] https://www.theben.de/service/fachpresse/hcba-gegen-energie-und-klimakrise/ (Accessed 01. March 2024).

[23] Https://www.proptechmap.com/ (Accessed 01. March 2024).

The future of the creative, construction, and real estate industry is diverse and will be characterized by individual planning approaches and lower planning tolerances in budgeting and implementation. High productivity gains in different dimensions will shape the division of labor approach of the construction industry with its own identity and its positive attitudes and methods.

Developers and real estate developers can use BIM, AI, metaverse, and AR to pursue innovative approaches to identify advantages and actively use them for their benefit. Through BIM, construction projects can be planned and executed more efficiently, which brings time and cost savings. Applying AI in the construction industry adds value by optimizing processes and detecting errors early. Metaverse offers the opportunity to create virtual spaces for 3D visualizations, which allows for improved planning and presentation.

AR technologies can be used to visualize construction processes directly on site and support workflows. Thus, workers on site can receive instructions or ask experienced remote experts for support. By combining these innovative technologies, developers and real estate developers can realize their projects faster, more cost-efficiently, and more sustainably. It is therefore advisable to actively use these tools to secure advantages in the competitive environment.

8.4 Transforming Urban Landscapes: Exploring the Impact of Web 3.0 on Smart, Net Zero-Compliant Cities. Revealing the crucial role of Construction Industry in the Digital Era. Aspects of Web 3.0 and its implications for smart, net-zero-compliant cities

Dr.-Ing. Dipl.-Ing. (FH) Bianca Weber-Lewerenz und Prof. Dr. Ingrid Vasiliu-Feltes

8.4.1 Introduction

The increased adoption of Web 3.0 is having a profound impact on the net-zero economy and represents a gateway to many frontier technologies. As the next generation of the World Wide Web (WWW), Web 3.0 brings a paradigm shift in how we interact with technology, data, and each other. The integration of technologies, such as AI, 5G or 6G, advanced computing, quantum computing, human–computer interfaces, and 3D printing, opens transformative advances across all industries. Current challenges facing urban areas, such as climate change, population growth and demographics, resource depletion, excessive energy consumption and increasing waste, lack of electromobility infrastructure, intelligent smart communication between buildings, transport and all digital networks involved, highlight the importance of defining innovative solutions.

One of the key drivers is the urgent need to achieve sustainable development goals and advance the transition to a net-zero economy. Climate change and environmental degradation require innovative solutions that promote sustainable practices and resource efficiency. Web 3.0 has the potential to support these goals and enable the creation of smart, resilient, and sustainable cities and infrastructure (Dashkevych and Portnov 2022). By harnessing the power of Web 3.0, cities can optimize energy consumption, improve transportation systems, waste management, and implement sustainable practices at scale.

In addition, Web 3.0 catalyzes the development of other frontier technologies and lays the foundation for the seamless integration and interoperability of these technologies. Web 3.0 enables the development of their full potential, e.g. AI portfolios can leverage the decentralized nature of Web 3.0 to securely access and analyze data complexity. The fast, low-latency connectivity of 5G or 6G networks facilitates real-time data exchange and improves the capabilities of AI applications. Advanced computing, quantum computing, and 3D printing benefit from the diversity of Web 3.0. They enable and strengthen collaborative design and sustainable manufacturing processes as part of Web 3.0's overall value chain (Fig. 8.11).

It is becoming apparent that human–computer interfaces are on the verge of being revolutionized by Web 3.0, enabling increasingly intuitive and immersive interactions with technology (Rowland et al. 2022). By integrating virtual and augmented reality, users can interact with digital content in more meaningful and interactive ways, increasing their productivity and creativity (Bastos et al. 2022)."

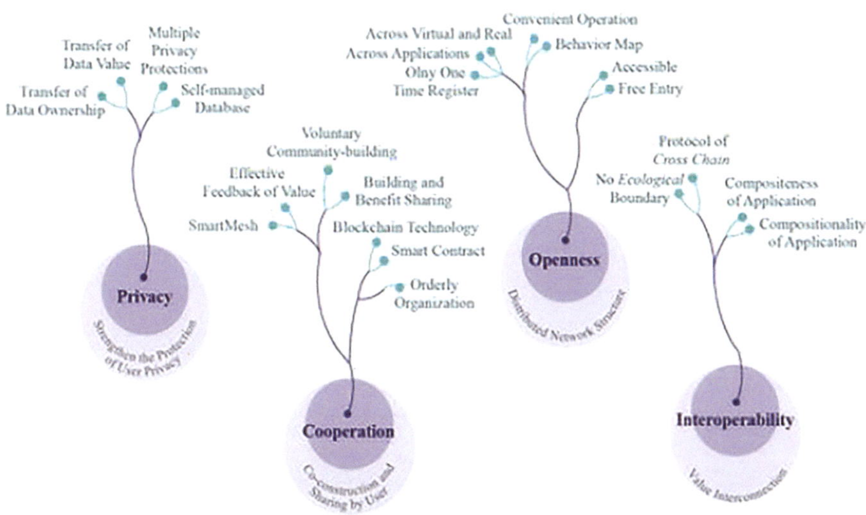

Fig. 8.11 Four Characteristics of Web 3.0. *Source* www.researchgate.net/figure/4-characteristics-of-Web-30_fig1_371347229

8.4.1.1 Leveraging Emerging and Frontier Technology

Web 3.0 and its converging technologies have a significant impact on the development of sustainable and climate-neutral cities. By integrating mature technologies, blockchains, spatial computing, virtual reality (VR), augmented reality (AR), extended reality (XR), edge artificial intelligence (AI), edge computing, and digital twins, cities can make their infrastructure and services more efficient and sustainable to achieve net-zero goals.

Intelligent technological solutions such as Web 3.0 help to respond to urban challenges, to involve customers and partners of cities from all walks of life and thus to promote the inclusion aspects of a city, to exploit opportunities and to achieve development goals. Smart digital infrastructure accelerates the development of smart cities and the potential to become a Web 3.0 hub. Thus, scientific research and literature sources increased significantly over past years (Fig. 8.12).

For example, blockchain technology is an essential part of Web 3.0. It plays a crucial role in creating transparency and security of technological systems for sustainable cities. By providing decentralized and immutable record-keeping, blockchain ensures the integrity and traceability of data related to energy consumption, waste management, and supply chains. This strengthens the trust of those involved, especially stakeholders, and makes it easier to implement decentralized energy trading, transparent supply chains, and efficient governance systems. This equally promotes sustainable practices.

In combination with VR, AR, and XR technologies, spatial computing enables immersive experiences and visualized urban environments (Rice 2022). This pioneering impact through a multi-sensorial and multi-modal experience enables urban planners and policymakers to make holistically thoughtful decisions for sustainable infrastructure design and resource allocation. These technologies simulate and analyze different scenarios help-

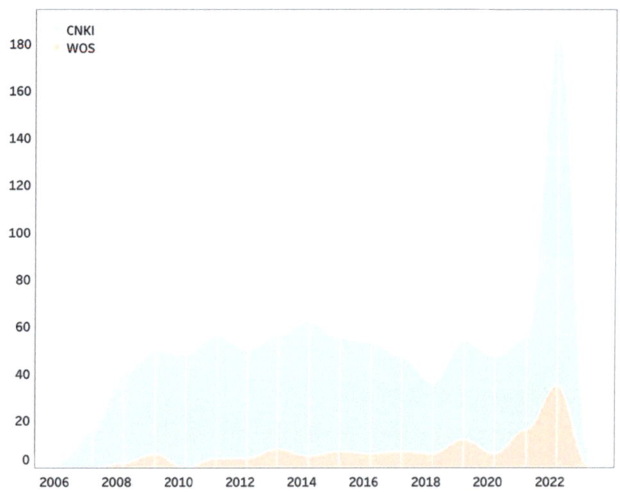

Fig. 8.12 Web-30-related-literature-publication-2003–2023. *Source* Ingrid Vasilu-Feltes

ing to optimize energy consumption, reduce waste, and improve the overall efficiency of urban systems.

Edge AI and edge computing are critical to the success of sustainable cities because they not only enable real-time data processing, but also significantly improved real-time decision-making at the network's edge. By applying these technologies latency can be reduced, responsiveness increased, and energy efficiency enhanced. They bring computational capacity closer to the data source. Edge computing has a transformative role: it transforms smart cities by enabling improved data processing, real-time insights, increased reliability and improved security. It enables cities to make data-driven decisions, optimize services and improve the overall quality of life of citizens. For example, edge AI and edge computing for the basis for real-time traffic management in intelligent transportation systems, reducing congestion, lowering emissions, and improving overall transportation efficiency (Bellini et al. 2022).

Digital twin technology combined with Web 3.0, offers high potential for sustainable and net-zero cities. Digital twins represent virtual replicas of physical assets, systems, and processes. These enable continuous monitoring, analysis, and optimization. Such combined capability makes predictive maintenance, urban planning, and resilience assessment possible, which in turn is crucial for sustainable infrastructure management. By leveraging real-time data from various sources, digital twins contribute to energy optimization, waste reduction, and the longevity of urban infrastructure (Antonios et al. 2023).

Furthermore, integrating cloud computing with edge computing complements sustainability efforts in cities. Cloud computing enables scalable and centralized resources for resource-intensive tasks like advanced analytics and machine learning. Cloud computing aggregates and analyzes data from multiple sources and, thus, enables holistic decision-making and optimization on all levels of smart city systems, enhancing sustainability, energy efficiency, and resource management (Haghshenas et al. 2022).

Web 3.0 architecture and its converging technologies offer a robust framework for developing sustainable and net-zero-centric cities. The implementation of blockchain, spatial computing, VR, AR, XR, edge AI, edge computing, and digital twin technology creates added value for cities by improving their infrastructure and services, thereby maximizing their share in achieving sustainable development goals (Zvarikova et al. 2022). The development of such technologies aims to create transparent transactions, immersive visualization, real-time data processing, predictive maintenance, and holistic decision-making. All of this represents the pillars for building resilient, sustainable, and net-zero cities of the future. Research in these technology fields demonstrate current trends when evaluating search of the keywords such as "blockchain", "engineering", "internet of things", "Network architecture", "AI" and "big data" in the Internet database (Fig. 8.13). These keywords also belong to the criteria used for selecting literature and sources for this study.

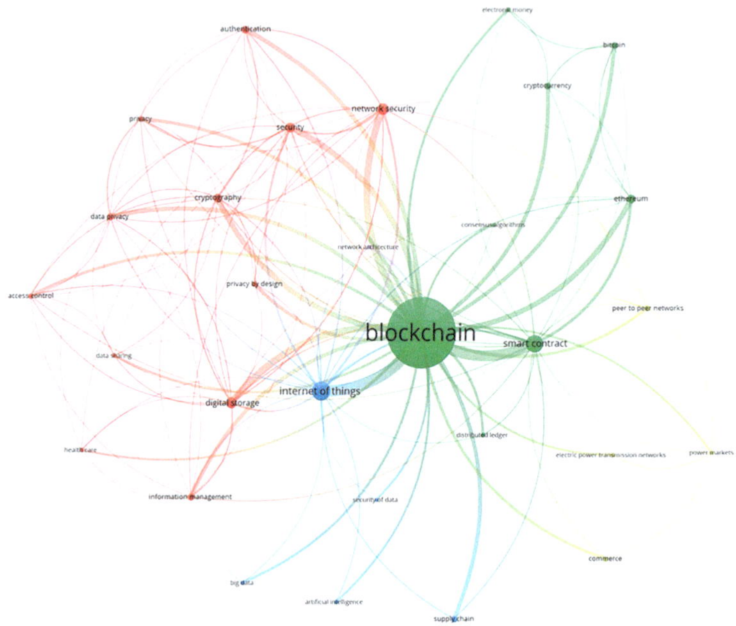

Fig. 8.13 Bibliometric map of the keyword's occurrence including "blockchain" in the "Engineering" field, in Scopus (2017–2022). *Source* Ingrid Vasiliu-Feltes

8.4.1.2 Spotlight-Impact of Web 3.0-Enablement in the Construction Industry

The transformation of cities into smart, resilient, and sustainable entities requires comprehensive efforts across various domains, including environment, energy, administration, education, economy, mobility, health, resilience, and safety (Deloitte 2023a). With the United Nations projecting that 70% of the global population will reside in urban areas by 2050 (Lai 2023a), it becomes imperative to address the resulting increase in emissions and energy consumption. This necessitates urgent action at all levels of decision-making to establish smarter urban transport networks, improve eco-friendly water disposal facilities, and design energy-efficient and secure smart buildings (Fabregue and Bogoni 2023). Given that most buildings are already constructed, retrofitting existing structures towards zero-emission construction is of utmost importance. Notably, energy communities, as exemplified by European Directives, promote cooperation among active and passive users to share locally produced energy through innovative management schemes, involving residential, industrial, commercial, and tertiary sectors (Trevisan et al. 2023). Software-defined networks (SDN), current trends in SDN-powered edge-enabled AI for IoT, edge computing-based IoT applications, edge computing to improve data network security of architectures and healthcare systems, strengthening against cyberattacks, improving simulation Energy consumption and smart city transportation, as well as AI-based technologies that facilitate consumer decisions, illustrate how these technologies are being used in construction projects to improve efficiency and sustainability. The high-tech smart cities showcase projects include Singapore, Amsterdam, Hamburg, Munich and Cologne. London scores with its technology and network infrastructure, for example with its very high number of 5G transmission masts. For the construction industry, this means ensuring a broadly developed cloud and network infrastructure that enables communication between construction machines, between intelligent buildings, remote building and city infrastructure maintenance, data transfer between all parties involved, data retrieval in real time, use of digital twins and AI.

8.4.1.3 Complexity of Technological Innovation and Human-Digital Transformation

This manuscript delves into the intricacies of technological innovation, the smart city ecosystem, and critical elements that underpin a harmonious human-digital transformation. It aims to guide key decision-makers in critically reflecting on their sustainability strategies with a social approach, thereby enhancing the quality of life. However, it should be noted that this manuscript represents a single facet of broader primary research on Corporate Digital Responsibility (CDR) in Construction 4.0. The research framework identifies, analyzes, and systematically evaluates essential factors for achieving sustainable digital transformation, particularly within the traditionally small-scale Construction Industry. Emphasis is placed on fostering a "healthy" human-technology interaction through human-centered engineering. In this regard, the "Excellence Initiative for Sustainable, Human-Led AI in Construction" was founded in 2020 to explore the potential of AI in standardized processes, data complexity management, cost and time efficiency, resource-saving construction, and the fulfillment of climate goals aligned with the UN Sustainable Development Goals (SDGs). As creators of the built environment, engineers, architects and construction companies assume crucial social and ethical responsibilities to achieve sustainable development goals. CDR focuses on the voluntary responsible development and use of technology by companies, while strengthening the responsible role of companies to positively shape the long-term social, environmental and economic impacts of digitalization and AI technologies with strong corporate ESG values. Such responsible behavior goes beyond legal regulations. The crucial question remains as to how humans can best design technologies that make sense and support human work and overall quality of life. AI streamlines planning, implementation, monitoring, and logistics processes, while digital twins and Building Information Modeling (BIM) enable visualization, simulation, risk assessment, environmental and social impact evaluation, and intelligent communication in Smart Cities. These methods offer a transparent and comprehensive basis for decision-making, paving the way for intelligent ecosystems. Education and awareness-raising efforts are integral components of the journey towards achieving these objectives.

A nuanced and differentiated understanding of the strategies for planning and implementing Smart City initiatives, as well as the factors contributing to their success, is crucial for achieving a harmonious balance between ecosystem growth, sustainability, and improved quality of life. Each domain of responsibility requires performance enhancement, a willingness to adapt (Deloitte 2023b), and targeted innovation and investment strategies to meet the increasing expectations of citizens and achieve Net-Zero standards. The German smart city market witnessed an average annual growth rate of 16.5% (in billion euros) from 2017 to 2022 (Arthur D. Little and eco—Association of Internet Industry 2017), with a projected doubling by 2030, as indicated by the Deloitte Manuscript (Fig. 8.14). The European Union leads globally in modern urban planning, investing significant resources in Smart City projects and initiatives (EU Parliament 2023a). The Internet of Things (IoT) plays a pivotal role in shaping the networked cities of the future, with mobile devices evolving into interconnected sensors and smart solutions. Intelligent and smart technologies offer opportunities to enhance environmental and energy efficiency, reduce citizens' stress levels, foster participation, and transparency, and improve connections to economic and social activities, even in rural regions. Moreover, smart city solutions optimize processes, value chains, and resource consumption, enabling CO_2-saving and climate-neutral effects, thereby enhancing the reputation, and expanding the value chain of the Construction industry (Fig. 8.15). Various communities worldwide have embraced strategies and initiatives aligned with this model, tailoring them to their unique geographic and infrastructural conditions. Effective administration acts as an enabler, putting the topic on the agenda and eliminating potential barriers, thereby facilitating the transition to smart cities or smart regions, and overcoming the challenges posed by rapid urbanization.

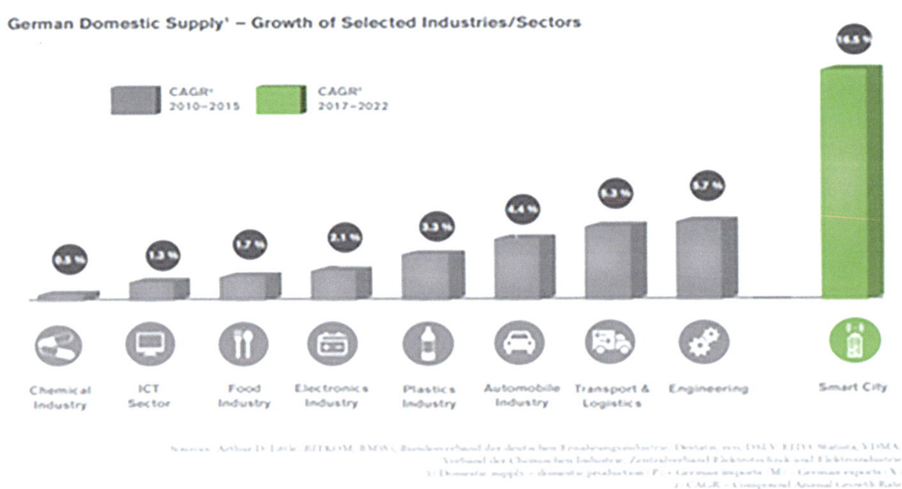

Fig. 8.14 German Domestic Supply—Growth of Selected Industries/Sectors. *Source* Arthur D. Little Study 2022

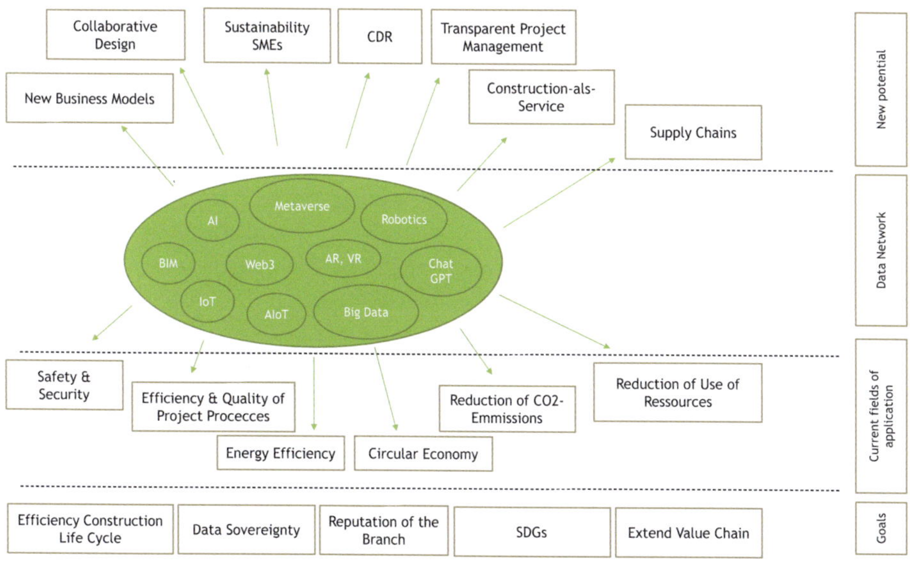

Fig. 8.15 Data Network in Construction. *Source* Bianca Weber-Lewerenz

8.4.2 Materials and Methods

This study applied a literature review, Internet and Database Research and the assessment of public discussions on media. It allows the scientific community to build on the published results. The criteria for selecting English language literature and sources include technologies' and smart cities' publications ranging from 2022 to date, which provide comprehensive knowledge of Web 3.0 ecosystems, current AI and edge computing-based technologies and socio-ethical implications.

The focus of the development of this paper is the adoption of social ethical principles, the promotion of interdisciplinary expert exchange and the expansion of multidisciplinary know-how to promote innovative approaches, responsibly use finite natural resources and optimize CDR in the construction industry. The study's research takes into account aspects of national and international strategies as well as the 17 Sustainable Development Goals (SDGs). Based on the fact that these new technologies are still at an early stage and are often limited to scientific research that lacks comprehensive practical experience, this study uses the hermeneutic approach as a qualitative research method to evaluate their results. It describes technological feasibility, examines its potential, effects and risks, finds connections and derives assumptions about innovative approaches for the sustainably efficient use of technologies.

8.4.3 Results

The development and implementation of Web 3.0 architecture and its converging technologies in smart cities present a host of challenges. To ensure successful integration and alignment with cybersecurity principles, ethics, legal and regulatory frameworks (Weber et al. 2023), interoperability, and performance measurement, critical factors must be addressed. This essay explores the challenges associated with these factors and emphasizes the need for comprehensive solutions in deploying Web 3.0 technologies for smart, resilient, and sustainable cities.

8.4.3.1 Cybersecurity Challenges

However, the increasingly complex interconnectivity of cities and the dependence on digital infrastructure simultaneously increase cybersecurity risks. While the decentralized nature of Web 3.0 offers higher transparency and resilience, it introduces new, additional vulnerabilities that require effective mitigation. Robust security measures, such as encryption, secure access controls, and continuous monitoring, are critical to protecting sensitive data and critical infrastructure from malicious actors. In the Construction Industry in particular, legal regulations adapted to new technologies are crucial for protecting contract, personal, user, building operational and overall project data that is communicated and exchanged between the complex participants in a construction project. It also plays a crucial role in ensuring that automated processes function safely and always remain accessible and controllable for humans. This remains a significant advantage for the construction industry when it has a Web 3.0 infrastructure but represents an intense challenge that still needs to be further scientifically and legally examined in relation to ethical aspects, secure cyber networks and data security.

8.4.3.2 Ethical Considerations

The vast collection and utilization of data in smart cities raises concerns regarding privacy, consent, and responsible data usage. Here, ethical frameworks adapted to technological feasibility such as Web 3.0 are mandatory. These are to ensure that individual rights are respected, inclusivity is promoted, and data misuse or abuse is prevented. The focus is on protecting privacy and handling personal data responsibly. In addition, ethical aspects represent an inner compass such as fairness, human and social values, moral orientation when deciding which technology makes sense to facilitate human work and which does not, which technology supports more efficient, productive and safe processes. The emphasis is on support. Technical feasibility does not necessarily have to be pursued solely based on technical progress and financial gains. Technical progress that recognizes human responsibility for society and the environment is the key to shaping and achieving sustainability in all areas of life and work for all future generations.

8.4.3.3 Legal and Regulatory

The existing legal frameworks must be adapted to the complexity and possible negative effects of decentralized systems and technologies. This cannot be achieved by companies respectively, for example by using CDR in line with CSR. Rather, policymakers and legislators need to update regulations and establish guidelines for the use, data ownership and accountability of Web 3.0 technologies. The development of international standards can facilitate interoperability and compatibility between systems, fostering seamless integration and collaboration between cities. This makes it all the more important to maintain secure data communication, ensure the security of smart and intelligent cities with a view to higher quality and more sustainable built environment ecosystems, and strengthen trust in and collaboration with new technologies. Digital twins and AI-based simulation methods only offer trustworthy support if they are maintained responsibly and used to expand the entire value chain.

8.4.3.4 Interoperability

Integrating diverse Web 3.0 technologies necessitates effective communication and data exchange between platforms. Standards and protocols for interoperability must be developed to achieve seamless integration, data sharing, and interoperability across applications and systems. Solving the challenges of interoperability would significantly increase the efficiency and effectiveness of smart city operations.

8.4.3.5 Customized Metrics and Performance Measurement

What is most essential for evaluating the performance and impact of Web 3.0 technologies in smart cities is the determination and creation of tailored metrics and Key Performance Indicators (KPIs). Customized metrics and KPIs should accurately assess the social, economic, and environmental aspects of smart cities. Implementing effective impact measurement enables stakeholders to monitor progress, identify areas for improvement, and ensure the fulfillment of sustainability goals.

8.4.3.6 The Role of Education

A crucial challenge lies in promoting education and awareness. The technical infrastructure and equipment of educational institutions and community buildings must adopt modern technologies to facilitate high-speed data transfer, communication, and storage. A smart city requires a society with a new culture of thinking, embracing innovative technology to act flexibly, environmentally friendly, resilient, and agile. Leading cities like London, Zurich, Singapore, Dubai, Amsterdam, Barcelona, Vienna, and Nanjing serve as best practice examples, demonstrating innovative technological architecture that reduces CO_2 emissions, minimizes traffic congestion, and enhances efficiency (Locke 2023a).

8.4.4 Discussion

8.4.4.1 Harmonization and Integration

Building Web 3.0-powered sustainable and resilient cities also presents many opportunities, including harmonizing and integrating net-zero metrics with other DEI (Diversity, Equity, and Inclusion), operational, or financial metrics. The combination of these metrics strongly supports cities to achieve better results and promote a holistic approach to urban development.

Web 3.0 provides a framework for seamless integration of different metrics and data sources. It also enables a comprehensive understanding of urban performance and progress, and how its sustainability goals, such as achieving net-zero carbon emissions, are aligned with other key metrics related to social equity, economic viability, and operational efficiency.

Only by harmonizing net-zero metrics with DEI metrics inclusive and equitable cities can be created. Sustainability initiatives recognize diverse factors, such as access to affordable housing, public transportation, and clean energy, and benefit all residents, regardless of their socio-economic background. In this context harmonization means ensuring the participation and access to digital ecosystems for all sections of the population, with and without physical or mental impairments and disabilities, without biases and without creation of disadvantages due to gender, origin and appearance. Harmonization requires equal treatment in urban environments and proactive inclusion measures for disadvantaged people by technical translating and recognizing their user behavior. Web 3.0 can aggregate and analyze diverse data sources. This technology enables cities to identify disparities, track progress, and implement targeted interventions to address both social and environmental inequalities.

Embedding net-zero metrics with operational and financial metrics enables comprehensive reporting on a city's performance in terms of sustainability and social and environmental responsibility. Assessing the combined data of energy consumption, waste management, and financial investment data enables cities to optimize resource allocation and identify cost-effective strategies to achieve sustainability goals.

This transparency fosters a conducive environment for investment, innovation, and collaboration, driving the transition to a sustainable future.

8.4.4.2 Hybrid Augmented Intelligence

Efforts to develop Smart Cities are based on the diverse fields of applying emerging and frontier technologies, optimized human–computer workflows, and hybrid augmented intelligence. The focus here lies on responsive adaptation to constantly changing external conditions and influencing factors and increased resilience. The basis for an efficient infrastructure and adaptable management at the municipal and regional level can be created by suitable digital solutions. Viewing recommendations provided by a streaming video service represents a daily example. Therefore, optimized human–machine interfaces are required to help users express their knowledge and retrieve information from difficult-to-use software.

As the central authority, citizens with their needs for a sustainable quality of life are the benchmark for further developments. However, economic power and climate protection are on an equal footing in the Smart City and are just as important in a livable city of the future. The new culture of re-thinking cities and smart, intelligent construction laying the foundation for generations to come requires the awareness to pro-actively design and protect the environment of work and life, knowing the potential of and applying innovative technologies. Technical progress means more societal responsibility in dealing with the unintended consequences of innovation and protecting values and democratic rights (Weber-Lewerenz 2022a, 2023a). Nevertheless, it also addresses the awareness of fully exploiting the potential of technical innovation to improve the protection of re-sources, environment, and climate and shape agile and resilient urban ecosystems as the caterpillars towards efficiency and productivity, but most of all, sustainability of the digital age (Weber-Lewerenz andTraverso 2022).

8.4.4.3 Continuous Optimization

Efficiency in waste management, reduction of CO_2 Emissions, regulation of urban circulation, and responsible material usage are pertinent challenges that require scientific investigations and research across multiple disciplines. The use of apps allows visualization of daily energy consumption compared to ideal values, simulation of the actions required to achieve the expected carbon footprint, comparability with other smart cities and possibly demonstration of best practices. These issues necessitate a comprehensive approach to all levels of responsibility to achieve holistic sustainability. This essay delves into the interplay between technological advancement and the societal and human common good, emphasizing the importance of a value-based foundation. Additionally, it explores the ethical considerations associated with artificial intelligence (AI) and emphasizes the need to preserve trust and prioritize the common good in human–machine interactions.

8.4.4.4 Technological Advancement and the Common Good

The continuous acquisition of new knowledge and the responsible use of experience ensures a profound understanding of the diverse and multiple interdependencies and interactions between technological progress and social well-being. In its March 2023 statement (German National Ethics Council 2023a), the German National Ethics Council, emphasizes that AI must not diminish human development. It highlights the importance of harmonious human–machine interaction, human intelligence, responsibility, and judgment. The preservation of trust and alignment with the common good remain paramount throughout this endeavor. These principles align with the European AI Strategy and the digital transformation framework, which promote individual participation, diversity, and inclusion. Education stands as the foundation for these endeavors.

8.4.4.5 Human Agency and Ethical Reflection

While discussions on technological feasibility, smarter tools, high-tech innovations, and artificial intelligence advancements proliferate, it is crucial to recognize the moral and ethical obligation to critically examine the multifaceted impact on human lives within these parallel virtual worlds. Human intelligence, the driving force behind technological progress, thrives on creativity and reflection. The progress made in the real world is inherently rooted in human ingenuity and thoughtful contemplation.

8.4.4.6 Governance

The rapid advancement of AI has raised concerns about the need for ethical reflection and responsible development. In March 2023, the Future of Life Institute issued an Open Letter (Future of Life Institute 2023a), signed by thousands of scientists, calling for a temporary pause in giant AI experiments to prevent uncontrolled competition and assess the implications of such developments. The letter emphasizes the importance of state regulation to ensure responsible AI development (Bucknall and Dori-Hacohen 2022; Bubeck 2023). However, addressing these challenges also requires the acquisition of new knowledge, skills, and qualifications, particularly in teaching and curriculum development. Human agencies play a pivotal role in determining the relevance of methods and technologies, optimizing human work, and monitoring and improving efficiency and productivity. Humans not only drive technical innovation but also exercise control and halt activities if they pose significant risks to sustainability (OECD 2023a).

8.4.4.7 Promoting Responsible AI Development

For even greater sustainability, it is crucial to promote innovations that are in line with human values and significantly support social well-being. AI systems should be subject to stricter security measures to ensure their safe operation throughout their lifespan. To minimize unreasonable safety hazards (OECD 2023c) continuous assessment and management of risk hazards are essential. A human-centered approach and inclusivity is the linchpin for the development of AI. This is the most feasible way to build trust and achieve broad acceptance in society.

8.4.4.8 European Initiatives for Trustworthy AI

Recognizing the need for a coordinated approach, the German Federal Association for AI launched the LEAM Initiative in collaboration with industry representatives and researchers (LEAM 2022). This initiative focuses on developing larger European AI models within a hyper-scale infrastructure, grounded in European values. It challenges political and regulatory stakeholders to establish a European Data Ecosystem that supports research and development efforts while ensuring trust and reliability. Governments have a critical role in supporting the development and accessibility of a digital ecosystem that fosters trustworthy AI.

8.4.4.9 Empowering Digital Twin Technologies in Web 3.0-Enabled Smart Cities

Web 3.0 offers a robust foundation for the digital transformation in the construction industry. As an integral part of Web 3.0 blockchain technology ensures secure and transparent transactional processes and counteracts the fraud of risks, counterfeit materials, and inefficient supply chains. Recording and validating the origin of materials as well as improved traceability via using blockchain have positive effects on building trust between those involved and streamlining the construction process.

The use of digital twins in combination with Web 3.0 offers unparalleled advantages in the fields of construction, infrastructure, and utility networks. Digital twins, virtual replicas of physical assets, systems, and processes, enable real-time urban infrastructure monitoring, analysis, and optimization. Integrating data from different sources using digital twins facilitates a comprehensive understanding of the built environment: Additionally, stakeholders can use reliable sources to make more informed decisions and optimize system performance.

Embedding Web 3.0 and digital twins into daily construction business and infrastructure design improves collaboration of all involved, not limited to stakeholders. Such team collaboration catalyzes the evaluation of designs, identification of potential problems, and optimization of construction processes through shared access to real-time data and immersive visualization. Such an innovative approach essentially improves communication, reduces errors, and accelerates project timelines, thereby improving efficiency and cost-effectiveness.

The resulting synergy effects also enable further technical progress regarding construction material sciences and nanotechnology. The use of these cutting-edge disciplines could revolutionize construction practices. The aim of such technologies is to ensure improved efficiency, sustainability, and safety. For example, the integration of nanomaterials in building components improves strength, durability, and thermal properties. Digital twins simulate and evaluate the performance of these advanced materials in the built environment, optimizing their use and reducing waste. The real-time simulation data provides basis for improved decision-making.

It is noteworthy that the high potential of Web 3.0 and digital twins in utility grids also lies in significantly improved energy management, distribution, and consumption. Real-time data collection, analysis, and predictive methods enable optimization of energy consumption, allocation of inefficiencies and improvement of the overall resilience of utility grids. This so-called intelligent monitoring and control of energy systems contributes to the development of sustainable, net-zero-compliant cities.

The strategic use of Web 3.0 architecture and mature digital twins offers transformative opportunities for the construction, infrastructure, and utility sectors. Responsible implementation of new technologies such as Web 3.0 could revolutionize construction practices and infrastructure design (Reis et al. 2022). Such an integrative approach lays a new foundation for leading cities towards a future characterized by greater sustainability, safety, and collaboration. It paves the way for smarter and more resilient urban environments.

8.4.5 Limitations

This manuscript focuses on the role of Web 3.0 and converging technologies in the construction sector in developing smart, resilient, and net-zero-centric cities. However, it is important to equally recognize that numerous additional factors come into play in shaping such urban developments. The technological limitations and potential biases or gaps examined represent important aspects to be considered when exploring Web 3.0 ecosystem scenarios. The challenges and opportunities discussed in this paper provide only a glimpse of the broader landscape of creating sustainable and resilient cities, simply reduced to the technological status quo at the time of writing.

Achieving smart, resilient, and net-zero-centric cities requires a multi-sectoral approach. In addition to construction, industries in sectors such as energy, transportation, waste management, and governance will play a crucial role in this transformation. Collaboration and careful orchestration of efforts across these industries will be essential for success.

There are a number of security risks associated with IoT/IoE, digital twins and blockchain and new challenges arise for protecting urban infrastructures and the data contained therein from failures and (cyber) attacks, the complexity of which has so far been little investigated. It is therefore necessary to discuss these more closely and implement new solutions for protection. The concept of smart city, concept of digital city belongs to national strategies to mitigate risks and ensure safety standards. The challenge consists in formulating national and internationally comparable safety and security standards recognizing the multiple aspects from legal, fire safety, protection from war and terrorism, police, operators of the safety–critical infrastructure.

We only mentioned challenges associated with implementing Web 3.0 and converging technologies in this manuscript. However, further research and collaborative efforts are required to design and deploy viable solutions for these barriers. Additionally, it is essential to recognize that each city has its unique context, challenges, and priorities. To design and implement smart, resilient, and net-zero-centric urban developments, it is even more important to understand and recognize the local context, cultural aspects, and community engagement. It is now clear that trustworthy, socially sustainable planning solutions tailored to the specific needs of a city require a deep understanding of local challenges, resources, and aspirations.

8.4.6 Conclusions

This paper introduces one of the pathways for sustainable innovation. Such innovation focuses on deploying converging emerging and frontier technologies and a complex Web 3.0-enabled infrastructure. Web 3.0 means a paradigm shift and a gateway to a new era. It requires corporate and governmental social and digital responsibilities in multiple fields. The integration of emerging technologies into planning and construction processes offers both a challenge and an opportunity. By leveraging Web 3.0 architecture and the associated development of emerging technologies, cities can promote inclusive development, optimize resource allocation, and enhance transparency. Integrating these metrics leads to sustainable outcomes while creating resilient and thriving urban environments. In these, the focus is on the well-being of residents, economic prosperity, and environmental protection.

By showing the courage to innovate and using profitable technologies such as Web 3.0, we shape a future that is not only geared towards net-zero, but also drives innovation and progress across all related industries and sectors. This approach ultimately benefits society, the environment, and sustainable digital ecosystems. The conclusions made provide actionable insights for policymakers, urban planners, and industry stakeholders. The highlighted recommendations and identified strategies for overcoming the challenges significantly advance the development of smart, resilient, and net-zero-centric cities. However, both ethical and sustainable practices in construction require continuous critical questioning in order to make work and decision-making processes more efficient, productive, safer and significantly easier for human work in order to address ESGs and achieve SDGs. Furthermore, such recommended practices not only provide actionable insights for policymakers, urban planners, and industry stakeholders, but for overcoming the challenges identified in the study and advancing the development of smart, resilient, and net-zero-centric cities. The more technical feasibility increases, the more impacts on human, society, environment increase, the more societal and environmental responsibility increase that companies need to assume. Here, the CDR approach represents an important catalyst and offers guidance to manage the new technologies and fully exploit their potential in a socially sustainable way. Further in-depth investigations into the diverse impacts of Web 3.0 technologies on urban governance, economic development, and social justice enrich the discussion in both applied sciences and research.

While this paper has shed light on the potential of Web 3.0 and AI technologies in the construction sector, it is clear that achieving the vision of smart, resilient, and net-zero-centric cities requires a comprehensive approach that spans various industries. Overcoming the limitations and challenges associated with this change, while addressing the specific needs and desires of each city, will be critical to creating sustainable and thriving urban ecosystems.

8.5 DARE Decentralized Autonomous Real Estate

Dipl.-Ing. Sandor Horvath

I would like to embark on a thought experiment with you. What you will read on the following pages is a self-developed concept, which does not yet exist in this form and can certainly only be manifested through healthy discussions in expert exchange circles. My concern is to rethink a property or a piece of real estate in a completely new way. I want to try with you to break through the principle of closed value creation silos along the life cycle considerations of buildings. Planning, building, operating, and redeveloping or demolishing, these are the four major sections of the built environment that keep repeating. At least when referring to the property, the unique, geographical assignment of a building. A meadow on the edge of the village becomes a cultivated field. If the settlement grows, there may be a development plan and the field becomes a new residential area, with a new parceling. Civil engineers come, lay house connections, roads are built, young families come, talk to financiers, look for specialist planners, construction companies, craftsmen, installers (Fig. 8.16). The planning is underway, official channels are pending, the construction execution shows every day that the dream of owning a home is getting closer. Construction acceptance, moving company, smarthome, operation, use, life. New heating law, appraiser, geothermal consultant, solar system distributor, replanning, fitters. Conversion, roof renovation, retirement, move out. Real estate evaluators, consultants, brokers, interested parties. Demolition, new construction, new beginning. What I want to say with this: A building lives, a place develops, but what never changes is its uniqueness of location in a limited environment. Many actors contribute to the history of this property.

Fig. 8.16 Many actors in construction—AI generated photo. *Source* Image Creator from Microsoft Designer

What they all have in common is the participation in shaping this place. But what is missing is to live this connection between the individuals, to gather experiences from each other, to pass on, to exchange information. Important information, important data. For example, what about the soil? What is its load-bearing capacity, can I use geothermal energy? Were bombs dropped in the area during World War II? What about the basement? Is there a lot of precipitation in the area, does the water seep away well? And groundwater? Can I build a well, any agricultural operations in the area? Nitrate contamination? What about the withdrawal fees? How high are the potable water discharge fees according to the municipal code and who do I go to to find out? And all this just from the perspective of the small house builder. And yet there are banks, insurers, city administration, public transport network operators and many others who come into contact with this little piece of earth in one way or another.

My concern here is to raise awareness for a new way of thinking about a property, whether it is already developed or still in the planning stage. Whether there is already a skyscraper at this location or fruit trees on the plantation with local fruits. As long as we as humans cultivate the land, settle down, expand and move on, we shape the environment. So there is untouched nature on the one hand and the environment where people intervene and shape on the other, referred to here as the built environment, which is what this is all about.

Ideally, the usage phases of real estate should last as long as possible, so we will not be able to ensure that the same participants involved in the development, planning, construction, and operation are also involved in the revitalization or demolition. Therefore, we need a continuous layer of information for optimal consistency of knowledge about the building. A data pool of the digital twin that ensures the best for everyone involved in the property in any way. The best possible transparency, calculation, the best living situation, and the highest level of safety during demolition, and so on.

So, what if we continue this thought experiment by considering the property as its own organ, organizational structure, or independent unit? However, such structures, like companies, are almost always centrally managed, and the best possible continuation of the organization depends on the wise decisions of management. This can lead to errors or imbalances in advantages and disadvantages. Centralized, hierarchical systems do not optimally meet such diverse and perhaps not yet fully defined needs, especially over such a long period.

New forms of distributed database structures are now producing solutions that do not require centrally controlled organizational structures. Organizational forms that can operate completely autonomously. We are talking about a concept called "DAO". DAOs are digitally supported, decentralized, autonomous organizations, in English: Decentralized Autonomous Organization. This is nothing less than a revolution in organizational structures.

This revolutionary concept originated in the world of blockchain technology. Blockchain is probably best known because of Bitcoin, a cryptocurrency that does not require a central identity. It does not need a mediator, a central trustee, a bank, or an intermediary to coordinate a transaction between two people in a tamper-proof and trustworthy manner. To briefly outline, this works by executing a certain logic on many computers distributed worldwide. The transactions, like in traditional banking, are recorded on digital tabs, nothing more than a simple database. But instead of individual banks communicating with each other to coordinate the coverage amount of a payment and the transfer from A to B, this happens in the Bitcoin blockchain through a globally spanning network. The tamper-proof validation of the transaction is checked, encrypted, and stored on all participating registers (nodes) by solving complicated computational tasks, which are done by so-called miners. The larger the number of nodes, the more secure the network. To hack the blockchain, you would need to crack at least half of all computers in the network. However, the transaction information is timestamped and elaborately encrypted, making it almost impossible to cheat this network.

In the Bitcoin network, the register only contains a few pieces of information, namely at what time A sent which sum to B and the verification whether A really had this sum in his account at this time. That's it. Other networks, like the Ethereum blockchain, can even execute small programs and thus create entirely new applications for this technology. Thanks to this blockchain, a large real estate transfer could be done without a trustee and notary. For example, a large real estate project worth several hundred million is to be transferred from a project developer to an investor. The costs for the notary and the entry in the land register amount to 1.5%, which in our case is a single or even double-digit million amount and can also take weeks until the process is completed. Real estate transfer taxes of 3.5% (Bavaria) to even 6.5% (NRW) are added on top. To avoid these taxes, such large projects are often put into their own companies and the shares in this, for example, GmbH are sold. In a so-called blockchain-based tokenization of the property, essentially the same thing happens. The packaging of the the project, however, is carried out digitally here and instead of the GmbH shares, the tokens that represent the property can now be transferred. Depending on the network chosen, this transfer then only takes a few minutes or even seconds and costs between a few euros or just a few cents. And all relevant information about the property, the value represented in the token and even the transaction conditions are stored in this small distributed computer program, the so-called Smart Contract.

But now back to our DAO.

The idea of Decentralized Autonomous Organizations was first outlined in 2013 by Daniel Larimer under the title of decentralized companies (Decentralized Autonomous Companies, or DAC for short). The breaking of hierarchical systems is not new. Cooperatives and associations, even corporations, can grant voting rights to members and shareholders, thus introducing a democratic element into rigid structures. However, this participation is usually limited to the annual general meeting and only a few functions. For example, one can decide on the discharge of the board or the appointment of management. The fortunes of the company are then delegated to those managers. A bit like the federal elections every four years. The election is democratic, but active and direct participation in the government is not granted to the ordinary citizen. Therefore, the idea of a DAO is much more profound.

So, let's imagine an organization without any hierarchy, supported by both the people and the previously determined rules of cooperation, which have been entered into the decentralized Smart Contract code of the blockchain. The first implementation of a DAO was built by developer Christoph Jentzsch in 2016 as one of the first employees of the Ethereum blockchain. During his studies, he was already fascinated by self-organizing systems in nature, such as ant colonies, beehives, or bird flocks. Systems in which there are different functions, such as a queen, but she does not distribute tasks in a hierarchical form. Rather, there is a natural predisposition as to how this people or this swarm positions itself to serve the community and thus itself. Christoph Jentzsch recognized the potential of digitization and the resulting globally networked communities and developed "The DAO" on the Ethereum blockchain. Network participants could buy into the DAO using their own cryptocurrency Ether and acquire shares of this digital organization in the form of tokens. For understanding: the money did not go into the pocket of Mr. Jentzsch or anyone else, but into the common pot of this DAO. The token served as a voting right, which was to decide what should happen with the money pot. The debates were then conducted via discussion forums, which were completely free of a centrally controlled committee. This concept excited so many people that within a month about 150 million dollars were put into the money pot, thus unexpectedly the largest crowdfunding of all time had taken place. Due to an error in a line of code, however, this visionary development was hacked and led to one of the most exciting cyber crimes, which can be read under the title "DAO WARS".

What remains, however, is the idea of decentralized, autonomously organized blockchain communities, which have now found many imitators. There are decentralized exchanges, aid organizations or various subculture interest groups based on this technology. So why not rethink our built environment (Fig. 8.17). A property could also be planned, bought, used and operated in a completely different way. If a building is to be rented, then the neighbors could join together as an investment community and collectively have a property developed that remains in their community ownership. The architect could bring in his visions and the intellectual property his design could be adopted by other real estate DAOs for a fee. A smart contract from the local utilities could have a say in the operation, in order to monitor waste disposal, sewage, electricity or heating, and thus coordinate the resource-saving optimum for the entire settlement, which in turn also acts as a DAO and defines and determines this optimum through a democratic process as a collective. And if various real estate, district, city or even state DAOs communicate with each other based on automated codes, i.e. the rules or digitally implemented legal regulations, errors could be avoided. The planning documents of the HVAC engineer could correspond directly with the requirements of a new heating law without human intervention. The smart contract then determines whether the planned boiler can be installed or not. And if the citizens do not like the law, they can participate in the legislation through the DAO of the heating engineers together with the DAO of sustainability research. Thus, the best could be extracted for a property from various perspectives. The property could be developed over the years with the decisions and plans without media disruption so that the best possible regulations are created for the respective use. Voting rights can of course also be transferred. The struc-

Fig. 8.17 The digital twin in the center of the decentralized network—AI generated photo. *Source* Image Creator from Microsoft Designer

tural engineer from the construction phase may not be able to have a say in a change of use, but the provided information about the load-bearing capacity of a certain wall should not be lost in order to avoid construction damage. All data can be assigned to the property in a tamper-proof manner. All decisions can thus be made decentrally for the benefit of all. This is the basic idea of DARE, the Decentralized Autonomous Real Estate.

8.6 Human-Centered Technology Design and Societal Perspectives in the Transformation of the Built Environment

Prof. Dr. Katharina Hölzle, Dr.-Ing. Steffen Braun, Dr. rer. nat. Moritz Maier and M.A. Michaela Friedrich

8.6.1 Technological Progress in the Urban Millennium

When it comes to the future of construction, it inevitably concerns the future of our cities as built environments and central living and economic areas of our society. The twenty-first century has rightly been described as the first urban millennium, at least since the 50% threshold of city dwellers in the total world population was exceeded in 2007. In the coming decades, despite planetary boundaries already being exceeded, the amount of urban space worldwide will double—a development that is hard to imagine with completely new requirements and consequences for the global planning, construction and operation of real estate and areas. Buildings and structures rarely arise or exist alone in a wide area, but are individual building blocks of larger, interconnected infrastructure and urban systems (Braun 2024). They are located either in public, semi-public or private spaces and are subject to different legal requirements or different functions and operator models depending on the type of use.

In view of this global paradigm shift, the understanding of technological progress at the level of urban development is correspondingly important: the built environment has always been influenced by technical innovations and is driven by them even more strongly in times of climate-neutral cities, smart cities or circular urban material flows. Previous research by the Fraunhofer Morgenstadt Initiative has shown that urban structures change only very slowly, at around 0.5% per year—so it takes mathematically 200 years for a city with all of its buildings and infrastructure to fundamentally renew itself. Digital technologies are diffusing much more quickly in the built environment: International construction projects address 'cognitive environments' or the relevance of 'Digital Urban Experience Design' as a new user-centered interface between the physical and virtual environments (Fig. 8.18). And not least in times of the "metaverse", the merge of virtual and real worlds, the question arises as to what role people and soci-

Fig. 8.18 Digital Twins in Construction with Laser Scan data in virtual realities. *Source* Fraunhofer IAO

ety play in technology development and transformation. Today and in the future, despite the increasing influence of digital technologies and the shift in many areas of life to virtuality, it is particularly important to create living environments that address the exchange and interaction of people with different needs and expectations.

8.6.2 Applied Research for the Human-Centered Transformation of the Built Environment

At the Fraunhofer Institute for Industrial Engineering and Organization IAO, applied research for the construction industry and urban system design represents a distinct cross-sectional topic, which began in 2007 with the joint research project FUCON (Future Construction), resulted in the Morgenstadt Initiative in 2011 and has since then, among other things, in the area of Virtual Environments' is continually being researched for the prototypical application of immersive planning and visualization tools based on Building Information (Wenzel et al. (2022) Modeling (BIM). Potential and scenario studies on AI applications in the construction industry (Giannakidis et al. 2021), strategies for the climate-neutral construction site in 2045 or on neighborhood development of the future (Braun 2023) offer scientifically based decision-making principles and recommendations for action for the construction industry—up to the real estate industry or the facility management of building complexes or campuses. Experiences were most recently gained in the EU H2020 research project BIMprove (Improving Building Information Modeling by Realtime Tracing of Construction Processes) with a European consortium for the combined use of drone-based building detection (Scan2BIM) until 2023 or in combining artificial intelligence with virtual reality, digital twins and 'Wearables' (Liinasuo et al. 2022). The high relevance of the right data for planning and execution is not only evident in these research projects. To date, availability, access and handling have been inadequate when it comes to the technical requirements for AI applications, for example. A fundamental understanding of ethics, diversity, data and personnel management is still an open issue in many trades. As recently as in autumn 2021, only 2.4% of construction companies surveyed said that staff could be replaced through automation and digitalization (HDB 2022). Despite and because of the lack of financing options for current construction projects due to the high construction costs and a simultaneous increasing demographic-related shortage of skilled workers, construction companies cannot avoid investing more in productivity-increasing measures in the future, such as series and prefabrication, serial renovation, 3D printing and BIM.

While labor productivity in manufacturing companies has almost doubled since the reference year 1991, in the construction industry it has fallen by four percent over the same period, according to calculations by the Federal Statistical Office (HDB 2022). For the future, construction needs to be radically and, above all, systemically rethought—this requires strategic future scenarios, comprehensive digitalization of processes, more diversity, creativity and new collaborations across industry boundaries.

8.6.3 How Do We Want to Live in the Future?—the Research Project. City. Country. Chances

In times of rising rents in many cities, vacancies and migration in parts of rural areas, changing social structures, increasing digitalization of life and new climate and environmental requirements, the question "How do we want to live in the future?" is a big one in Germany. Since answering this question should not only be the responsibility of experts, in the research project "City.Country.Chances." (Fig. 8.19) (German "Stadt.Land.Chancen") of the Center for Responsible Research & Innovation (CeRRI) at Fraunhofer IAO, 8,787 citizens were asked about their wishes and needs with regard to the building and living of the future (Maier 2021).

Before the survey, the participants were presented with different hypotheses for the future in a scrollytelling format, which are briefly described below.

1. Building Information Modeling digitally records building data in order to meet the increasing demands on building planning.
2. Flexible living arrangements and modular interior design enable adaptation to various phases of life.
3. Circular economy in the construction industry improves the ecological balance and contributes to achieving sustainability goals.
4. Integration of digital technologies transforms living spaces into smart homes.

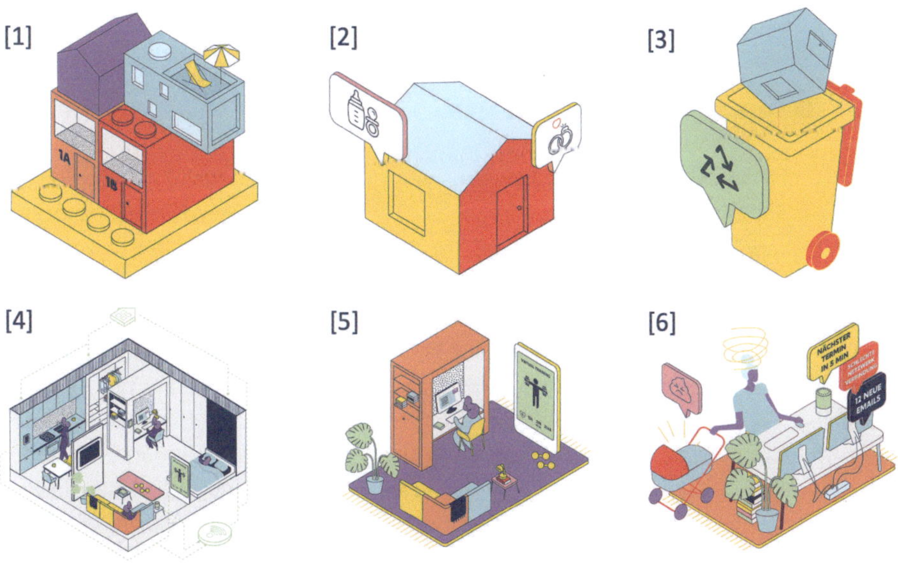

Fig. 8.19 Visualized future theses. Research project Stadt.Land.Chancen. *Source* Fraunhofer IAO

5. Increasing demand for smaller living space, caused by single households and high rental prices, is leading to new concepts such as tiny houses and micro apartments.
6. Mobile working outside of offices is becoming more and more common, which places new demands on living spaces.

The surveys revealed some key principles that should guide the design of the built environment in the coming decades. There is interest in ecologically sustainable construction in the sense of a circular economy in both urban and rural areas. At the same time, city residents and women in particular are concerned about rising housing costs. It is therefore important to further research the compatibility of sustainability and cost efficiency in building materials and to make existing processes better known. City dwellers, low-income people and women welcome more flexible living arrangements, whereas rural residents, high-income people and men are more skeptical. In the interests of equal opportunities in the housing market, which responds flexibly to changing needs over the course of life, flexible living arrangements can be an important lever for needs-oriented design. To attract rural and higher-income population groups as well as men to flexible forms of living, the associated opportunities and sustainability aspects should be highlighted. Furthermore, respondents expressed clear concerns and wishes regarding leisure and social behavior, particularly in urban areas. There are fears of a lack of local recreational areas, which particularly affects lower-income people, who are therefore particularly concerned about this. Overall, the results show that future building and living environments require a significantly more differentiated and diverse offering, which should be designed through targeted diversity research and knowledge transfer from science.

8.6.4 Diversity in Applied Research

Diversity and equal opportunities between men and women, particularly the increasing proportion of women in leadership positions, is a goal set by politicians and those involved in applied research and is itself the subject of research. From 2003 to 2006, the Fraunhofer Society carried out the "Discover Gender" project in an interdisciplinary team. Around the same time, Fraunhofer ISI was involved in the BMBF-funded project "Gender Aspects in Research". Fraunhofer IAO has been conducting research in various EU-funded projects since 2009, including "STAGES—Structural Transformation to Achieve Gender Equality in Science" or "JERRI—Joining Efforts for Responsible Research and Innovation" on the relevance of a balanced relationship between genders and cultures. Diversity dimensions play an increasingly important role, particularly in technology design and transformation, e.g. in research into the development of generative AI technologies. On the one hand, it is important to ensure that the AI is trained with data and uses it accordingly that reflects the reality of life in the sense of diversity and heterogeneity. On the other hand, those who research and develop AI need different cultures, experience, and knowledge backgrounds in order to address the potential and risks of these technologies.

Diversity dimensions are also becoming more important as our built environment transforms. In addition to cultural diversity and the increasing mobility of the young generation worldwide, accessibility or social integration through the creation of living arrangements that bring people of different ethnic, religious, and social backgrounds together are a strategic location factor for attractive cities (Dienes et al. 2022). Public spaces such as parks, squares and streets should be designed in such a way that they convey a feeling of security on the one hand and are attractive to female and male city residents on the other. This can be done, for example, through lighting, various play and leisure equipment or seating. The Swedish city of Umeå should be mentioned as a model city, which has gained an international reputation for its comprehensive work on gender mainstreaming by integrating gender equality into its urban planning since the 1980s (Smart City Sweden 2020).

8.6.5 Outlook

Human-centered technology design in the construction industry and the transformation of the built environment between digitalization and climate protection is characterized by a variety of influencing factors. The current convergence of technology fields such as AI, 5G or digital twins in construction is highly dynamic, which must, on the one hand, be based on common standards such as data management and, on the other hand, consider the plural needs of tomorrow's society. At the same time, it is important to sharpen our role as Germany as a business and innovation location between globalization, cultural values, technological sovereignty and new markets in the active shaping of these fields—especially for the next generations and the cities of tomorrow. Winston Churchill put this aptly in 1943: "First we shape our buildings; thereafter they shape us."

8.7 I can... Heide Ackerbauer

Dr.-Ing. Heide Ackerbauer

"And you, what do you want to become?" When I was asked this question as a child, I usually answered with "I am good at calculating and I like to doodle*".
About 25 years later, I am asked if I would like to write a book contribution on the topic of diversity in construction. About my own story…

I became a civil engineer (Fig. 8.20), earned a doctorate in building physics, worked for a long time as a research assistant at a university and simultaneously in the construction industry, and now head a department in the field of materials testing in construction. What

Fig. 8.20 Author 2023. *Source* Heide Ackerbauer

happened in between? How did I get there? I have thought long and hard about what I could tell about my journey so far. I'm not that old yet. But what could be my resume so far.

I could tell about my childhood. About growing up in the countryside. Very close to nature in a forestry household. After school, it was off to the forest, to the forest workers, fishing or to the high seat, when my sister and I were not just scheduled for piano lessons or tennis. Or we helped with the harvest around the house, in the garden or the meadows around it, processing it into jam, juice or cake. I fondly remember how my sister and I got the old moped, a barn find, running again after days with the help of my father, circled the village, brought the most beautiful flowers home, as proof of how far we had driven. At the same time, we sewed, from today's perspective, terrible clothes. We thought it was fantastic to show them off in grandma's high heels, having converted the small path in the garden into a catwalk. The variety of tasks, which I did not perceive as a duty, was like a huge playground for me. Without pressure, a wide range of skills and possibilities for their development were probably unconsciously created. For my sister and me, it was completely normal that we changed the tires on our car ourselves, helped with the construction of our parents' house and the farmer next door or with hunting. Just as normal as helping in the kitchen and around the house. Nothing was mutually exclusive. My father could also paint watercolors as well as cook, mend holes or build cellars and roofs (he can still do it today, just slower). My mother is also equally creative and musical, as well as handy. Interjection: No bragging. Really not. They just tried or did everything, even if they were not the best at it.

I still notice this today when my parents equally support me in sewing my wedding dress, in my private renovation projects, such as how to apply the composite seal most skillfully, in concreting or electrical installations.

Why am I telling this, if it is so normal? Because what was completely everyday for me back then and until slowly growing up, I would retrospectively describe today as a childhood without gender- or character-specific selection or restriction of trainable skills. And for me, this was an essential ingredient for my development. Not to lose confidence on the way there and beyond, what I have now become. Especially because my family has always shown me by example that you can just try everything and nothing is impossible if you just try.

I could continue to tell that like all teenagers who are faced with the choice, I was eventually able to answer the question of career aspiration concretely for myself. From "I like to doodle" I deduced that my interests in natural sciences and mathematical technology, coupled with a healthy dose of creativity and interest in sustainable construction methods, eventually led me to the construction industry.

I could tell you that my parents, despite their positive encouragement and support in all directions, had kept one thing from me about the big wide world. I once overheard them talking, saying that there might be a problem. That it might be a world where I would encounter difficulties, for which I might not be suited. Strictly speaking, there were two things. One was obvious, because I was a girl and, physically speaking, quite small. The other thing was something I needed even more. Because of the first thing. For a world that, in my parents' view, is associated with self-confidence and assertiveness, I might not have had the best character traits. I was shy, not fond of being in the front row, not a leading child. Not like children usually are, but one that was afraid to make phone calls, afraid of strangers and even turtles, who couldn't even say "yes" when the sausage seller at the supermarket counter offered me a slice of salami. I remember very well when my mother once showed me a newspaper article about shyness when I was about 14 years old. According to the main message, this trait is innate and one never gets rid of it throughout life. Naturally, this subsided in familiar surroundings, but is always there again in unfamiliar situations. My sister didn't have these problems, but that's just how I was. I was aware of this as a child and as a teenager. I had discarded my alternative plan to become a sports journalist for these reasons, but in a profession where I could hide behind my "numbers" and logics, let them speak, I initially didn't foresee any problems. What, if not a clear result to the right of the equals sign, could show my good performance and talent? Besides, I was equipped with my craft experience and the joy of everything to do with building and self-building. So I stuck with my decision.

I could tell you what this decision meant for me. I used every conceivable pressure situation that I recognized as such to actively train myself out of my shy behavior. Whether it was about preparing and presenting separate presentations at school, presenting project group results at university, doing student assistant jobs at the institute I favored, or taking semesters abroad. Or whether it was about seeking situations in the first years of my professional life where I had to speak in front of people, in construction meetings or in front of colleagues. Every time it was a huge effort. To this day. You can imagine it like a balloon that needs to be inflated over and over again to grow or even just to maintain its shape. I believe that this very conscious process of training myself to be brave and self-confident, which I would have had to go through anyway on the way to becoming an adult, also specifically helped me to move in the world of construction. If you are already striving out of your own natural character comfort zone, you can go even further.

Upon entering the professional world after my studies, the construction industry seemed to me to fulfill every cliché. A strongly masculine power cascade but also a predominantly male workforce in breadth, a rigid corporate and working culture, all rather non-diverse, characteristically uniform. Very straightforward, with elbows but still lovable. To fit in here, especially in the beginning, I left a lot of what I suspected could be interpreted as typically female or modern, and therefore rather inappropriate, at home when I went to work. It wasn't particularly difficult for me, I had enough other intersections with colleagues and superiors. Enjoyment of work and with the environment that was offered to me. Nevertheless, it sometimes felt as if you were on the move with a visor down. You observe, take perspective changes, adapt iteratively, depending on the reaction. it affects you and your environment. Years after our collaboration, a colleague told me that she put on armor every morning on her way to the office. It took a lot of energy from her, energy she would have liked to use for work. Changing that someday was my dream at the time. For me, that was the key to improving work quality by focusing on the essentials, because then you wouldn't have to focus on any armor or adaptation processes. But first, you need a good reputation, otherwise people won't listen to you. How you are perceived is essential for your reputation.

In this context, I could of course also talk about the classics that have certainly not only happened to me, but to many other women.

That I was often perceived as a secretary during construction site inspections and still am. That as soon as words like drill, concrete, grinding, cutting etc. are mentioned, people like me suddenly become invisible. Even when a male assistant from the engineering office accompanied me in my expert activities, he was naturally addressed as the main contact person on site by the participants. Every time, I first had to overcome the hurdle of professional recognition through a short knowledge comparison, let's call it nicely. To date. By the way, the doctorate title sometimes helps not to have to jump quite so high. Because even though I don't care much about it, but prejudice also exists in the other direction and then it can be helpful to make yourself heard, as banal and stupid as it may sound.

I could tell about formative experiences, like one where I was asked to move to another office. When I asked why I should do that, I was told that my office partner had complained to the institute director that I would wear shorts and skirts in the summer and he could not concentrate because of my visible legs. Logically, they assigned me a single office and moved me away so that my colleague could concentrate on his scientific work again.

I could also tell the story that after years of joint team work with my male university and office colleagues and really good joint successes, the founding of a new engineering office was suddenly agreed upon, which was thought of without my participation and initially even kept secret from me. When I asked why I was left out, the answer was that I had not yet finished my doctorate and was in my mid-30s (!!). Of course, the same circumstances were not a problem for my male colleagues.

That after asking for a salary adjustment (this was by no means excessive, but was only about raising the hourly wage I received in the engineering office to the level of an equivalent employment level in the public service), I was called "money-grubbing".

The more visible I became in terms of scientific expertise, the more frequent the before described experiences became. The examples are as diverse as they are striking, always following the same pattern. I am sure that some readers will recognize themselves in such examples. They are not made up and even if you don't want to see them, they are there. Every day and everywhere. Often told. Therefore, I don't want to bring up more. To classify and evaluate such stories, it certainly helps to keep in mind that the world of science is often characterized by strong competition. Of course, it is largely about gaining knowledge, which brings us all progress and improvement in the aggregate. This makes science a sovereign and enriching task, which also only works through cohesion and common goals of the researchers. For me, scientific work was and is the most meaningful thing I could experience in the course of my professional occupation. Despite all ambition to honor the common sense and purpose also with a team performance, the external structures in which scientists move provide enough basis to develop competitive behavior. Fixed-term contracts, Individual pressure to succeed, strong dependencies on well-disposed scientific superiors and evaluators, and above all the ever-present need to demarcate one's own research project from that of others, create a sometimes highly competitive environment. Often enough, it is more about personal liking and conformity rather than professional measurement and progress. I have often wondered why obvious, also perceived by companions, injustice, unequal treatment or not being seen for the professional depth and significance, was not somehow punished by them, but completely omitted. The most blatant of my examples, the shorts action, everyone noticed and shook their heads in conversation with me about it, but no one officially, that is, with my boss who decided my implementation, took sides for me and my legs. Then and again and again, when something like this happens, I have asked myself what I did wrong, that the presumed sense of justice in my fellow human beings did not visibly report for me. Considering the described competitive climate and very own individual goals of each individual, it is easier for me today to not take the events personally. On the contrary, I began to classify it as a somehow natural structurally conditioned survival action.

The thoughts I had at the described experiences, I do not want to elaborate at this point. What it has done to me until today, all the more. I also like to share this on my Instagram profile "@frauackerbauer", XING and LinkedIn.

Because, what I could also tell, is that besides this, a kind of natural selection of those who I know as supportive companions, especially during the intensive time until the completion of my doctorate and until today at my side. It is indeed the second look of sponsors that has helped my research ideas to take their first steps. With their support, I then later crossed the finish line and hopefully will continue to do so. A sudden being seen provides enormous impulses. As individual and idiosyncratic as we tech people are, without a team, common discourse, rubbing against each other and mutual "challenging" with sparring partners, without building a network, no idea, no matter how good, can be successful and contribute to progress. Especially in long-term projects like my research project on predicting microbial growth on exterior walls, regular feedback in the team is essential. Naturally, doubts about the right path taken arise. You don't immediately get answers to the research questions and theses from the project. It is exclusively the team that provides the necessary support and the necessary professional self-confidence. In our case, several awards for the research work confirmed the approach, but you don't know that beforehand and alone, at least for me, it would have been unthinkable to just go through with it. The power of role models in this context should not be underestimated. They can really captivate one and inspire.

Now I have written a lot about burden, adaptation processes, also suffering, injustice and effort and about what I could tell in this context. Surely there is more to mention here. But for me, what is crucial is what task I can derive from all these experiences for myself.

If, as my parents apparently rightly feared at the time, this world of construction is associated with very uniform attributes such as roughness, assertiveness, self-confidence and strongly heteronormative role models, and it is therefore difficult for children, like I was one, to exist in this world, although I brought the professional aptitude, then it is from my point of view the task to change this association. From within, from oneself. My great wish for all children is that they can answer talent- and ability-guided, what they would like to become and the answer to that is as simple as the answer to "what am I good at?", "what do I feel comfortable with", without having to ask at the same time the question of fitting into this world. If it doesn't matter whether one is hard enough, big enough, loud or quiet enough, the right gender, the right sexual identity possesses, but the professional suitability is far in the foreground, no talent will be deterred anymore. At the same time, this is also my wish for the technical professions themselves, which isolate themselves strongly through the above-mentioned associations and thus leave many talents behind. The selection or recruitment principle must be "best skills" instead of "best fit".

Whoever follows this principle will automatically get a diverse personality structure. Diversity is in this case not a end in itself, but in addition to content-related enhancement of work quality through the bundling of people with the best skills for the corresponding tasks, the most valuable and an automatic by-product. Impulses and creativity do not arise from equality, but through the interaction of different character traits, perspectives, work styles, as well as social and cultural backgrounds. Diversity, very importantly, is not defined by what one is, but how one is.

I have decided that in my current position I can derive a task from the experiences made and contribute to improvement. The goal for me is to create conditions where everyone leaves their armor at home, enjoys work and the colleagues, thereby creating creative solutions in the matter, without using their energy on side issues. Create a positive mindset, form a team that, despite all heterogeneity, has the same set of values. Looking in the same direction. But how? What can I do concretely?

I can be a role model. The power of role models I have already addressed. I believe that I would have felt encouraged back then if a female civil engineer had told me that she once was afraid of everything possible and still did what she was good at and that she is going her own way and everything is not bad at all. Therefore I think it is important to tell my path and thoughts about it. Often I can also make connections in personal conversations with my employees. Also during my teaching activity I have made corresponding mental excursions. I will always remember how at my graduation ceremony three students came to me. They said that I might not know them, but they stayed because of me and my encouraging stories about the world of construction at the study. I am still embarrassed today that I really didn't know the three. But in my defense I may say that there were almost 400 students in my building physics lecture. Nevertheless, I do not forget the positive feedback.

When I notice that employees, for example, do not trust themselves to take on a task, admittedly, a very human typical impulse often arises in me. I think the leap into the cold water would help. After all, I had to do it too. Experience has shown, however, that this does not always create a positive experience for those who have jumped. So I usually decide to accompany first and then let go. This approach also always offers me the opportunity to "stretch" my own balloon again and provides the employees with a suggestion based on my approach. Usually, this results in new ideas for future comparable situations, from which I learn again.

I can be close and approachable. The cognitive diversity that results in diverse teams, to promote individually is for me the key and my solution approach for the door to progress according to the "best skills" principle. Anything else could cost us dearly. To find out the individual strengths of each person and to use them accordingly, requires careful observation, getting to know each other, close collaboration, a lot of feedback, empathy and understanding for also personal circumstances. Therefore, I intensively exchange with my colleagues and usually we find out together, how and what is the optimal deployment situation for each and every one is. It is particularly important to me that the feedback and openness is not only bilateral between my employees and me, but multilateral in the entire team. Ideally, everyone looks, helps or gives each other, sharing even private circumstances relevant to work.

I can be corrected, show weakness, seek reassurance or ask for advice. Everyone has a perception distorted from pure objectivity due to their personal history, imprinting, and socialization. Me too. It would be naive to believe that one can judge everything perfectly objectively. Therefore, the use of opinions and advice is absolutely necessary for me. Regularly, I consult with my team about my decisions or future paths to take. Pursuing my goal also requires the recognition and appreciation of different personality types and work styles. For example: I know that I would fall into the so-called similarity bias if I were to select new employees alone. That I develop an automatic sympathy for people who stand shyly in the second or third row. This would contradict my goal of automatic diversity through professional selection criteria. So, I have gotten into the habit of inviting potential new applicants for a mutual get-to-know or trial day at our company. Here too, the team is actively involved in the evaluation. The job interview is still the formally essential act for me, but I have long placed my weight on this one day of getting to know each other.

I can be powerful and steadfast. Especially in the public service, my current work environment, where the innovation-braking effect of a missing view beyond the horizon is motivated by great job security on the one hand and on the other hand also by missing mechanisms such as reward possibilities is even more intensified, it is particularly important to persistently stand up for new ideas. Things that are already proven elsewhere, to enforce, to transfer. Starting with the necessity of digitization, its advantages, proving that mobile working or home office works more than loss-free, or simply changing, optimizing grown work processes, willpower and perseverance are required. Even if I do not yet know the course or success of the new paths I have taken, I must dare to go them, because otherwise I cannot provide the counter-evidence.

For example, it is a great concern of mine and for me the central approach, to no longer leave great talents lying around, but to lift the treasure of skilled workers, to lift the stigmata for women, by no longer just talking about making the compatibility of family and career possible for women. For this to work, it must become real for male colleagues to be active fathers, to be able to share parenting times equally with their partner at best. This is only possible if they no longer have to fear a career break because of it. If they dare to ask for work time models that allow equal participation in family life. Only then will the stigmatization of women be able to end. If care work is thus distributed more equally, the construction industry will automatically become more attractive and can also benefit from many more female professionals. I myself have only benefited from the home office work of my forester dad and my mom (of course, it was not called home office back then, but it worked just the same). Therefore, I actively encourage male colleagues in my team to work with me to develop work time models that allow them intensive family participation without decreasing their professional development opportunities. And, who would have thought, they thank the entire team with a balanced work atmosphere, goodwill, they are simply positively minded. A thought experiment on this: If, who wants to, without worrying about their own professional advancement, set up their work time model deviating from the grown 40-h presence, i.e., reduce it, then I could automatically hire more employees mathematically. More heads, more ideas, same costs, more people for whom a job in construction would be an option, because we no longer drive on the old 40-h tracks. Automatically also more women. It's worth thinking about. Convincing of these and other thought approaches, I see as one of the great tasks of our time and therefore also as mine.

We can have fun with it. It's relatively easy. In the end, it's about wanting success for my team and myself. I can only achieve this if everyone and each individual feels comfortable at work, is motivated. Has good associations. Likes to work. Of course, also earns money, but not through compulsion, but through joy. Perhaps also wants to waste one more thought on solutions after work. The result of all aspects I have described is enjoyment at work. And that is the engine. For all of us. As heterogeneous as we are.

By the way, showing leg is definitely allowed in my team.

8.8 What Does Our Heart Beat for? Implementation of Smart Cities in the Age of Industry 5.0

Dr.-Ing. Dipl.-Ing. (FH) Bianca Weber-Lewerenz

The design of intelligent, resilient, and sustainable smart cities deals with the transformation of all their areas: environment, energy, administration, education, economy, mobility, health, resilience, and security (Deloitte 2023c). Furthermore, it sets new political standards for an ethical and responsible smart city development. The United Nations' forecast states that by 2050, 70% of the world's population will live in cities and urban areas (Lai 2023b), leading to a significant increase in emissions and energy consumption. Therefore, there is an urgent need at all levels of decision-makers to install smarter urban transport networks, improve environmentally friendly water disposal and treatment plants and design intelligent buildings with high energy efficiency and security (Fabregue and Bogoni 2023).

Given the fact that most buildings are existing buildings and new constructions should be reduced as much as possible to avoid unnecessary consumption of natural resources and a significant increase in CO_2 emissions, there is an urgent need to retrofit the existing structure to a zero-emission construction. Numerous role models present concrete solutions, e.g. defined energy communities in European guidelines, to strengthen the cooperation between active and passive users and to share locally generated energy according to innovative management systems. Energy communities can consist of various participants, e.g. residential, industrial, commercial and service sectors, to develop within city districts or in suburban environments (Trevisan et al. 2023). Singapore is among the best practices of globally leading Smart Intelligent Cities and a Green City[24] (Fig. 8.21).

This contribution aims to explain to the readers the complexity of technological innovation, the smart city ecosystem, and the key elements of a harmonious human and digital transformation and to raise awareness. It particularly aims to strengthen urban decision-makers' commitment to improving the quality of life by critically reflecting on their individual sustainability strategy with a socio-ethical approach. The contribution

[24] https://www.businesstraveller.de/ziele/singapur-die-loewenstadt-auf-dem-weg-in-eine-gruene-zukunft/ (Accessed 17. May 2023).

Fig. 8.21 Singapore: Green City and one of the globally leading Smart Cities. *Source* https://www.businesstraveller.de/ziele/singapur-die-loewenstadt-auf-dem-weg-in-eine-gruene-zukunft

illustrates the interplay of the diverse influencing factors and links technical innovation with human, social sustainability.

The contribution in your hands is a small excerpt from a large-scale primary research on a corporate responsible approach to digitization and AI in construction (English: Corporate Digital Responsibility CDR). The CDR concept designs a political framework by identifying, analyzing, and systematically evaluating key factors of a sustainable digital transformation in the traditionally fragmented, highly conservative construction industry. The focus is on a "healthy" human-technology interaction with a human-centered, value-based engineering focus. To give this approach face and name, the author founded the "Excellence Initiative for Sustainable, Human-Led AI in Construction" in 2020 (Fig. 8.22). The need for clarification and orientation is high, as are the expectations of AI: standardized processes executed by intelligent technologies, structured data complexity, increased cost and time efficiency, resource-saving construction, monitoring and fulfillment of climate goals, and safer, more efficient and productive processes for implementing the UN goals for sustainable development (SDGs).

The central question remains how we humans want to shape technologies, where they make sense and in which areas they are useful to best support human work and quality of life. AI simplifies processes in planning, implementation, monitoring, and logistics. Digital twins and Building Information Modeling (BIM) visualize the performance promise as a planned end product with its technical equipment and simulate user behavior, technical building operation, energy consumption, possible risks, environmental and social impacts, and intelligent communication, e.g., in Smart Cities. Such methods pro-

Fig. 8.22 Author in discussion round "AI in Construction" at the 25th anniversary celebration of the existence of the Engineering Chamber of Saxony, Oschatz, 2023. *Source* Engineering Chamber of Saxony

vide transparent and holistic decision-making bases. Digital twins thus enable a world of intelligent ecosystems. Education and sensitization are part of the path to achievement. The concept of the digital twin helps companies master the challenges of their own digital transformation and their positioning in the ecosystem. In addition, the digital twin not only leads to more transparency in the presentation, planning, and implementation of Smart Cities in the age of Industry 5.0, but also enables further innovations. The European Parliament calls for a social and inclusive approach in the transition process to Smart Cities (EU Parliament 2023b).

A more nuanced understanding of the strategies for planning and implementing Smart City initiatives and the success factors is essential to align the increase of ecosystems, sustainability, and quality of life. Each task area requires performance, willingness to change, and a targeted innovation and investment strategy to meet both the rising expectations of citizens and the net-zero standards. It is reasonable to assume that from 2021 to 2026 the average annual growth rate of the German Smart City markets will increase by 17.1% (in billion euros) (Arthur and eco 2021–26) (Fig. 8.23).

The European Union is considered a pioneer in modern city planning. It spends most worldwide and invests in Smart City projects and initiatives. The Internet of Things, or IoT, plays a key role in the connected city of the future.

Mobile devices are increasing in the form of connected sensors and smart solutions. Intelligent and smart technologies can significantly contribute to making life in the city more environmentally and energy-efficient and less burdensome for citizens, or more participatory and transparent. Even in rural areas, there is the opportunity for improved

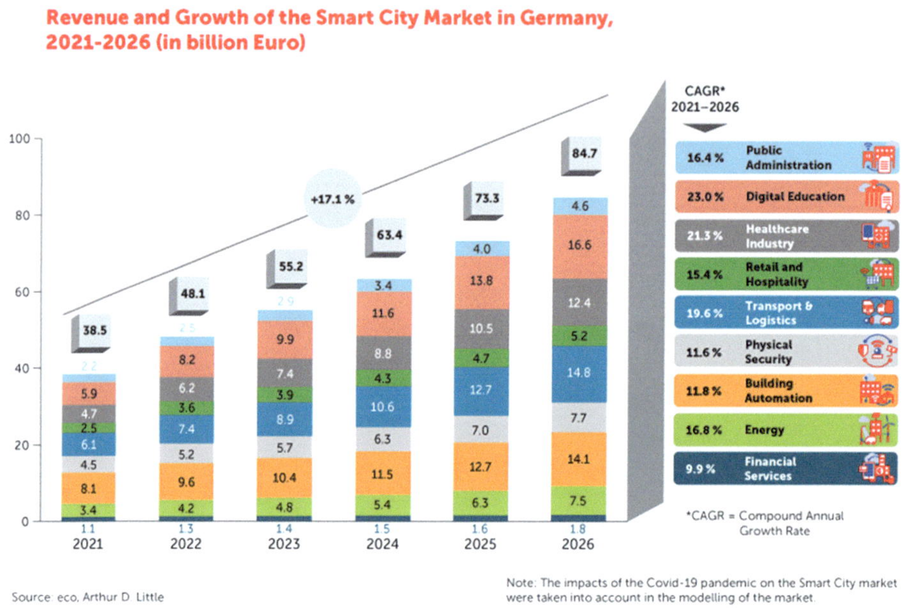

Fig. 8.23 Average growth rate of the German Smart City market 2021–26 *Source: Arthur D. Little and eco—Facts and Figures.* https://www.eco.de/en/topics/internet-of-things/studies/study-the-smart-city-market-in-germany-2021-2026/#download*)*

connection to economic and social activities. In addition, Smart City solutions have the potential to optimize process and value chains and minimize resource consumption. In this way, CO_2 -saving or climate-neutral effects can be achieved, in addition to enhancing the reputation of the construction industry and expanding the value chain (Fig. 8.24).

The term "Smart City" refers to a concept in which information and communication systems are connected with other physical devices and interact with the IoT network. The goal is to optimize the efficiency, quality, and service portfolio of an urban operation, ensuring data connectivity between citizens, urban infrastructure, and administration, and at the same time reducing cost and resource consumption. Numerous communities around the world have launched strategies and initiatives to follow this model. However, new strategies are based on the individual geographical and infrastructural conditions of a city. The administration is challenged to act as an "enabler", to put the topic on the agenda and remove possible hurdles. Because only in this way can the transition to a Smart City or Smart Region be successful and only in this way can the increasing challenges for cities due to increasing urbanization be managed.

The efforts to develop intelligent cities are based both on the diverse areas of application of digital technologies such as IoT, AI, cloud computing, and AI technologies, as well as on human intelligence, to drive sustainable environments. The focus is on the responsive adaptation to increasingly complex external conditions and influencing fac-

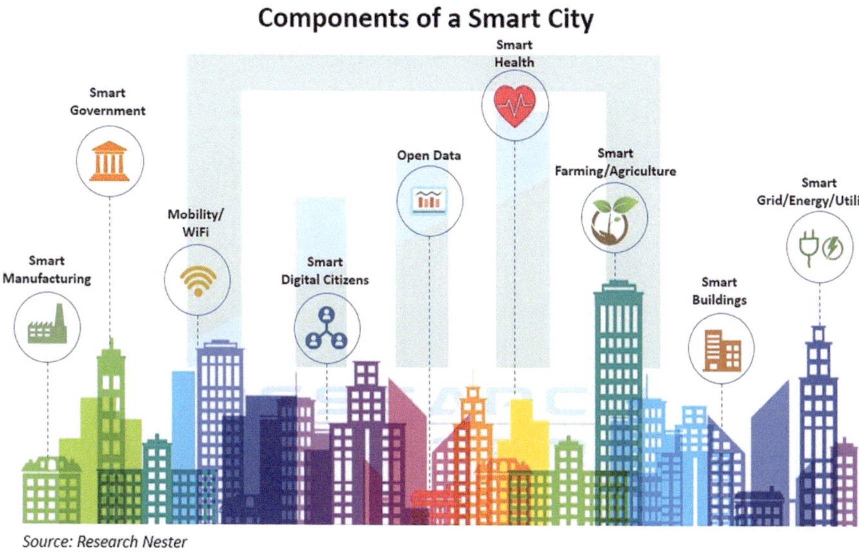

Fig. 8.24 Network of components of a Smart City *Source:* https://www.researchnester.com/reports/smart-cities-market/2156

tors, as well as increasing resilience. Through suitable digital solutions, the foundation for an efficient infrastructure and adaptable management at municipal and regional level can be created.

As a central entity, the citizens with their needs for sustainable quality of life are the benchmark for further development. However, economic power and climate protection are equally important in the Smart City and just as important in a livable city of the future. The new culture of new thinking of the cities of tomorrow and of intelligent, sustainable construction, which lays the foundation for future generations, requires awareness for proactive design, as well as the protection of the working and living environment, and the continuously new knowledge to be generated about the potential of innovative technologies and their application.

Technological progress means more societal responsibility in dealing with the unintended consequences of innovations as well as in protecting values and democratic rights (Weber-Lewerenz 2022).(Weber-Lewerenz 2023b, c). But it is also about the awareness of recognizing the potential of technical innovations and fully exploiting them to improve the protection of resources, environment, and climate, and to design agile and resilient urban ecosystems as catalysts for efficiency and productivity, but above all sustainability of the digital age (Weber-Lewerenz, B. 2022, 2023a). The question is more about the MUST, less about COULD or SHOULD.

Fig. 8.25 Paris: Host for the 2024 Summer Olympics *Source:* https://www.ecowoman.de/freizeit/natur/paris-2050-stadtentwicklung-grossstadtgaerten-nachhaltig-gegen-klimawandel-3786

How can waste management work more efficiently? How can CO_2 emissions be reduced? How can urban circulation be efficiently regulated, monitored, and controlled? What technologies and what further development of new innovations do we need to reduce high energy consumption, inefficient material use up to the irresponsible disposal of unused material? How does a city like Paris tackle this challenge, now on the way to becoming the host of the 2024 Summer Olympics (Fig. 8.25)? These are current questions that scientific investigations and research in several areas deal with. These are the fields of action that need to be pursued at all levels of responsibility in order to achieve holistic sustainability. The city of Paris has been pursuing the Paris Smart City 2050—urban development strategy for some time[25].

New insights and experiences contribute to an understanding of the diverse relationships between technological progress and societal and human common good, which can only thrive on a value-based foundation. When the German National Ethics Council announced in its statement in March 2023 (German National Ethics Council 2023b) that artificial intelligence must not diminish human development, the Council emphasized the harmonious human–machine interaction, which is based on the fact that AI cannot replace human intelligence, responsibility, and judgment. It is important to maintain trust and focus on the common good. This declaration ties in with the milestones of the measures within the framework of the European AI strategy and the digital transformation,

[25] https://www.ecowoman.de/freizeit/natur/paris-2050-stadtentwicklung-grossstadtgaerten-nachhaltig-gegen-klimawandel-3786 (Accessed 17. May 2023).

which promote the participation of each individual and Strengthen diversity and inclusion. All statements have one thing in common: Education is the key!

Regardless of whether discussions about technological feasibility, the development of even smarter tools, innovative high-tech, and improved approaches to artificial intelligence are being conducted, there is a human, moral, and ethical obligation to critically question the complexity and human impacts of such technologies, which actually create parallel virtual worlds. It is beyond question that human intelligence in a real world is the one that enables all technical progress. A meaningful, socially sustainable progress that should develop on the basis of human creativity and human reflection.

In order to be able to think about the latest AI developments and to prevent the technical competition from getting out of control, the Future of Life Institute calls in its open letter in March 2023 to pause huge AI experiments for at least six months (Future of Life Institute 2023b). This letter was signed by thousands of scientists and IT experts and is intended to set a strong signal against the out-of-control race of AI development for the sake of competition alone (Bucknall and Dori-Hacohen 2022; Bubeck 2023). The letter calls for state regulation as a follow-up measure if the branch does not follow the recommendation of the letter.

The digital age is historical and offers diverse opportunities. However, there is a lack of the necessary new knowledge, new skills, and qualifications, both in terms of teaching competencies and new curricula, to train new generations of qualified professionals. These areas require our urgent attention and active implementation.

It is important to understand that it is the human who determines which method or technology is meaningful, who defines which best supports human work, who continuously monitors, adjusts, and improves efficient productive work methods and processes. It is the human who develops technical innovations and their applications, continuously monitors, controls, and also stops when serious risks are predicted, impacts on people, society, and the environment are not in line with sustainability. But it is also the human who can show new ways through the development of innovative technologies that lead to the achievement of sustainability (OECD 2023b). AI systems must function robustly and safely throughout their entire lifespan, and potential risks should be continuously assessed and managed and must not pose a disproportionate security risk (OECD 2023d). AI development is seen as an integrative, participatory process and promotes a human-centered approach to strengthen trust in and the adoption of AI in society.

In addition, AI and other new technologies provide tools for efficient and productive management of challenges in the health sector and demographic change, which requires more hospitals, health centers, and thus municipal efforts to ensure the quality of life for both the young and older generation. In the education sector, the technical infrastructure and equipment of schools and community buildings require modern technologies for fast data transmission and communication as well as high data storage capacity, e.g. for digital Archives of cities, daily digital municipal services, public administration, building automation, energy management, e-mobility, and public transport and logistics.

A Smart City requires a society with a new thinking culture, driven by the constant effort to find new ways and to use innovative technologies to act flexibly, environmentally friendly, resilient, and agile. London, Zurich, Singapore, Dubai, Amsterdam, Barcelona, Vienna, and Nanjing represent best practices with their lead in Smart City platforms and their diversity of innovative technological architecture, leading to measurable CO_2 reduction, less congestion, and faster traffic (Locke 2023b).

In the context of legal handling of data, numerous responsibilities arise: order competence, copyrights, multi-party contracts, warranty, disruptions in BIM projects, data protection (Weber et al. 2023), cybersecurity, building data in facility management, data in technical building equipment and building users, data from intelligent building communication and Smart Cities. Since the regulations by laws and standards vary from country to country and from country to country, only an overview of the main fields of action is highlighted.

The Federal Association of AI has launched its LEAM initiative[26] and cooperates with industry representatives and researchers. The initiative aims at larger European AI models towards a hyperscale infrastructure for the development of methods based on European values. However, it questions research and development as a political and regulatory level to build a European data ecosystem. Governments should promote the development and access to a digital ecosystem for trustworthy AI.

This contribution concludes that such a proposed ecosystem represents the next step of sustainable innovation by including digital technologies and infrastructure as well as potential mechanisms for the exchange of AI knowledge. In this context, governments should consider promoting mechanisms like data trustees to support the safe, fair, legal, and ethical exchange of data. This is all the more important when discussing how Web 3.0 can be used to accelerate smart, net-zero-compliant cities in the age of Industry 5.0.

Let us all, architects, engineers, all those involved in construction and co-decision-makers, always pull out the inner compass from the pocket in the face of all technical feasibility and rapid progress and ask the question: What does our heart beat for? And what is needed in the meaningful and sustainable implementation of smart, intelligent cities?

8.9 Diversity—the Opportunity for More Innovation in Timber Construction

M.Sc. Niki Karatza

The construction industry is a broad field of different sectors. With some detours I ended up in the field of timber construction. Without any training or direct connection to the timber industry except for my interest in the material wood, I decided to study wood technology in Rosenheim. As a more practical and visual type, the theoretical parts of the

[26] See LEAM—Large European AI Models. https://leam.ai/ (Accessed 17. May 2023).

engineering studies were no walk in the park for me. However, the many internships and wood-specific modules gave me a broad insight into the wood processing and machining processes and machines, thus providing a good balance to the theory. While many fellow students focused more on the furniture and interior design sector, I was more drawn to wood processing and timber construction during my studies, so I chose my internships and project work accordingly. I also learned about process optimization and workplace design through additional training.

As common habit in my studies, I carried out my bachelor's thesis in parallel with my job and thus found my way into a medium-sized timber construction company, where the house elements are prefabricated in the production hall. With the introduction of a continuous improvement process in production, I got to know the manufacturing processes intensively and dealt with the topics of lean management. My main focus was on waste avoidance, which further deepened my understanding of optimal workplace and process design.

My time in production taught me one thing overall: many processes in timber construction are outdated and are not questioned. In addition, most productions, especially of small and medium-sized enterprises, are characterized by manual activities and hardly use automation. This realization led me back to Rosenheim University of Applied Sciences, where I took up my master's degree in Applied Research and Development while working in the research and development department. As part of a research project, I developed a concept for a production cell together with a project team, which manufactures timber construction elements not in the traditional way lying down, but standing up. The processes were also to be carried out fully automated by robots. During this time at the university, I expanded my knowledge in the field of timber construction processes and specialized in particular on topics of digitization, automation and robotics.

After the completion of the research project, I agreed with a round of project colleagues that we wanted to continue pursuing our common vision: to make construction processes—especially in timber construction—more efficient through digitization and automation. Thus, we jointly founded holzbau.tech GmbH. Of the various business ideas that were supposed to shape the timber construction industry in different ways, the offer of services and consulting stood out for us. Over the past years, we have been able to win numerous customers, whom we were able to support with our in-depth knowledge and wealth of experience in the areas of automation and digitization in the new and further development or optimization of their productions. With our background and our connection to the university, our work is characterized by scientific approaches and structured analytical procedures. Our goal is to create customer-specific and manufacturer-independent concepts that are comprehensible for our customers. Our derivations form the basis for further planning and execution. Since we distinguish ourselves from comparable consulting companies both in our approach and in the topics we cover, we have found a niche in which we can support timber constructors. In this setting, I now found my role as Head of Consulting in our company.

My desire to contribute to advancing timber construction stems primarily from the increasing relevance of this construction method. Although the demand for timber buildings has been steadily increasing for years, it seems the general understanding of it outside the industry still seems to be very patchy. When people talk about "building", many still primarily think of solid constructions. On the other hand, when people talk about "wooden buildings", many people think of log or half-timbered houses. Much more clarification should be given on this.

8.9.1 Diversity in Timber Construction for Me

Diversity in construction includes the timber construction method in terms of different construction methods. But also within timber construction itself, there are various possibilities of manufacturing. The most common today can probably be reduced to the timber frame or panel construction and solid timber construction. In Germany and many other countries, these construction methods have in common that they consist of individual flat elements that are prefabricated in the factory. At the core of these elements are statically load-bearing wooden structures, which are usually stiffened and insulated with panel and insulation materials. In timber frame construction, we speak of a framework, which is composed of individual wooden rods, a solid wood element on the other hand, consists of connected solid wood parts (Fig. 8.26). The flat elements can be assembled directly on the construction site into a building or in the module construction method still be assembled into room modules in the factory.

Further diversity in timber construction can also be found in the constructive details and the choice of materials. If preferably wood materials are used, the sustainability of the buildings is increased. By choosing about glue-free materials and detachable fasteners such as screws, the dismantling of the elements is also made possible, which benefits the circular economy.

8.9.2 Opportunities of Prefabrication in Timber Construction

The degree of prefabrication also varies from manufacturer to manufacturer. So some companies only deliver semi-finished elements, which are completed on the construction site, while others pre-assemble the elements including windows and finished facades.

Prefabrication of the buildings in the factory brings many advantages. A main aspect is quality assurance, which is optimized in the hall in many ways. On the one hand, both raw materials as well as semi-finished and finished parts are protected from weather influences and damage. Prefabrication also allows for a quick assembly of the shell and significantly shortens construction times, especially compared to solid construction, so that the materials and elements are not exposed to external influences for long.

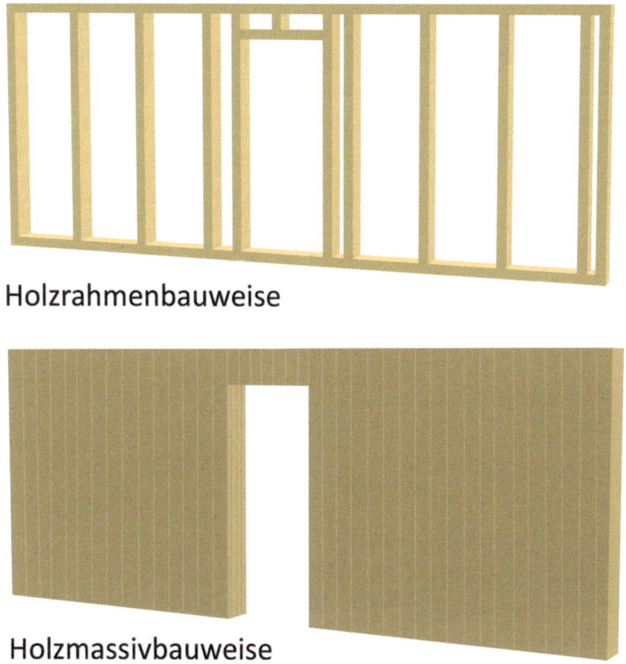

Fig. 8.26 Static cores of the flat wooden building elements in timber frame construction (top) and Solid wood construction (down)

On the other hand, the monitoring of the processes in production is possible to avoid errors and thus rework. Here, the ongoing shortage of skilled workers plays a major role. Due to the difficult search for workers, companies are increasingly employing non-specialist workers to respond to growing sales numbers or personnel gaps.

In addition to the improved supervision of personnel within the factory it makes sense to rely on automation. This is possible within the production halls, but not to the same extent for construction site activities. With the use of machines and equipment, workers are supported by partial automation in their actions or completely relieved in fully automated solutions. Since the scope of tasks in this case moves from skilled work to general machine operation, the requirements for the training and background knowledge of the workers also change. This allows companies to continue to scale their capacities independently of personnel.

Another advantage of prefabrication in timber construction is the detailing of planning and construction. While exactly this property is often negatively associated as complex and laborious, especially by non-specialists, it opens up at the same time opportunities for digitization. The topic of BIM is not yet widely spread in timber construction, but would be particularly beneficial in collaboration. be of great benefit with subcontractors and expansion trades. Because above all, the interfaces between the many participants often lead to ambiguities and errors in both planning and execution. The reason for this is usually insufficient communication and ignorance about the respective requirements of adjacent trades, so that work handovers can often not be optimally prepared. Collaboration in a unified planning model could reduce many complications. However, the very manually dominated timber construction industry is still in the early stages of digitization. There is a lack of knowledge and experience. Many companies are not clear about how data can be used or that usable data is already available. Here it is necessary to provide more clarification both in timber construction and in the adjacent trades.

8.9.3 Where I Wish More Diversity

The construction and especially the timber construction industry holds great potential for the future, not least due to the growing importance of sustainability and ecology. The possibilities of constructions, which are continuously being developed, allow the realization of increasingly complex building types in timber construction. At the same time, the timber construction industry is a very traditionally shaped area that slowly accepts innovations. Despite all changes and developments in the areas of construction, automation and digitization, there is a need for knowledge dissemination at this point.

A lack of diversity, as in many construction areas, can also be seen in the gender distribution in timber construction. In the male-dominated field, I was often the only woman in the round during my studies and still am in my collaboration with customers. This can sometimes be difficult and may deter some women who are interested in the industry. Events and programs specifically designed for women, such as a mentoring program, are quite useful for making contacts with women in similar professional fields. In such an environment, insecure women can be encouraged and valuable contacts can be made.

Both professionally and privately, the right companions can favor one's own development. I was lucky to have found a mentor and good friend who supported and promoted me in my professional career. In addition, in my opinion, professional meetings such as trade fairs and congresses are particularly valuable. Here you can not only exchange news about the industry, but also make and maintain contacts.

I too had to struggle with doubts at the beginning as to whether the chosen field of a male-dominated STEM profession was right for me. But my interest in the field helped me overcome my initial difficulties with some perseverance. After years of study and professional experience, I am now glad to have taken this path. As one of the still relatively few women in the industry, a consultant and co-founder of a company, I have found a role in which I feel comfortable and can drive something forward. In my work as a consultant for timber construction companies, I am also fortunate to be able to continuously learn and build a network of exciting and valuable contacts. I recommend to all those who are interested in STEM professions not to be deterred by external circumstances, as this field offers many fascinating opportunities.

8.10 Reconstruction of the Cathédrale Notre Dame in Paris

Dr.-Ing. Dipl.-Ing. (FH) Bianca Weber-Lewerenz

The start of the reconstruction of the Notre Dame Cathedral Paris[27], after the fire four years ago in 2019, also marked the opening of a scientific construction site[28]. The National Center for Scientific Research (CNRS), mentioned by the Ministry of Culture, coordinates a scientific construction site—one that holds new exciting discoveries parallel to the restoration site. 50 research teams and laboratories, 175 researchers—including historians, art historians, archaeologists, architects, engineers, physicists, chemists, computer scientists, anthropologists—expand the knowledge of this unique cultural heritage (Figs. 8.27, 8.28 and 8.32).

The construction industry has a high signal and role model effect, as far as innovation and use of the potentials of technical progress are concerned: it offers a newly created virtual duplicate of the cathedral, an innovative "monumental" information system.

Fig. 8.27 A restorer is working on the famous choir screen from the early fourteenth century. (Die Tagespost 16.Dec2022, Source Photo: Patrick Zachmann—Magnum Photos)

[27] https://notre-dame-de-paris.culture.gouv.fr/fr (Accessed 21. March 2024).

[28] www.culture.gouv.fr/de/Thematisch/Wissenschaft-des-kulturellen-Erbes/Themen-der-Forschung/Die-Baustelle-von-Notre-Dame-de-Paris (Accessed 21. March 2024).

Fig. 8.28 Construction of an arch scaffold, 2019 (Ministère de la Culture. www.culture.gouv.fr/de/Thematisch/Wissenschaft-des-kulturellen-Erbes/Themen-der-Forschung/Die-Baustelle-von-Notre-Dame-de-Paris)

Fig. 8.29 Renovation of Notre Dame roof truss. (Die Tagespost 16.Dec2022, Source Photo: Patrick Zachmann—Magnum Photos)

Fig. 8.30 Construction of the individual parts of the wooden roof truss. *Source* Neue Zürcher Zeitung, April 16, 2023. https://www.nzz.ch/feuilleton/notre-dame-von-paris-die-wiedereroeffnung-2024-scheint-moeglich-ld.1732217)

Fig. 8.31 Construction of the spire in the workshop of Le Bras Frères in Val de Briey. Here: Lateral roof truss, which connects the tower with the transept. *Source* Neue Zürcher Zeitung, April 16, 2023. https://www.nzz.ch/feuilleton

Fig. 8.32 A glazier manually copies the head of the Prophet Isaiah onto a new piece of glass. (Die Tagespost, December 17, 2022, Photo source: Olivier Caillard)

Fig. 8.33 View from below of the place where the spire is being built. *Source* Neue Zürcher Zeitung, April 16, 2023. https://www.nzz.ch/feuilleton/notre-dame-von-paris-die-wiedereroeffnung-2024-scheint-moeglich-ld.1732217

Fig. 8.34 Notre-Dame de Paris in December 2022: The scaffolding of the crossing will have reached a height of 100 m by the time the crossing tower is rebuilt. (Die Tagespost, December 18, 2024; Photo: davidbordes.com)

Despite digital plans to resurrect the roof of Notre Dame with medieval craftsmanship, carpenters, including one of the best in his field in the country, who has repaired roofs of famous Loire castles, use an ancient technique in woodworking: hand-hewn, later perfectly fitting beams (Fig. 8.29 and 8.30). Many of the individual parts were numbered and driven from the Loire by truck to Paris and installed there on site with mechanical cranes. About 2000 oaks were felled for the roof truss and the spire throughout the country, partly also on private land[29]. In ateliers, workshops and factories all over the country, parts for the reconstruction of Notre-Dame Paris are being built and processed (Fig. 8.31 and 8.33).

A glassmaker from Cologne is involved (Fig. 8.32).

The organ with 8,000 pipes is being cleaned in three workshops. As of March 2024, the roof truss has already been restored and the workers celebrated the completion of the reconstruction of the roof truss of Notre Dame Cathedral (Fig. 8.35). Proud craftsmanship, value creation and appreciation.

Overall, the scientific construction site is divided into eight thematic working groups: Wood and scaffolding (construction of frames, analysis of the remains of the so-called "forest" due to the number and density of wooden beams), Acoustics (song of the organs and the masters and choirs; sound environments), Metal (use of iron and lead in the

[29] https://www.nzz.ch/feuilleton/notre-dame-von-paris-die-wiedereroeffnung-2024-scheint-moeglich-ld.1732217 (Accessed 21. March 2024).

Fig. 8.35 Inauguration of the completed reconstructed roof truss of Notre-Dame Paris in January 2024. *Source* IMAGO/MAXPPP. https://www.mdr.de/nachrichten/welt/panorama/notre-dame-brand-wiederaufbau-dachstuhl-paris-100.html

building), Glass (glass surfaces from the 13th to twentieth century), Stone mortar and monumental decoration (vaults, walls, columns), Structures (skeleton of the cathedral), Digital (digital ecosystem of the cathedral) and Emotions and mobilizations (handling of collective ideas by anthropologists and sociologists).

Due to its history, complexity, diversity and innovation, the construction site is also referred to as the "construction site of the century". (Fig. 8.36).

With a virtual reality show about the history of the cathedral[30,31], the so-called "Eternelle Notre-Dame", they want to provide tourists and visitors with a new "visit" experience

---

[30] https://katholisch.de/artikel/43948-zeitpunkt-der-wiedereroeffnung-von-notre-dame-bekanntgegeben (Accessed 21. March 2024).

[31] https://parisjetaime.com/ger/kultur/eternelle-notre-dame-p3662 (Accessed 21. March 2024).

Fig. 8.36 View of the cathedral, 2024. *Source* © KNA/Corinne Simon (Archive image), https://katholisch.de/artikel/43948-zeitpunkt-der-wiedereroeffnung-von-notre-dame-bekannt-gegeben

Fig. 8.37 Immersive Tour "Eternal Notre Dame Paris" *Source* @Orange-Emissive. https://www.en-vols.com/en/inspirations-en/culture-en/eternal-notre-dame-paris

of Notre-Dame for all senses (Fig. 8.37, 8.38 and 8.39). At the same time, it is a journey through time that spans centuries. The entire range of technical feasibility in artificial intelligence and virtual reality supports the triad of history—architecture–art.

The opening is planned for December 2024, originally it was timed to coincide with the 2024 Summer Olympics. As early as 2021, Jean-Louis Georgelin, who was commis-

Fig. 8.38 Immersive Tour "Eternal Notre Dame Paris". *Source* @Orange-Emissive. https://www.en-vols.com/en/inspirations-en/culture-en/eternal-notre-dame-paris

Fig. 8.39 Immersive Tour "Eternal Notre Dame Paris". *Source* @Orange-Emissive. https://www.en-vols.com/en/inspirations-en/culture-en/eternal-notre-dame-paris/notre-dame-of-paris-the-reopening-2024-seems-possible-ld.1732217

Fig. 8.40 View from one of the construction cranes. *Source* Neue Züricher Zeitung, 16.April 2023. https://www.nzz.ch/feuilleton/notre-dame-of-paris-the-reopening-2024-seems-possible-ld.1732217

sioned by Emanuel Macron with the construction, called for tenders « *Craftsmen and companies with expertise in monument preservation are called upon to mobilize and apply for the community of builders who with enthusiasm and determination are working to reopen the cathedral for services and visitors by 2024, as planned by the President of the Republic. The restoration of the crossing tower and the north and south transepts as well as the completion of the interior restoration, two symbolic measures of the restoration project, require a high level of expertise in the field of timber construction, lead roofing, scaffolding, but also in the restoration of ashlar and masonry, carpentry, painting and gilding, blacksmithing as well as the restoration of stained glass windows and sculptures*». The supervision of the restoration is carried out by architect Philippe Villeneuve (Fig. 8.40).

A Te Deum was planned for April 15, 2024, exactly five years after the fire[32].

[32] https://parisjetaime.com/ger/artikel/notre-dame-die-aktuelle-lage-a741#:~:text=Ein%20Te%20Deum%20ist%20f%C3%BCr,wieder%20aufzubauen%2C%20einschlie%C3%9Flich%20der%20Turmspitze (Accessed 21. March 2024).

References

Antonios, P.; Konstantinos, K.; Christos, G. (2023). A systematic review on semantic interoperability in the IoE-enabled smart cities. Internet of Things, p. 100754sqq.

Arthur D. Little and eco – Association of Internet Industry (2017). The German Smart City Market 2017–2022. Facts and Figures 2017. https://www.eco.de/wp-content/uploads/2018/01/Manuscript_German_Smart_City_Market_Web.pdf (Accessed 11 December 2023).

Arthur D. Little and eco. Facts and Figures. Durchschnittliche Wachstumsrate des deutschen Smart City Markt 2021–26. https://www.eco.de/themen/internet-of-things/studien/studie-der-smart-city-markt-in-deutschland-2021-2026/#download (Accessed 17. May 2023).

Bastos, D.; Fernández-Caballero, A.; Pereira, A.; Rocha, N. P. (2022). Smart City Applications to Promote Citizen Participation in City Management and Governance: A Systematic Review. Informatics, 9(4), p. 89sqq. https://doi.org/10.3390/informatics9040089.

Bauen aktuell (2021). Automatisierung in der Baubranche: Welches Potenzial Roboter bieten. (WIN-Verlag GmbH & Co. KG). https://www.bauen-aktuell.eu/automatisierung-in-der-baubranche-welches-potenzial-roboter-bieten/ (Accessed 12. January 2024).

BauNetz (2024). Architektur aus dem Drucker. Baustoffvielfalt mit additiver Fertigung. (BauNetz) https://www.baunetz-campus.de/focus/architektur-aus-dem-drucker-8016618 (Accessed 12. February 2024).

Bellini, P.; Nesi, P.; Pantaleo, G. (2022). IoT-enabled smart cities: A review of concepts, frameworks and key technologies. Applied Sciences, 12(3), p. 1607sqq. https://doi.org/10.3390/app12031607.

Braun, S. (2023). #futuredistricts2053. Strategische Zukunftsszenarien und Handlungsanforderungen für die Quartiersentwicklung von heute bis übermorgen. Eine Veröffentlichung im Rahmen der Future District Alliance. https://doi.org/10.24406/publica-1932.

Braun, S. (2024). Mustersprache für Innovationsdiffusion in urbanen Systemen - Urban Innovation Pattern Language. Dissertation. Beiträge zum Stuttgarter Maschinenbau; Band 24.

Bryde, D., Broquetas, M., & Volm, J. (2013). The project benefits of Building Information Modelling (BIM). International Journal of Project Management.

Bubeck, S. et al. (2023). Sparks of Artificial General Intelligence: Early experiments with GPT-4. Journal for Computation and Language, https://doi.org/10.48550/arXiv.2303.12712.

Bucknall, B. S. & Dori-Hacohen, S. (2022). Current and Near-Term AI as a Potential Existential Risk Factor. In: Proceedings of the 2022 AAAI/ACM Conference on AI, Ethics, and Society, 2022, pp. 1–7. https://doi.org/10.1145/3514094.3534146.

Dashkevych, O. & Portnov, B. A. (2022). Criteria for Smart City Identification: A Systematic Literature Review. Sustainability 14(8), p. 4448 sqq. https://doi.org/10.3390/su14084448.

Deloitte (2023a). Smart Cities and Regions: Shaping transformation. Start-City-Manuscript by Deloitte and Fraunhofer IOSB-INA 2023. https://www2.deloitte.com/de/de/pages/public-sector/articles/smarte-staedte-und-regionen-studie.html?id=de:2ps:3gl:4gps-smarte-st%C3%A4dte-sicherheit-und-resilienz:5:6ps:20230312::&gclid=CjwKCAjw0N6hBhAUEiwAXab-TZSOnSCKomLca8-SQL3_Ai1JJEk_otj_oHNEoQogwSWEfjKmFXqAWxoCV0cQAvD_BwE (Accessed 01. April 2024).

Deloitte (2023b). Smart Cities and Regions: Shaping transformation. Resilience and Safety. Start-City-Manuscript by Deloitte and Fraunhofer IOSB-INA 2023.

Deloitte (2023c). Smart Cities and Regions: Shaping transformation. Start-City-Study by Deloitte and Fraunhofer IOSB-INA. https://www2.deloitte.com/de/de/pages/public-sector/articles/smarte-staedte-und-regionen-studie.html?id=de:2ps:3gl:4gps-smarte-st%C3%A4dte-sicherheit-und-resilienz:5:6ps:20230312::&gclid=CjwKCAjw0N6hBhAUEiwAXab-TZSOnSCKomLca8-SQL3_Ai1JJEk_otj_oHNEoQogwSWEfjKmFXqAWxoCV0cQAvD_BwE (Accessed 17. May 2023).

Dienes, K. et al. (2022). Ökosysteme im Wandel – Zukunftsszenarien für Business Events im Zeitalter grenzenloser Kommunikation. Innovationsverbund »Future Meeting Space«. Fraunhofer IAO, GCB German Convention Bureau e. V., 01/2022.

EU Parliament (2023a). Social Approach to the Transition to Smart Cities. Manuscript by STOA Panel for the Future of Science and Technology 2023.https://www.europarl.europa.eu/RegData/etudes/STUD/2023/737128/EPRS_STU(2023)737128_EN.pdf (Accessed 11. December 2023).

EU Parliament (2023b). Social Approach to the Transition to Smart Cities. Study by STOA Panel for the Future of Science and Technologyhttps://www.europarl.europa.eu/RegData/etudes/STUD/2023/737128/EPRS_STU(2023)737128_EN.pdf (Accessed 17. May 2023).

Fabregue, B. and Bogoni, SA. (2023). Privacy and Security Concerns in the Smart City. Journal ofd Smart Cities. MDPI Publishing. Vol. 6, Iss. 1, p. 586–613. https://doi.org/10.3390/smartcities6010027.

Future of Life Institute (2023a). Pause Giant AI Experiments: An Open Letter 2023. www.futureoflife.org/open-letter/pause-giant-ai-experiments/ (Accessed 11. December 2023).

Future of Life Institute (2023b). Pause Giant AI Experiments: An Open Letter. www.futureoflife.org/open-lette.r/pause-giant-ai-experiments/ (Accessed 17. May 2023).

German National Ethics Council (2023a). Humans and Machines – Challenges of Artificial Intelligence 2023. https://www.ethikrat.org/en/press-releases/press-releases/2023/ethics-council-artificial-intelligence-must-not-diminish-human-flourishing/?cookieLevel=not-set (Accessed 11. December 2023).

German National Ethics Council (2023b). Humans and Machines – Challenges of Artificial Intelligence. https://www.ethikrat.org/en/press-releases/press-releases/2023/ethics-council-artificial-intelligence-must-not-diminish-human-flourishing/?cookieLevel=not-set (Accessed 17. May 2023).

Giannakidis, A. et al. (2021). KI in der Bauwirtschaft. Einsatzmöglichkeiten für Planung, Realisierung und Betrieb von Bauwerken. Fraunhofer Studie. https://doi.org/10.13140/RG.2.2.15557.99041.

Haghshenas, S. S.; Guido, G.; Vitale, A.; Ghoushchi, S. J. (2022). Quantitative and Qualitative Analysis of Internet of Things (IoT) in Smart Cities and its Applications. In: 2022 IEEE Intl Conf on Dependable, Autonomic and Secure Computing, Intl Conf on Pervasive Intelligence and Computing, Intl Conf on Cloud and Big Data Computing, Intl Conf on Cyber Science and Technology Congress (DASC/PiCom/CBDCom/CyberSciTech) (pp. 1–6). IEEE. (2022, September).

HDB (2022). Produktivität im Bau(haupt-)gewerbe – ein statistischer Befund. https://www.bauindustrie.de/zahlen-fakten/auf-den-punkt-gebracht/produktivitaet-im-bauhauptgewerbe (Accessed 15.March 2024).

Jarosch, M. (2023). Vermessung im Bauwesen: Eine Einführung für Bauingenieure und Architekten. Wiesbaden: Springer Fachmedien.

Lai, O. (2023a). Top 7 Smart Cities in the World in 2023. https://earth.org/top-7-smart-cities-in-the-world/ (Accessed 01. April 2024).

Lai, O. (2023b). Top 7 Smart Cities in the World in 2023. https://earth.org/top-7-smart-cities-in-the-world/ (Accessed 17. May 2023).

LEAM (2022). Large European AI Models. https://leam.ai/ (Accessed 11. December 2023)

Liinasuo, M. et al. (2022). BIMprove - Building's digital twin model shown as first-person and third-person views. IOP Conf. Ser.: Earth Environ. Sci. https://doi.org/10.1088/1755-1315/1101/8/082018.

Locke, J. (2023a). Top 10 Most Futuristic Smart Cities in the World 2023. Digi International 2023. https://www.digi.com/blog/post/smart-cities-in-the-world (Accessed 11. December 2023).

Locke, J. (2023b). Top 10 Most Futuristic Smart Cities in the World 2023. Digi International. https://www.digi.com/blog/post/smart-cities-in-the-world (Zugriff 17. Mai 2023).

Lorenz, M. (2018). Digitale Führungskompetenz. Was Führungskräfte von morgen heute wissen sollten. Wiesbaden: Springer Gabler.

Maier, M. J. et al. (2021). STADT.LAND.CHANCEN. Wünsche und Sorgen von Bürgerinnen und Bürgern in Stadt und Land. Ergebnisse der Onlinebefragung (acatech KOOPERATION), München. https://doi.org/10.48669/aca_2021-1.

Meßner, S. (2021). Handwerk+Bau. Anwendungsgebiete und Herausforderungen. https://www.handwerkundbau.at/bauen/ki-am-bau-anwendungsgebiete-und-herausforderungen-17266 (Accessed 12.January 2024).

Mittelstand-Digital Zentrum Bau (2021). IoT im Bauwesen https://www.digitalzentrumbau.de/kos/WNetz?art=News.show&id=954 (Accessed 12.February 2024).

OECD (2023a). OECD AI Principles - Inclusive growth, sustainable development and well-being (Principle 1.1) 2023. www.oecd.ai/en/dashboards/ai-principles/P5 (Accessed 11. December 2023).

OECD (2023b). OECD AI Principles - Inclusive growth, sustainable development and well-being (Principle 1.1). www.oecd.ai/en/dashboards/ai-principles/P5 (Accessed 17. May 2023).

OECD (2023c). OECD AI Principles - Robustness, security and safety (Principle 1.4). www.oecd.ai/en/dashboards/ai-principles/P8 (Accessed 11. December 2023).

OECD (2023d). OECD AI Principles - Robustness, security and safety (Principle 1.4). www.oecd.ai/en/dashboards/ai-principles/P8 (Accessed 17. May 2023).

Reis, J.; Marques, P. A.; Marques, P. C. (2022). Where are smart cities heading? A meta-review and guidelines for future research. Applied Sciences, 12(16), p. 8328sqq. https://doi.org/10.3390/app12168328.

Rice, L. (2022). Digital twins of Smart cities: Spatial data visualization tools, monitoring and sensing technologies, and virtual simulation modeling. Geopolitics, History, and International Relations, 14(1), pp. 26–42.

Ries, S. (2018). Beginn einer neuen Ära. Automatisierung war früher. In B. Volkens, & K. Anderson, Digital Human. Der Mensch im Mittelpunkt der Digitalisierung (S. 46). Frankfurt am Main: Campus Verlag.

Rowland, Z.; Cug, J.; Nica, E. (2022). The Geopolitics of Smart City Digital Twins: Urban Sensing and Immersive Virtual Technologies, Spatio-Temporal Fusion Algorithms, and Visualization Modeling Tools. Geopolitics, History and International Relations, 14(2), pp. 56–71.

Smart City Sweden (2020). Umeå builds for everyone and attracts gender equality tourism. https://smartcitysweden.com/umea-builds-for-everyone-and-attracts-gender-equality-tourism/ (Accessed 01. March 2024)

Trevisan, R. et al. (2023). Renewable Energy Communities in Positive Energy Districts: A Governance and Realisation Framework in Compliance with the Italian Regulation. Journal of Smart Cities, 6(1), pp. 563–585. https://doi.org/10.3390/smartcities6010026.

von Hattburg, A., & Troxler, C. (2020). Gesunde Digitalisierung heißt: Der Mensch steht im Mittelpunkt. In SRH Fernhochschule (Hrsg.), Gesundheit – Arbeit – Prävention. Tagungsband zum 3. Kongress für Betriebliches Gesundheitsmanagement (S. 33–48). Wiesbaden: Springer.

Wagner, D. (2018). Digital Leadership. Kompetenzen - Führungsverhalten - Umsetzungsempfehlungen. Wiesbaden: Springer Gabler.

Weber, B. et al. (2023). Rechtskonformes Datenteilen im Bauprozess – Anforderungen des Data Governance Act an Common Data Environments. (English: Legal Compliance for data sharing in the Construction process - Requirements of the Data Governance Act for Common Data Environments). German Bauingenieur Journal, Vol. 98, Iss. 3. https://doi.org/10.37544/0005-6650-2023-03-66.

Weber-Lewerenz, B. & Traverso, M. (2022). CDR and Best Practices in AI in Construction - Catalysts and key to sustainability in the Digital Era. In: S. A. Pillay, et al. (Eds.), SACAIR 2022. Proceedings of the third South African Conference for Artificial Intelligence Research (pp. 290 sqq.). SACAIR 2022 Publishing, South Africa. https://2022.sacair.org.za/proceedings/ (Accessed 11. December 2023).

Weber-Lewerenz, B. & Traverso, M. (2023). Best Practices in Construction 4.0 – Catalysts of digital innovations (Part II). Journal of Architectural Environment & Structural Engineering Research (JAESER), 6(2), pp. 1–21. https://doi.org/10.30564/jaeser.v6i2.5408.

Weber-Lewerenz, B. (2022a). Accents of Added Value in Construction 4.0 – Ethical Observations in Dealing with Digitization and AI. Springer Nature Publishing. https://doi.org/10.1007/978-3-658-39407-3.

Weber-Lewerenz, B. (2022b). Unternehmerische Compliance im digitalen Transformationsprozess Teil 2. Vol. 2 (2022). S. 63 ff Fachjournal Risk, Fraud and Compliance (ZRFC), Erich Schmidt Verlag GmbH & Co. KG Verlag, Berlin. https://doi.org/10.37307/j.1867-8394.2022.02.

Weber-Lewerenz, B. (2022c). Wertakzente im Bauwesen 4.0 - Ethische Beobachtungen im Umgang mit der Digitalisierung und KI, Springer Nature Publishing, Wiesbaden. https://doi.org/10.1007/978-3-658-38238-4

Weber-Lewerenz, B. (2023a). Digitalisierung und KI am Bau. Schwung für die Branche. Teil 1 – Bauen neu denken. Im Interview mit Andreas Grosz im März 2023. KAP FORUM ARCHITECTURE & URBAN DEVELOPMENT. Veröffentlichung unter: https://www.kap-forum.de/ki-am-bau-teil-1/.

Weber-Lewerenz, B. (2023b). Thinking otherwise: Integrating Existing Buildings and Monument Protection in Smart Cities -Experience shared from user practice, In: Impact of Digital Twins in Smart Cities Development. In: Special Edition by Ingrid Vasilliu-Feltes, IGI Global Publishing. https://doi.org/10.4018/978-1-6684-3833-6.

Weber-Lewerenz, B. (2023c). Wertschöpfungskette Datensicherheit und CyberSecurity für das Bauwesen, Vol. 4, Fachjournal Risk, Fraud and Compliance (ZRFC), Erich Schmidt Verlag GmbH & Co. KG Verlag, Berlin. https://doi.org/10.37307/j.1867-8394.2023.04.05.

Wenzel, G. et al. (2022). Szenarioprozess "Bauen 2030". Kurzbericht: Zukunftsszenarien für das Planen, Bauen, Betreiben und Rückbauen im Jahr 2030. Fraunhofer IAO. https://doi.org/10.24406/publica-fhg-417722.

Zvarikova, K.; Horak, J.; Downs, S. (2022). Digital twin algorithms, smart city technologies, and 3D spatio-temporal simulations in virtual urban environments. Geopolitics, History and International Relations,14(1), pp. 139–154.

Part IV

KNOWLEDGE

International Law and Human Rights: European Politics Including Diversity and Inclusion

9

From Innovation to Smart Innovation in the Construction Sector Appreciation of Social Value, ESG Business Approach and 'Smart by Design'

Virginie Martins de Nobrega

Virginie MARTINS de NOBREGA (EMBA)
International law and human rights lawyer and Adviser for Ethics, legal, societal, and business AI (EMBA), France

9.1 Introduction

The construction sector is not the sector that first comes to mind to best assess the risks, impact and opportunities of artificial intelligence (AI) and emerging technologies on international law, human rights and value business creation from a sustainability point of view—sustainability being understood as 'meeting the needs of the present without compromising the ability of future generations to meet their own needs' following the definition of United Nations (UN) Brundtland Commission. In this article, those needs encompass the ethical, human rights, social, environmental, and societal implications based on the conviction—and hopefully the growing realization and integration—that with artificial intelligence and emerging technologies 'it is impossible to separate economic development issues from environment', ethical, human rights and social issues (United Nations Brundtland Commission 1987). And the conviction that the increasing interlocking crises faced by every sector, industry and actors should be dealt with a comprehensive approach, integrating all those dimensions and implications to political agendas and governance, public and private decision-making and, to business strategy and processes.

Furthermore, the Architecture, Engineering, and Construction (AEC) industry is 'one of the most influential markets globally' (Liang et al. 2023) due to its demographic, economic and financial importance and its environmental, social, and societal influence

B. C. Weber-Lewerenz (ed.), *Diversity in Construction—The Gamechangers*,
https://doi.org/10.1007/978-3-658-45308-4_9

and impact on individuals, public spaces, politics, and the res publica. 'The AEC industry has been continually growing over the last decades' (Liang et al. 2023) and whatever the level of development of countries, it is 'one of the major industries in any national economy' (Ayassamy and Pellerin 2023) representing 'around 7% of the world's total workforce' while 'real estate represents almost 60% of the value of all global assets [USD 217 trillion]' (Institute for Human Rights and Business 2019).

The construction industry is also 'responsible for substantial material and resource consumption and [has a major] impact on climate change' being 'responsible for 40% of global greenhouse gas emissions, and 36% of final energy use (...). With a lack of consideration for waste management and waste reduction in the early phases of projects, there tends to be waste generated by construction and demolition through the life cycle of buildings, with a remarkable impact of 50% at the end of the life of a project. It also has a reputation for its high consumption rate of natural resources, which generates between two and three billion tonnes of building waste per year' (Ayassamy and Pellerin 2023). According to the United Nations World Urbanization prospects, 'cities only cover 2% of the world's surface, yet they consume 78% of the world's energy and produce more than 60% of greenhouse gas emissions'.

The AEC industry has also an impact on the building and shaping of cities, hence impact and influence individuals and society especially since 'around 55% of the world's population [resides in cities] with this figure reaching a peak of 85% in countries like Australia, the UK, and the Netherlands' (Ziosi et al. 2022).. Global estimates highlight a continuous increase of global urbanization with 'rural-to-urban migration expected to reach (...) 68% [of the world's population]' by 2050 (Givens and Lam 2020). The role and impact of cities, urbanism and architecture on its population and society whether to protect, influence, control or/and surveille is not new and has long been a consideration of urban planning with Baron Haussmann 'acknowledging the military value of having broader and straighter streets to dispatch riots quickly' (Ziosi et al. 2022). Such considerations have reached another level of impact and influence when policy making, cybernetics, data and computing started to be combined to respond to an increased concern of states and military 'about the 'unruliness' of cities and the possibility of both military action by non-state combatants (such as terrorism) or conventional Military Operations in Urban Terrain' (Wood 2015)—such concern amplified following the 9/11 terrorist attacks in the United States.

That urge to protect, control and/or surveille has coincided with technological breakthroughs and a push by the technology industry to penetrate new markets with their products, services and solutions using as sales arguments the power of data to better understand, run and shape cities in alignment with the need for results-based management within the public sector to harness urbanization and its challenges, to innovate and to protect the populations, while facing budget restrictions, poor or deteriorating infrastructures, and rampant inequalities. 'IBM', for example, promised a "smarter planet and a new strategic agenda for progress and growth (...) Cisco solutions claim[ed] that it encapsulates a new way of thinking about how communities are designed, built, managed, and renewed to achieve social, economic, and environmental sustainability (...)

Start-ups, small and medium-sized enterprises (SMEs), and consultancies naturally joined, each offering their own solutions and analysis to support cities. It is this confluence of urgency and expectation in cities, and the increased ability with industry-led solutions and optimism, that birthed smart cities (…) estimated to be worth USD 2.4 trillion by 2025" (Givens and Lam 2020). This techno-solutionism approach has taken different shapes, from improving the efficiency of current services by collecting data to address their shortcomings, creating new optimized services to building new modern neighborhoods or sites from the "internet up" like the Songdo International Business District in South Korea and the Toronto Quayside project by Sidewalk Labs, a subsidiary of Google's parent company Alphabet.

Despite those trends, the construction sector remains one of the least digitalized industry due to many factors such as projects' complexity, resistance to change, project management complexity, 'lack of clear demand from clients and stakeholders' (Royal Institution of Chartered Surveyors 2020), as well as 'high initial costs of deploying AI-based solutions, talent shortage, computing power [and], internet connectivity' (Abioya et al. 2021).. AI-based solutions could be implemented throughout the project life-cycle to improve the management of projects, optimize the construction phase, and decrease the environmental cost of buildings and construction though.

At the same time, the questioning of the role and impact of smart cities on human rights, of mass surveillance and surveillance capitalism, of the facial recognition society and of the issue of privacy continues to both unveil new dynamics with a privatization of public spaces, an increase of public–private shared decision-making to the detriment of the common good, increased inequalities between digitally smart citizens and marginalized digitally illiterate populations, reduction of the role of public spaces and buildings as hubs for social innovation, cultural exchange, community participation and, the urgent need to abide by and promote the application of international law and international human rights standards.

9.2 Smart Construction By Design—Innovation Vs. Smart Innovation By Design

Like every industry evolving in a competitive and high-demand global environment, the AEC sector has to adapt to new demands in a context of constant evolution having to deal with multiple crises (climate change, demographic explosion in urban cities, rarefication of natural resources, reputational risks and crises, new consumers' interests and needs, pressure from civil society and policy makers) while maximizing their business model, preserving their margins—or at least their bottom line—investing in new technologies to improve their business and production processes and, abiding by their legal obligations to limit their liability and legal responsibilities, which implies to mainstream and operationalize international human rights standards and obligations within their business and processes.

Not only the AEC needs to invest and scale up AI-applications to maintain their comparative advantage and meet those different constraints, but it should do it through a virtuous circle to both generate profits and create and bring value to all stakeholders impacted, hence respecting the social role it plays in cities and societies.

9.2.1 Overview of AI-Applications Applied to Construction

Over the years, 'there has been a surge in the global digitization of corporate processes and concepts such as digital technology development which is growing at such a quick pace that the construction industry is struggling to catch up with latest developments' (Egwim et al. 2024) despite AI applications proving to be able to open opportunities throughout the project-life cycle (design, project management, construction, building maintenance and management) to improve cost-efficiency and risks analyses, reduce the production time, improve workers' safety and, improve the quality and sustainability approach towards the sustainable development goals (SDGs). The UN SDG 11 relates indeed to 'making cities (…) inclusive, safe, resilient and sustainable' emphasizing that 'we need new, intelligent urban planning that creates safe, affordable, and resilient cities with green and culturally inspiring living conditions' such as waste and energy management (SDG 11, target 11.6). That implies all actors of the built environment cycle (planners, investors and developers, owners, architects and designers, construction, and engineering) at international, national, and local level.

Research shows that most of AI technologies (machine learning, robotics, quantum computing, blockchain) has been explored and tested at the exception of language model (LM) and large language model (LLM). Those AI technological advancements combined with big data, the internet (e.g., the internet of things (IoT)) and advancements in technologies such as 3D printing, 'sensors, actuators, wireless technologies, cloud computing' (Abioya et al. 2021) have been supporting the transformation of the construction sector thanks to smart environments and interconnected systems. It is worth noting that such connectiveness can be unreliable on sites due to lack of power, lack of internet connectivity or lack of telecommunications, which might hamper the scaling up of such technologies especially since 'the danger of inefficiency and risk, which eventually contribute to project costs and delays, grows geometrically with the project's scale' and its complexity (Egwim et al. 2024).

Machine learning (ML) can be used to improve risks prediction, logistics and supply chain processes. Robotics can support with heavy tasks (bricklaying, assembly, or demolition) preserving the safety and well-being of workers, assemble pre-designed modular structures by autonomously identifying the prismatic building components (Pillai and Marus 2019) and, conduct site monitoring inspections and performance evaluation when coupled with drones and sensors (Holzmann and Lechiara 2022). Algorithmic decision-making can support procurement, tender evaluation, contract, and administrative tasks. Blockchain can help manage data safely throughout the building's lifecycle in accordance with the OCDE AI guiding principles and the European guidelines for trustworthy AI, thus limiting and mitigating the digital risks e.g., data leak, confidentiality breaches, disclosure of sensitive information such as security system information, disclosure of intellectual property, malware, phishing.

Two other illustrations that are more relevant when put into perspective with the imperatives of international law and human rights standards and, with the ethical and human rights risks, are the Building Information Modelling (BIM) and the building automation and management systems (BAMs) used in connection with AI. 'BIM [plus the technology of] IoT is the formula for most of the AI applications in the construction processes' enabling the collect, share and storage of a large quantity of data on the building components leading to big data mining and analysis creating a more comprehensive and centralized source of information to analyze (Pillai and Marus 2019) and even conduct autonomous data analysis for operation optimization. For example, machine learning (ML) and BIM can support cost estimation, risks predictions and project management by enhancing the so called "iron triangle" or "triple constraints" optimizing the analysis and triangulation of information related to 'schedule, cost, scope, and quality management, jointly with project risk management' (Holzmann and Lechiara 2022). It can also 'achieve real-time visibility and traceability (…) and generate early warnings and alarms as dynamical safety barriers for hazard energy on underground construction sites' (Abioya et al. 2021). Coupled with AI, BIM has also an important leverage effect on value-driven services shifting the construction industry from a management and mitigation approach towards a data-driven and data-analytics approach. For example, the construction industry can shift from a waste management approach towards a waste analytics approach to reduce and minimize waste during the design and construction stages, thus improve its environmental impact. Similarly, BAMs use computational and data capacities to optimize and automate building operations and best manage them making on-time decisions to improve buildings' performances (heating ventilation, air conditioning, energy and water management, security systems). For example, it has 'recently emerged as key solutions for predicting building occupancy, forecasting thermal comfort, boosting energy saving, and enabling demand-side response' (Himeur et al. 2023). Coupled with behavioral analytics, it can also predict and forecast demands and needs to adjust the management and maintenance accordingly providing, for example, energy on-demand, saving energy, provide demand-side response, understand consumers demand patterns. 'Moreover, AI-based big data analytics are contributing to building automation by making BAMs self-learning, self-configuring and self-diagnosing, and self-commissioning. Additionally, using AI-based analytics can adapt existing building systems to promote the deployment of BAMs with fewer investments from building owners' (Himeur et al. 2023).

Though quite late in developing and integrating AI-based solutions, the construction industry has the peculiarity to follow the pattern of 'innovation spill-over', (Pillai and Marus 2019) which could speed-up its digitalization and the deployment of big data analytics projects. In following its digitalization, the construction industry should continue to implement its whole-life and whole-asset thinking by embracing the value-driven approach across the asset life cycle and measuring its social value to continue its shift to a performance and outcomes-centric regime (Royal Institution of Chartered Surveyors 2023) to not only innovate, but innovate smartly by design.

9.2.2 Smart Innovation by Design: Embedding Human Rights and Dignity by Design to Increase Social Value

As highlighted by the 2020 RICS report, impact measurement metrics for infrastructure projects has long shifted from a mere economic argument towards a social value and social benefits approach centred around economic prosperity, wellbeing, and quality of life. To that regard, social value is defined as 'the impacts that an intervention, policy, or project [i.e. all types of assets] has on society and the value that these impacts have, both positive and negative (…) The social value of a project is the net value generated to society (net of negative impacts). Social value actually captures all types of impact: on the economy, the environment and society more widely because they all affect people in society'. A similar approach as long being adopted by the United Nations Human Rights and Business Guidelines based on the international law principles, which states that states have the obligation to respect, protect and fulfill human rights and fundamental freedoms and that companies across industries are required to comply with all applicable laws and to respect human rights.

In the context of digitalization, big data and the ever-increasing use of AI-applications, the construction sector also has to follow three tiers of international law and human rights obligations: the current general international and regional frameworks in relation to international law and human rights e.g., Universal Declaration of Human Rights, the International Covenant on Civil and Political Rights and the International Covenant on Economic, Social and Cultural Rights, the International Labor Organization conventions; the specific AI related general international and regional frameworks e.g., UNESCO recommendation on the Ethics of Artificial Intelligence, the OCDE AI principles, the Ethics Guidelines for Trustworthy Artificial Intelligence and its ALTAI assessment list, the newly adopted AU AI Act, as well as more specific regulations in relation with specific technologies, for example, the Council of Europe 108+Convention on data and its guidelines on facial recognition. To that you add, the laws related to the construction sector at national and local levels, to which most actors are used to depending of their realm of operation. The general or sector specific international instruments (charters, conventions, guidelines United Nations recommendations (UN, 2011) are good indicators and a solid ground to build or adapt a strong human-rights and ethical-based strategic positioning –also referred as human rights by design, ethics by design—enabling all actors, depending of their scope of action and level of responsibilities, to define and steer their strategic and governance policy, as well as operationalize and mainstream best practices, ethical safeguards and guidelines grounded on international law and human rights throughout the AI-life cycle projects and throughout the built environment cycle not only in their due diligence, but as integral part of the cost–benefit analyses and the measurement of the projects' social value. To that end, the social life-cycle assessment could be of interest (Ayassamy and Pellerin 2023).

As in every industry, the strategic approach to the data ecosystem should also be of high importance not only to protect the data, ensure transparency in its use and re-use, but also to guarantee that the data-driven processes and the data feeding the AI-applications are unbiased, have sufficient granularity and, rightly represent the impacted populations with a particular attention to be given to known disadvantaged groups, such as women, minorities, indigenous groups and vulnerable groups (Martins de Nobrega 2022). Needless to say that such business approach will be adapted to the peculiarities of the field of operation (geography, culture, context, scope of actions, level of responsibilities). This comprehensive approach of an operationalized international law and an ESG-oriented business strategy should be grounded on the larger risk and needs analyses, gaps analyses and SWOT integrating the ethical and international law aspects from day one enabling companies to decide ab initio about their ethical and human rights priorities, criteria, objectives, expected outputs and outcomes and their KPIs (Martins de Nobrega 2018). Such strategic assessments should involve multistakeholder processes to enhance the collaboration and maximize the efforts done throughout the construction or built environment life-cycle including with local populations in due time and depending of the project.

The strategic thinking process of mainstreaming and operationalizing smart innovation by design should be:

Step 1: Clarify your leadership positioning regarding international law and human rights within your ESG approach.

Step 2: Have clear mandate-driven priorities and projects.

Step 3: Conduct relevant strategic business assessments including adapted parameters in relation to your human rights and ethics by design approach.

Step 4: Map out the relevant sector specific, geographic specific or any other relevant legal and regulatory instruments, policy guidelines available (e.g., RICS 2020 and 2023 recommendations, Institute for Human Rights and Business Principles for Dignity in the Built Environment).

Step 5: Define, test, and control the processes.

Step 6: Define clear chain of responsibilities and accountability.

Step 7: Measure.

Thinking of the conception, development, scaling up and integration of AI-applications in the construction industry is leading to a profound transformation of the sector encouraged and supported by authorities, professionals, and technological companies. From a business point of view, it presents risks and opportunities, which requires to shift towards a 'smart by design' approach. The later should also be applied—as well as be the lens of analysis—to the interactions between buildings and its direct (cities) and indirect environment (individuals, groups, communities, society).

9.3 Smarter 'Smart Cities' with Human Rights by Design

As highlighted by the Mass Design Group, 'Architecture is not neutral; it either helps or hurts. Architecture is a mechanism that projects its values far beyond a building's walls and into the lives of communities and people. To acknowledge that architecture has this kind of agency and power is to acknowledge that buildings, and the industry that erects them, are as accountable for social injustices as they are critical levers to improve the communities they serve' (Institute for Human Rights and Business 2019). From monitoring behaviors and patterns of consumption to both services on-demand and management at-distance, without properly informing nor obtaining the consent of inhabitants, to a generalization of surveillance in neighborhoods and cities leading to mass surveillance, the data analytics approach to urbanism and construction is increasing the role, and therefore the responsibility—whether legal or ethical even moral—of the stakeholders involved in the construction industry (planners, investors and developers, owners, architects and designers, construction and engineering). Technology and data are not neutral either. The business model of technological companies, the treatment of data (collect, storage, use, re-purposing, treatment, and interpretation) and the power given to both data and the data holders are creating new paradigms that needs to be thought through the lens of international law, human rights, and ethics, as well as shaped according to those.

The tension between new technologies and human rights is not new as Warren and Brandeis underlined the invasion of social privacy by photography in violation of the right to privacy as early as 1890. However, the dual-nature of AI and emerging technologies, the pervasiveness of AI technologies, scale, speed, real-time monitoring, 'unprecedented granularity of data' (Ziosi et al. 2022) frequence and numbers impacted makes it a political, structural, economic, and societal dilemma that needs to be addressed and tackled effectively and efficiently (Martins de Nobrega 2022).

9.3.1 Ethical and Human Rights Risks in Smart Cities

Understanding thoroughly the design, planning, development, and operations of smart cities to integrate and mainstream a 'smart by design' approach would certainly require to have a consensus on the definition of smart cities also referred to over time as digital cities, tech cities, wired cities, ubiquitous cities depending on the level of interoperability and connectiveness and the importance of data analytics. All refers to a technocentric, techno-optimistic and technology driven vision mentioned before. In this article, we englobe three forms of cities: the optimization of services of traditional cities through technological additions; the creation of new services through data-analytics to collect, monitor and decide, which can refer to data-driven cities e.g., new ICT; the creation of new cities and neighborhoods from the 'ground up' (Ziosi et al. 2022). We also envision 'smart' as a multidimensional concept englobing the ethical, human rights, social, environmental, and societal implications echoing the European Parliament's 'six dimensions of smart economy, smart governance, smart living, smart people, smart environment, and smart mobility' (Ziosi et al. 2022). Concretely, following Albino's research, we agree on 'four prevailing aspects that enable one to identify a city as "smart": the presence of a city's networked infrastructure that enables political efficiency and social and cultural development, an emphasis on business-led urban development and creative activities for the promotion of urban growth, social inclusion of various urban residents and social capital in urban development, and the natural environment as a strategic component for the future' (Ziosi et al. 2022).

In relation to such understanding of smart cities, the possible upcoming trend of increased datafication in the construction industry means it will be facing, increasing, and possibly generalizing the current ethical and human rights risks identified, as well as unintended consequences. The first risk is a cybersecurity risk proportionally increased by the reliance on big data, virtuality, internet, and the integration of external data, such as with Level 3 BIM. The risks can be a cyberattack on the infrastructure or of the Intelligent Operation Center (also called Control Rooms) that centralize the flow of data in real-time. It can also be linked to a data breach, a data leak or hack, a misuse, unauthorized or/and untransparent re-use of data and a lack of anonymization. In that respect, the increase use of 5G high speed transmission, bad storage habits (e.g., home office, local storage, lack of security of data centers), multiplicity of users, synchronization of different software, data sharing and the cloud would increase the risks (Liang et al. 2023).

The second risk is the risk on fundamental freedoms due to a mass surveillance. Indeed, this end-to-end integrated system approach has a direct ripple effect on the level of control and surveillance of citizens and populations in a general context where the trade-off between security and human rights continues to be unquestioned and no real alternatives is presented either leading to 'everyday surveillance [being endemic] to modern societies'. This 'unbroken trend' (Liang et al. 2023) varies depending of the political and geographic contexts with countries like China, Egypt, Nigeria, India, Korea, Brazil, Tunisia and most countries of Central Asia, South Caucasus, Eastern Europe, Turkey having largely adopting it. This trend has been steered outside China by the high-market penetration rate of Chinese companies specialized in facial recognition (Givens and Lam 2020). Commonly, even in European cities like London, the tendency went from urban planning to an increased deployment of ICT and cameras linked to predictive analytics, facial recognition systems or algorithmic cameras trained to detect suspicious behaviors (e.g., thief, terrorist attacks…), radars, GPS localization, sensors through public spaces morphing the optimization of cities' systems into a network of surveillance leading to mass surveillance as the new normal. The citizen is now seen as a commodity whose behavior could and should be influenced. That influence is facilitated by the citizens' ubiquitous trackability rendered possible by their wearable devices, increased transnational collaboration between private actors and the shift of perception from a disciplinarian society for governing purposes to a 'surveillant assemblage' and controlled society for consumerism purposes (to construct and monitor consumption patterns, constructing consumer profiles) through 'data-double'. 'Social control today is, thus, decentralised and shape-shifting—it is not focused just on collecting information but on decoding and recoding, sorting, altering, circulating and re-playing information' without the citizens knowing nor should be made aware of (Galič et al. 2017). An underlying unmeasurable risk to date is the emotional impact and psychological harm done on individuals with the pervasiveness of such technologies. The second risk is therefore the violations of the right to privacy, in private and public spaces, with seven threats being identified: 'identification, localization and tracking, profiling, privacy-violating interaction and presentation, lifecycle transitions, inventory attack and finally, linkage' (Ziosi et al. 2022) without the citizen being informed, asked for consent or knowing clearly and precisely the future use of its data and whether they will be re-purpose in violations with international and regional legal instruments.

Another risk is linked to algorithmic bias and the lack of transparency in algorithms (the black-box effect) leading to possible discriminations and inequalities whether reinforcing old stigmas and patterns or creating new ones based on the 'average citizen' standard selected to train the algorithm and interpret the data (Martins de Nobrega 2022). Another risk is the exclusion or limitation in the exercise of human rights, such as the right of peaceful assembly and the right of freedom of expression, when internet access and connectivity are tempered with e.g., throttling, blocking of applications or websites. The smart city system even permits to profile and track individuals leading to possible human rights violations based on political or religious beliefs. Such trends are already prevalent in some countries especially since the exceptions to international law on the ground of national security are broadly used leading to companies being requested—if not obligated—to transfer their data to the states.

9.4 Smarter Smart Cities with International Law and Human Rights Safeguards by Design

Those risks could seem quite remote from a construction company even more if not involved in the discussions with political authorities and decision-makers. However, in the new global context and focus on social value and impact, as well as the questioning of states' monopoly on the regulation of their territory due to an increased privatization and corporatization of public spaces and an increase of public–private decision-making processes due to the privatization of data and infrastructure, such risks need to be addressed and taken into consideration mainstreaming an ethical and human-rights based approach, thus building smarter 'smart cities' beneficial for the 5Ps (people, planet, prosperity, peace, and partnerships) of the UN SDGs. Private actors should also be aware of their responsibilities to mitigate their cost–benefit analysis by the social value factor when selecting AI-systems and products. For example, to date, the best facial recognition technologies in terms of value for money are Chinese companies that not only do develop their products according to Chinese ethical and legal framework less focused on privacy and human rights, but also must transmit all data to the government when not state-owned. Therefore, when measuring the return on investment, companies should include the social value and impact to avoid mainstreaming technologies that are not 'human rights by design'. That could prove difficult since there is not enough measurement yet in the construction sector, but would constitute a good practice of anticipatory and active risk management approach. It is even more needed in a context where there is an increasing role of cities.[1]

In terms of international and human rights legal frameworks, all international human rights texts apply (e.g., 1966 International covenants, articles 12, 17, 19, 20, Universal declaration of Human Rights) with the United Nations constantly reaffirming than digital rights are human rights. Various international guidelines and indicators can also be referred to (UNESCO internet universality indicators, United Nations Human Rights and Business Guidelines, OHCHR reports such as A/HRC/51/17 on privacy in the digital age, the New Urban Agenda), as well as regional (e.g., article 8 of the European Charter on Human Rights) or sector-specific (e.g., The European Charter for the safeguarding of human rights in the city, Institute for Human Rights and Business Principles for Dignity in the Built Environment) legal instruments as mentioned previously. When acting locally, it is of interest to look at the Cities Coalition for Digital Rights, the Human rights cities declarations, and charters and, the Gwangju guiding principles for a human rights city.

Regarding the issue of data security in a context of increase private–public decision making, some authors have already pointed out few best practices with little effect on innovation, such as: 'adjust[ing] the default parameters of the system so that they are as well-balanced as they are feasible. Using the smart meter as an example once more, the goal is to adjust the default read frequency to a certain interval of minutes which satisfies both the information gathering needs of the operator and the user's privacy; individual's permission to modify the system's default settings; aggregate data processing.[2]

The strategic thinking process of mainstreaming human rights by design could be:

Step 1: Map out your outputs and outcomes integrating the ethical and human rights dimensions

Step 2: Conduct the relevant assessments and tests including international law tests od necessity and proportionnality

Step 3: Map out the relevant sector specific, geographic specific or any other relevant legal and regulatory instruments, policy guidelines

Step 4: Define, test, and control

Step 5: Define clear chain of responsibilities and accountability

Step 6: Monitor, define best practices and measure

---

[1] (Kempin Reuter 2020).

[2] (Fabrègue and Bogoni 2023).

Step 7: Think about remedies and redress mechanisms

9.5 Conclusion

Increased interconnectedness in a global economy facing intricated crises call for a global systemic thinking, commitment, and actions of all stakeholders to design, plan, build, implement, deploy, and manage smart innovation bringing social value for the common good and individuals. The possibilities offered by artificial intelligence and emerging technologies can be harnessed for such purpose-driven innovations or can be deployed rapidly for the sake of its efficiency and cost-reduction promises without thinking ahead of the ethical, legal, and social implications. Judging from the current challenges, the increased inequalities, the digital divide between the north and the south as well as between groups within countries, the multiplications of conflicts and human rights violations, it is not guaranteed that such a choice is still a real viable option. We have a collective responsibility to act at our level of action to be smarter and more focus on our impact and our outcomes.

The question is not whether AI-applications will be deployed but when, how and for whom. Societal challenges are complex and challenging thus calling for smarter approaches and smarter businesses, which involves ethical and human rights design approaches.

References

Abioya, S. et al. (2021). Artificial intelligence in the construction industry: A review of present status, opportunities, and future challenges. Journal of Building Engineering 44 (2021) 103,299. https://doi.org/10.1016/j.jobe.2021.103299.

Ayassamy P, Pellerin R. Social Life-Cycle Assessment in the Construction Industry: A Review of Characteristics, Limitations, and Challenges of S-LCA through Case Studies. Sustainability. 2023; 15(19):14,569. https://doi.org/10.3390/su151914569.

Egwim, C.N. et al. (2024) Artificial Intelligence in the Construction Industry: A Systematic Review of the Entire Construction Value Chain Lifecycle. Energies 2024, 17, 182. https://doi.org/10.3390/en17010182.

Fabrègue, B.F.G.; Bogoni, A. Privacy and Security Concerns in the Smart City. Smart Cities 2023, 6, 586–613. https://doi.org/10.3390/ smartcities6010027.

Galič, M., Timan, T. & Koops, BJ. Bentham, Deleuze and Beyond: An Overview of Surveillance Theories from the Panopticon to Participation. Philos. Technol. 30, 9–37 (2017). https://doi.org/10.1007/s13347-016-0219-1.

Givens J.W. & Lam, D. (2020). Smarter Cities or Bigger Brother? How the Race for Smart Cities Could Determine the Future of China. Democracy, and Privacy, 47 Fordham Urban Law Journal 829. https://ir.lawnet.fordham.edu/ulj/vol47/iss4/2. (Accessed 16. March 2024).

Holzmann, V., & Lechiara, M. (2022). Artificial Intelligence in Construction Projects: An Explorative Study of Professionals' Expectations. European Journal of Business and Management Research, 7(3), 151–162. https://doi.org/10.24018/ejbmr.2022.7.3.1432.

Himeur, Y. et al. (2023). AI big data analytics for building automation and management systems: a survey, actual challenges, and future perspectives. Artificial Intelligence Review (2023) 56:4929–5021. https://doi.org/10.1007/s10462-022-10286-2.

Institute for Human Rights and Business. (2019). Dignity by Design: Human Rights and the Built Environment Lifecycle. https://www.ihrb.org/focus-areas/built-environment/report-dignity-by-design-human-rights-and-the-built-environment-life. (Accessed 15. January 2024).

Kempin Reuter, T. (2020). Smart City Visions and Human Rights: Do They Go Together? Carr Center Discussion Paper Series Understanding the impact of technology on urban life. Spring 2020 issue.

Liang, C. J., Le, T. H., Ham, Y., Mantha, B. R., Cheng, M. H., & Lin, J. J. (2023). Ethics of Artificial Intelligence and Robotics in the Architecture, Engineering, and Construction Industry. arXiv preprint arXiv:2310.05414.

Martins de Nobrega, V., (2018). A.I and innovation for good in the nexus of human rights, humanitarian, and development. Real opportunity for a greater impact of policies, advocacy, and human rights? [Unpublished Executive Master in Management Dissertation]. SDA Bocconi School of Management, Milan, Italy.

Martins de Nobrega, V. (2022). A Diverse (AI) World: How to Make Sure That the Digital World Reflects the Richness and Diversity of Our World. In I. Vasiliu-Feltes (Ed.), Impact of Women's Empowerment on SDGs in the Digital Era (pp. 180–204). IGI Global. https://doi.org/10.4018/978-1-6684-3637-0.ch008.

Pillai V. S. & Marus K. (2019). Regulation of AI Technologies in the Construction Industry. HKUST IEMS Working Paper No. 2019–65. http://dx.doi.org/10.2139/ssrn.3439826.

Royal Institution of Chartered Surveyors. (2020, September 2020). Measuring social value in infrastructure projects: insights from the public sector).

Wood, D. M (2015). Smart City, Surveillance City. https://www.scl.org/3405-smart-city-surveillance-city/ (Accessed 15. January 2024).

United Nations. (2011). Guiding Principles on Business and Human Rights: Implementing the United Nations "Protect, Respect and Remedy" Framework.

United Nations Brundtland Commission. (1987). Report of the World Commission on Environment and Development: Our Common Future. https://www.un-documents.net/our-common-future.pdf. (Zugriff 15. Januar 2024).

Ziosi, M., Hewitt, B., Juneja, P. et al. (2022). Smart cities: reviewing the debate about their ethical implications. AI & Society. https://doi.org/10.1007/s00146-022-01558-0.

New Thinking and Learning Culture 10

10.1 Diversity For Managing Complexity in Construction Based on System Thinking

Dipl.-Ing. M.-Eng. LL.M. Michael Frahm

10.1.1 Introduction

This article provides an overview of the management of complexity in construction based on system thinking. To this end, the term system thinking, complexity and diversity are presented from a systemic perspective. The approaches are then transferred to the management of complexity in construction. Diversity is viewed systemically and cybernetically in the sense of diversity of action. There are similarities to diversity, which describes the variety of differences between people and their recognition. What diversity and diversity of action have in common is that they are based on the concept of diversity. Both concepts are effective in dealing with complexity. This article is about understanding the diversity of actions in organizations to better deal with complexity.

Why System Thinking? If the question, when dealing with today's challenges, is "complexity," then the answer is "system thinking." Many management methods come and go, promising simple solutions to complex problems but failing to deliver. System thinking, on the other hand, uses a multi-method approach to offer ways of thinking and methods that, contrary to the linear cause-effect principle, offer approaches that can overcome the challenges of complexity. System thinking has its origins in general systems theory, cybernetics and complexity sciences (Jackson 2019), i.e. at the origins of thinking about complexity. Due to the interdisciplinary approach to the concept, there is currently no generally accepted definition of system thinking. For this article, reference

B. C. Weber-Lewerenz (ed.), *Diversity in Construction—The Gamechangers*,
https://doi.org/10.1007/978-3-658-45308-4_10

is made to the methodological canon of SCiO—System (SCiO 2023) and Complexity in Organization and the SoSM—System of System Methodology by Mike C. Jackson (Jackson 2019) and an understanding of the term based on this is used. Methods for dealing with complexity that are attributed to system thinking include the Viable System Model, Soft Systems Methodology, System Dynamics, Interactive Planning, Syntegration and Critical System Heuristics, to name just a few approaches. But not only the use of suitable methods is important, but these should be accompanied by an adaptation of the way we think about systems. For example, Hovestädt (Hovestädt et al. 2022) considers the recognition and understanding of emergence in systems to be a central and important competence. Emergence means that the whole, through the interaction of the individual parts, is more than the sum of its parts. The properties of the whole cannot therefore be derived from the individual properties of the elements and often cannot be explained. This view contradicts the reductionist focus of individual considerations. The aforementioned approaches and conceptual concepts explicitly take complexity into account and take into account the resulting properties of systems such as emergence, nonlinearity, unpredictability, openness, self-organization and rapid phase transitions.

10.1.2 About Complexity

There is also no uniform definition of complexity. This can also be largely attributed to interdisciplinarity. Understanding and dealing with complexity is a highly relevant topic today in all disciplines and scientific areas. However, each field defines complexity differently. In computer science, complexity essentially means computing power (Frahm and Roll 2022) (ref. e.g. to Kolmogorov complexity). Complexity or complex systems are characterized by special properties compared to simple or ordinary systems. These are among others:

- Emergence (as already mentioned)
- Nonlinearity (sensitivity of the initial conditions with regard to the initial conditions—also known via the metaphor: butterfly effect)
- Self-organization
- System openness
- Unpredictability

Almost regardless of which discipline or actor is consulted, the concept of complexity is associated with fundamental properties. According to Frahm (Frahm and Rahebi 2021), a minimal definition that includes these basic properties is as follows:

> *„in simple terms, complexity means that a system has many elements, relationships and states that change over time in different dynamics."*

the time factor is decisive for complexity and, thus, the ability to change over time and, for example, the rapid phase transitions from stable states to unstable states, which appear to occur out of nowhere. The ability to change over time is also the main difference to complicated systems (e.g. technical systems such as a clock or an engine or a finished building), which do not have this dimension. This article is based on an understanding of complexity that has already found its way into the context of project management through approaches such as Bennet and Fines (Bennett and Fine 1980), Baccarinis (Baccarini 1996) or Morris (Morris 1997). This does not change the basic properties of complexity, as described above; it essentially creates a structuring or classification to generally sort the complexity into drawers and allocate it to work areas. The amount of complexity always remains the same, it is just cut into pieces like a cake. In project management, which is also relevant to the construction industry, there are the following classifications according to Table 10.1.

Accordingly, complexity is classified into:

- Structural/Organizational
- Technical
- Directional/Environment/Stakeholder
- Temporally

For specific articles, dealing with structural/organizational complexity will be further considered. To get an idea of what is meant by structural/organizational complexity, Damayanti, Hartono and Wijaya (Damayanti et al. 2021) are used. For you, complexity is characterized by the size of the internal and external organizations and the quantity and quality of the interrelationships. Construction projects and especially large-scale con-

Table 10.1 Classification of complexity in Project Management & Large projects. *Source* Frahm

Complexity	Remington and Pollack (2007)	Shenhar and Dvir (2007)	Hertogh and Westerweld (2010)	Bosch-Rekveldt et.al. (2011)	ICCPM (2012)	Giezen (2012)
Structural/ Organizational	x		x	x	x	x
Technical	x	x	x	x	x	x
Directional/ Environmental/ Stakeholders	x	x	x	x	x	x
In terms of time	x	x	x		x	x

struction projects have a very high structural/organizational complexity. In order to manage complexity in the construction industry, various actors such as He, Luo, Hu, Chan (He et al. 2015), Hoffmann (Hoffmann 2018), Nyarirangwe & Babatunde (Nyarirangwe and Babatunde 2021) and Jia, Xiang, Chen (Jia et al. 2023) recommend the use of methods such as systems thinking or alternative system approaches.

10.1.3 About Diversity from a Systemic Perspective

For diversity and the management of diversity from a systems perspective, Ashby's variety theorem, also known as "Ashby's law of requisite variety", is used (Ashby 1956). Variety stands for variety of actions. Ashby's law of required variety is considered as important among management sciences as Einstein's theory of relativity about the equivalence of mass and energy with E = m * c2 is among physicists (Pfiffner 2020). Ashby's law fundamentally contributes to understanding how to deal with complexity, it says:

"Only variety can destroy variety "

or

"Only variety can absorb variety"

So only variety can absorb variety, or only variety can cope with variety.

In the remainder of the article, the term diversity is used instead of variety, as both terms can be viewed as synonymous.

When applied, Ashby's Law means that the diversity that a system or organization has to cope with in a particular environment or system environment must correspond to the diversity of actions of the system or organization. If this is the case, there are stable relationships between the system and the environment. If there are unstable conditions and therefore no balance, this can almost always be explained by a disruption in the management of diversity. Figure 10.1 shows Ashby's theorem transferred to the management paradigm (Frahm and Roll 2022).

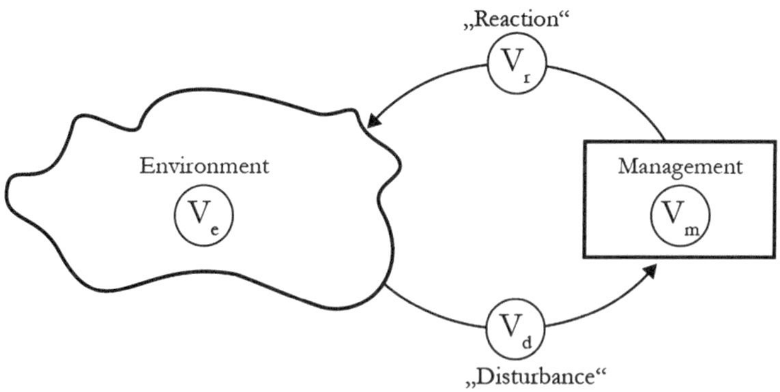

Fig. 10.1 Ashbys Law 1

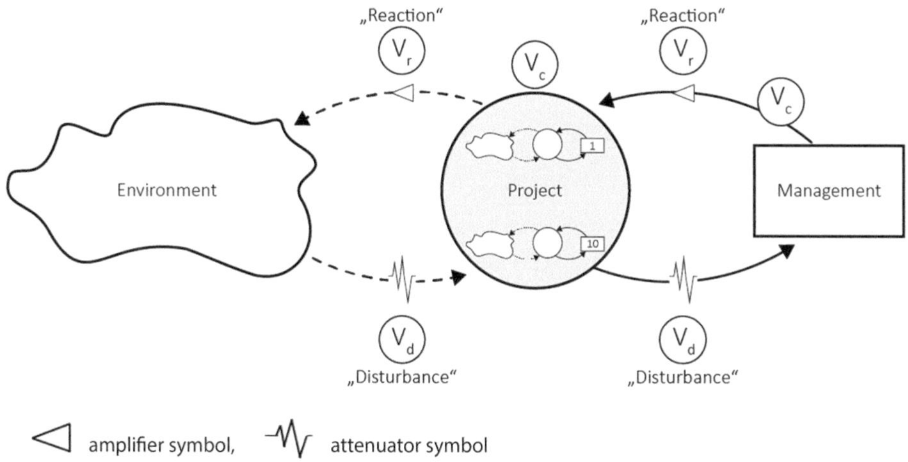

Fig. 10.2 Ashbys Law 2

Environmental diversity (Ve) far exceeds that of management (Vm) (Eq. 10.1). For construction projects, it can be concluded that the environmental diversity (Ve) from e.g. customer requirements, regulatory requirements, suppliers, the public, etc. is generally greater than the behavioral repertoire of an organization's construction management.

$$V_e >>> V_m \tag{Eq. 10.1}$$

Ashby uses diversity as a measure of complexity and bases his equilibrium theorem, which represents the diversity of consequences (Vc) as the quotient of the diversity of disturbances (Vd) and the diversity of system reactions (Vr) (Fig. 10.2). If this equation is balanced, then stability is present (Eq. 10.2). If the variety of consequences (Vc) is larger or much larger, ultrastability exists and the system can compensate for changes without a crisis and remains stable (Eq. 10.3).

$$V_c \geq \frac{V_d}{V_r} \text{ stable} \tag{Eq. 10.2}$$

$$V_c < \frac{V_d}{V_r} \text{ ultra stable} \tag{Eq. 10.3}$$

A system collapses when the disturbances are significantly greater than the system reactions. The variety theorem then becomes unstable (Eq. 10.4).

$$V_c < \frac{V_d}{V_r} \text{ not stable} \tag{Eq. 10.4}$$

According to Markus Schwanninger (Schwanninger 2006), Emeritus Management Professor at the University of St. Gallen, diversity can be calculated as follows (Eqs. 10.5 and 10.6):

Relationship-oriented equation

$$V = m * \frac{n * (n - 1)}{2} \quad \text{(Eq. 10.5)}$$

Element-oriented equation

$$V = z^n \quad \text{(Eq. 10.6)}$$

Codes:

V: Variety/ Diversity
m: Number of relationships between each pair of elements
n: Number of elements
z: Number of possible states for each element

It doesn't matter whether the relationship-oriented Eq. 10.5 or the element-oriented Eq. 10.6 is used as a basis, the diversity always follows the power law. This means it grows exponentially with the number of elements and the number of relationships, as shown in Eq. 10.5, or with the potential or number of states, as shown in Eq. 10.6. This means that it is not necessary to calculate the diversity for practical application. But it is important to know that diversity is growing exponentially and therefore very quickly and that this must be given particular attention when managing diversity and thus managing complexity.

10.1.4 Application Concept of Diversity with the Viable System Model

The possible applications of the previously made considerations are presented below based on Ashby's Law and its requirement for sufficient diversity in systems. To consider complexity, structural/organizational complexity was selected for this article. According to the SOSM—System of Systems Methodology (Jackson 2019), the Viable System Model is the approach of choice as a management model for organizational complexity. The Viable System Model (Beer 1972, 1979, 1985) is the theoretical and practical application of Ashby's law and depicts the mathematical equilibrium theorem with its system structure, its system levels, its communication channels and its information channels as a homeostat. As a model, it represents the flow of diversity within and outside an organization. It can be used both as a design and as a diagnostic tool (Beer 1985).

The VSM consists in six horizontal System levels (Fig. 10.3):

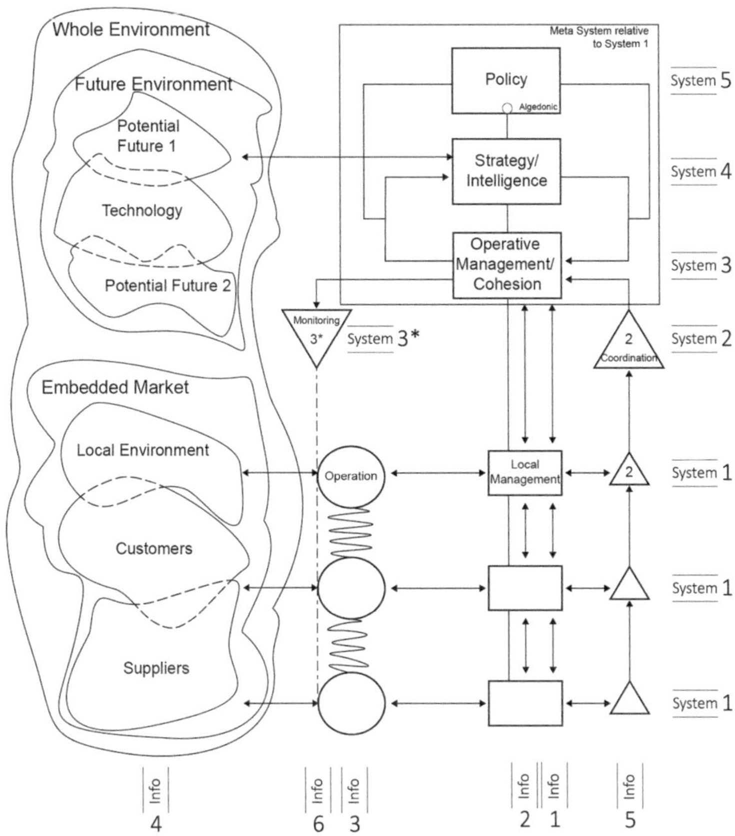

Fig. 10.3 Viable System Model (Frahm and Roll 2022)

- System 1 Production
- System 2 Coordination
- System 3 Operative Management
- System 3* Monitoring
- System 4 Strategic Management
- System 5 Normative Management

and six vertical Information channels

- Channel 1 Intervention & Regulation
- Channel 2 Allocation of resources
- Channel 3 Interrelationships of production
- Channel 4 Interrelationships of the environment
- Channel 5 Coordination = System 2
- Channel 6 Monitoring = System 3*

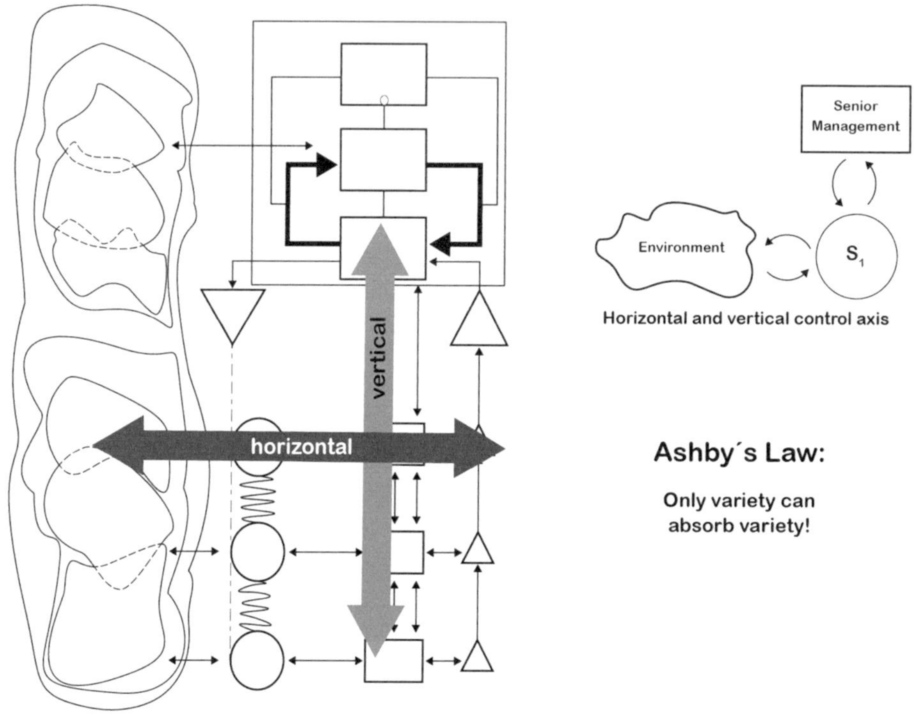

Fig. 10.4 Horizontal and vertical diversity (Frahm and Roll 2022)

The organization must be able to cope with the diversity that flows into the organizational system from outside, i.e. from the environment. To achieve this, diversity within the organization must be able to be dealt with sufficiently effectively on both the horizontal levels and the vertical levels (Frahm and Pfiffner 2023) (Fig. 10.4).

As described above, the flow of diversity is to be understood mathematically in order to obtain stable or ultra-stable system states. If an analogy about the flow of diversity within organizational systems had to be used, weather maps with high pressure and low pressure areas that flow into one another and depict different states of order and disorder would seem suitable. The five critical equilibrium states (Fig. 10.5) in the organization, which should be in equilibrium, create a deeper understanding of the flow of diversity in the organization based on the Viable System Model. These are:

1. Workload/workload
2. Coordination of production/line balancing
3. Autonomy vs. cohesion
4. Rate of change
5. Change vs. Status Quo

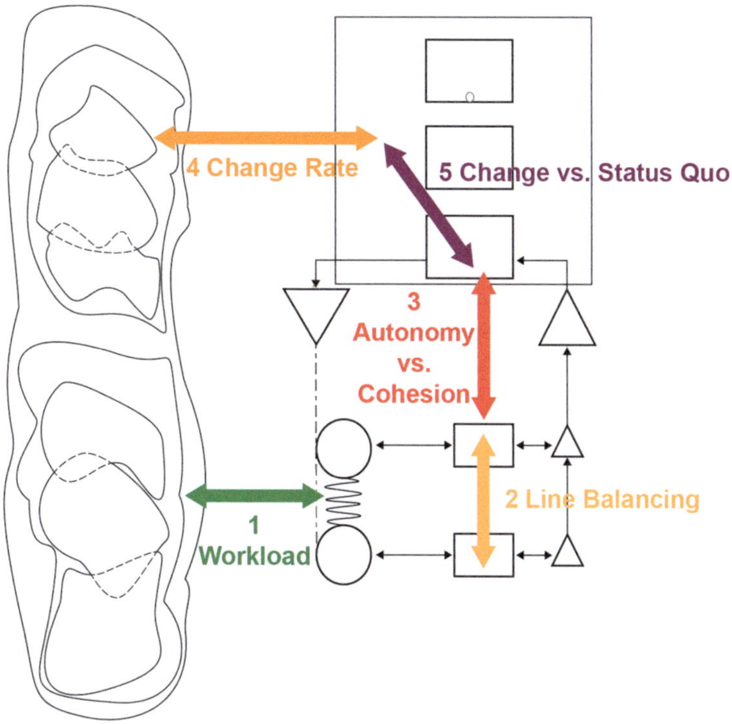

Fig. 10.5 Five critical Equilibrium states (Frahm and Roll 2022)

This results in the following areas of action for construction organizations:

Workload/workload—Equilibrium State 1:
Equilibrium state 1 shows the critical relationship between production and its operational environment (i.e. market, customer). At this level of the production system, the question can be asked: What are our complexity drivers? And how can we manage the diversity that arises through the relationship with the environment?

Coordination of Production/line Balancing—State of Equilibrium 2:
This state of equilibrium involves the coordination of the production systems by the coordination system. This equilibrium state is one of the strongest for adapting vertical complexity. The importance of this balance (coordination system) of diversity is often chronically underdeveloped.

Autonomy Versus Cohesion—State of Equilibrium 3:
At equilibrium state 3, two areas can be considered that should compete with each other.

1. Resources Versus Performance:
   First, it is considered whether the cycle between resource provision and result requirements is closed. This means that there are enough resources in production for the expected performance and that the monitoring of resource allocation and performance measurement are adequately designed?
2. Autonomy Versus Cohesion:
   This element of the state of equilibrium considers whether the operational areas can decide and act independently and therefore have a sufficient degree of autonomy. Furthermore, it can be considered whether the overall system as a whole can work strategically coherently, even though the operational areas have been granted a certain degree of autonomy.

Rate of Change—Equilibrium State 4:
Equilibrium state 4 is about the known and unknown future of the entire organization. The greater the diversity of these environments, the greater the diversity of the strategic area must be. Through this area, strategic planning becomes the framework for operational planning. However, this often does not work in companies. For this to work, the operational and strategic areas must be equally strong.

Change Versus Status Quo—State of Equilibrium 5:
Equilibrium state 5 is, on the one hand, about the identity of the organization and, on the other hand, about a balanced decision-making balance between the operational and strategic areas for sufficient strategy development. This means, for example, the correct and timely redirection of human and financial resources from the old to the new business, which is a difficult and at the same time important task.

The diversity in a construction project can be represented as follows (Fig. 10.6) based on the viable system model, the system assignments and under the approach that the diversity flows through a construction organization in different forms or strengths. The classification was made using a Likert scale. In this case, the starting point and perspective of this example is the construction site. It is easy to see how far a construction site is structurally and organizationally anchored in a construction organization and what different forms of diversity exist at different levels. Obviously, if the perspective is changed, the expression and view of diversity, for example, also changes.

Based on Fig. 10.6, an example of the responsibilities and activities that may arise in relation to the systems and information channels is shown below.

System Levels:
System 5—Normative Management: Managing Director, Branch Manager.
Examples are: Corporate values and principles of a group. These principles of the organization are reflected in all subsystems.

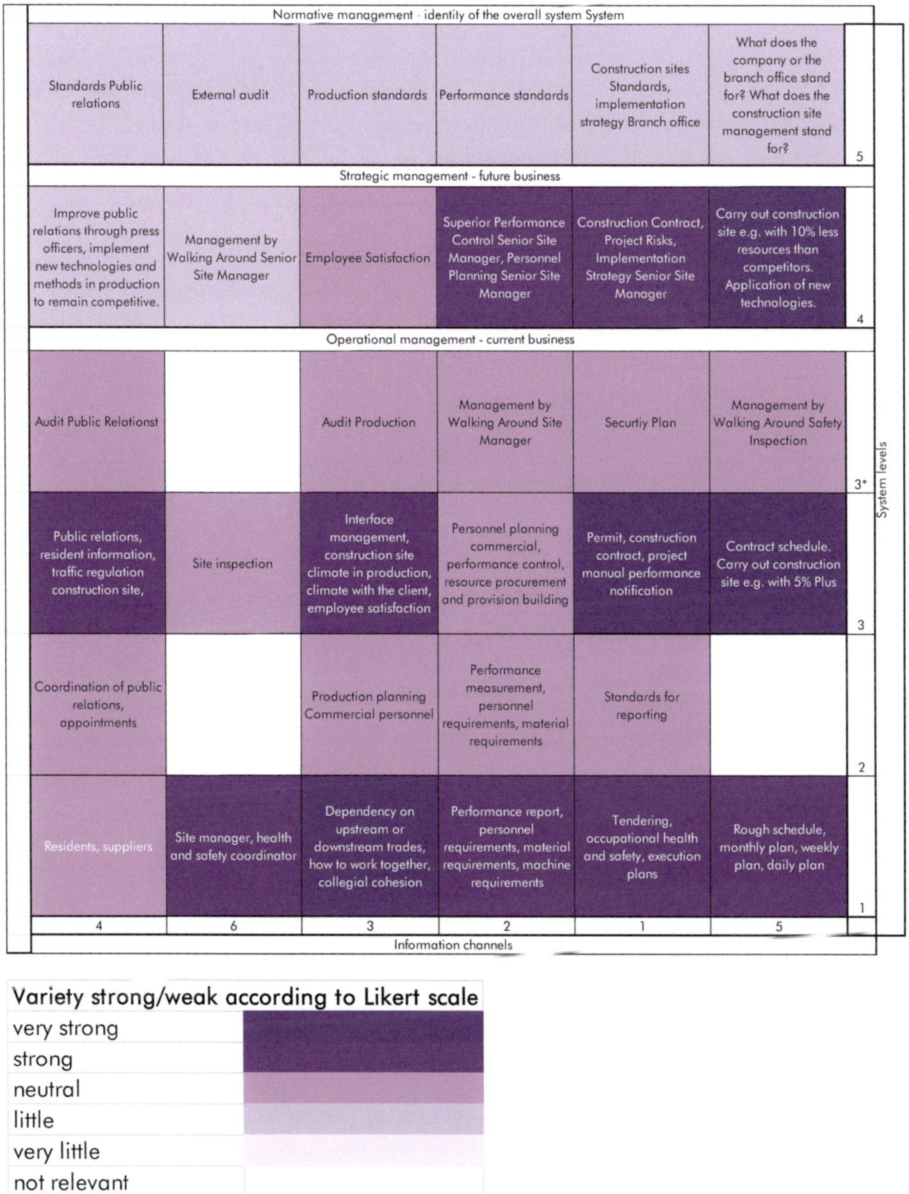

Fig. 10.6 Diversity Construction site

System 4—Strategic Management: Managing Director, Branch Manager, Senior Construction Manager.

Examples of System 4 are strategic purchasing or the development/development of new business areas and technologies such as a BIM strategy for the entire company.

System 3*—Monitoring/ Audit: Senior construction manager, construction manager, SiGeKo, etc.
An example of this system is a regular walk by the management staff around the construction site and conversations with the foremen and the industrial workers in order to get an additional picture of what is happening in production, or a project audit with an outside view of/about the project.

System 3—Operative Management: Construction manager.
Examples of a System 3 instance are: project, construction, team, department leader, or a managing director.

System 2—Control and coordination: construction manager, office staff.
Examples are daily short, informal meetings (see also Daily Scrum/Lean), operational and production plans of any kind, Scrum/Kanban boards, project management, Six Sigma, the Last Planner System or cycle planning.

System 1—Production: Foreman, commercial staff.
Examples of an S1 are producing units such as construction crews and construction companies.

Remark: People with certain roles, such as construction managers, can occupy different functions in the Viable System at the same time in different capacities, depending on the size of the company. A solo self-employed person, for example, initially fills all functions themselves.

Information Channels:
Channel 1—Intervention and Rules:
Central project specifications can be found here: e.g. project manual, construction contracts, engineering contracts, building permits. Question: What is expected and what are the rules?

Channel 2—Resource Allocation:
This involves the overall allocation of resources (e.g. personnel planning, resource planning, contractual regulations). Question: What is needed, according to which rules is it distributed and how does one have to justify it?

Channel 3—Interrelationships of production:
Channel of internal relationships, which is depicted materially and informally.

Channel 4—Interrelationships between the external and internal environments:
Management and producing units are associated with different environments. These environments, in turn, are connected to each other in various ways. These are environments that are relevant for the present and environments that are relevant for the

future (e.g. stakeholders, public, politics). Question: How do environments interact with each other and what consequences arise for the system?

Channel 5—Coordination:
See system 2.

Channel 6—Monitoring/Audit:
See system 3*

10.1.5 Résumé

The following article provides an insight into the understanding of complexity based on a sub-area of systems thinking. In contrast to various management fashions, the concept presented therein is based on a basic understanding of complexity and how this diversity flows from the system environment into organizational systems and can be managed. For this purpose, "System Thinking" was first introduced as a multi-method approach, which comes from dealing with complexity and thus represents a counter-model to cause-and-effect thinking. Furthermore, what is meant by complexity in the context of project implementation was introduced. This was followed by the presentation of Ashby's Law, which can be seen as elementary in dealing with complexity through diversity. There was also a brief look at the Viable System Model (VSM), which as a systems thinking approach is the method of choice for dealing with structural and organizational complexity. Finally, an example of the possible use of the VSM from the perspective of a construction site was presented. The aim of the article is to create a deeper understanding of how to deal with complexity in the construction system. If you want to deal with complexity well and effectively, you need suitable concepts. This article presented an alternative perspective on the concept of diversity and how effective and powerful understanding and effective management of diversity is.

10.2 Innovative Female Civil Engineers Since 1928

Dipl.-Ing. Natalia Bienkowski and Univ.-Prof. Dr.-Ing. Dipl.-Wirt.-Ing. Katharina Kleinschrot

10.2.1 Introduction to the "Vision2028" Project

The BMBF-supported project "Vision2028" (FKZ 01FP21011) has researched the pioneering contributions of women in civil engineering between 1928 and 2020. The starting point was the year 1928, marked by the career entry of Martha Schneider-Bürger, Germany's first female civil engineer. The project's goal was to capture the innovative achievements and ideas of these women and to present them in both a technical and historical context. This meticulous research filled a significant gap. Until now, only two pioneers were documented in the technical literature, such as in Klaus Stiglat's "Civil Engineers and Their Work".

The project aimed to make the achievements of these women accessible to a broad public through sophisticated visual communication strategies and to anchor them on the website bauingenieurinnen.de. These measures were intended not only to increase the visibility of the women, but also to serve as inspiration and role model for young women. By publishing on various social media platforms, the project aimed to particularly appeal to, motivate and encourage young women to actively participate in the professional discourse. Furthermore, the project developed specific strategies to encourage young women through a network to participate in shaping the future of civil engineering and to strengthen their presence in this field in the long term.

10.2.2 Selected Research Results

10.2.2.1 Patents

In the patent classes relevant to civil engineering, a total of about 206,000 patents were identified in the research for the period from 1928 to 2021. The detailed analysis of these patents was carried out via the first names of the inventors. From this extensive research, it was found that approximately 2,860 patents were filed by women, which only represents 1.39% of the total number of patents in civil engineering. To better illustrate the development of patent applications by women in civil engineering since 1928, the data was divided into decades and analyzed accordingly.

Figure 10.7 shows both the patent applications of all patents (blue bars), as well as the patent applications of female inventors (yellow bars). The diagram shows that the patent applications in general, with the exception of a drop between 1940–1949 and a small drop between 1980–1989, have steadily increased from 1928 to 2009. In the last decade, between 2010 and 2019, the number of patent applications had been declining for the first time. The patent applications from female inventors have been continuously increasing after a drop between 1940–1949. Thus, the proportion of patent applications from female inventors has increased from 0.2% (1940–1949) to now 3.2% (2010–2019).

10.2.2.2 Academic Positions

The search for information on scientific staff in civil engineering was carried out by evaluating the staff directories of the TU9 since 1928. Since these directories are only partially digitized, the historical staff directories had to be physically examined in the German National Library (DNB) in Leipzig and Frankfurt, Germany. This detailed analysis led to the identification of Ms. Dipl.-Ing. Natalie Robens, who, according to current research, is considered the first scientific staff member in civil engineering in Germany and has been working at the Institute for contributed to the steel construction of Technical University TU Braunschweig, Germany.

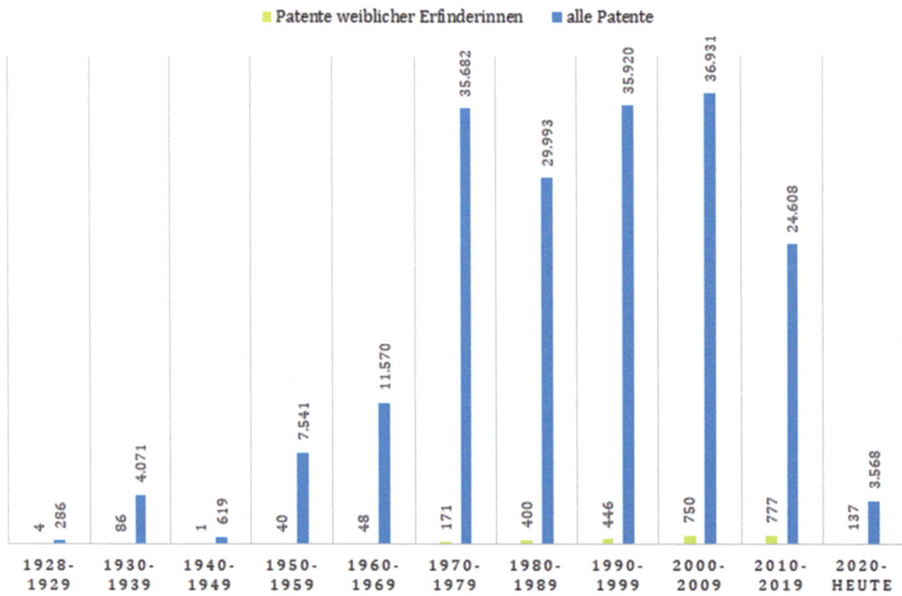

Fig. 10.7 Development of patent applications since 1928 until the end of 2020

Another investigation concerned dissertations and habilitations in civil engineering, which were written by women, and resulted in an Excel list with about 600 such works since 1928. The first verified dissertation of a female civil engineer was submitted in 1957 by Else Hartmann at the Technical Engineering College TH Stuttgart, who dealt with the topic "*On the Effect of frost and de-icing salts on concrete without and with air-entraining additives*".

In 1995, Prof. Dr.-Ing. Ulrike Kuhlmann achieved a milestone by becoming the first woman to hold the professorship for steel and timber construction as well as the leadership of the Institute for Design and Drafting at the University of Stuttgart. This made her the first female professor in the field of civil engineering at the German TU9 universities. Two years later, in 1997, Martina Schnellenbach-Held became the first professor in the field of solid construction at the TU Darmstadt (Fig. 10.8).

10.2.3 Network "Female Civil Engineers"

The final event of the research project "Vision2028" in September 2023 served as the starting signal for the network "Female Civil Engineers", which started its activities immediately after this event. The reactions to this initiative and the increased visibility were particularly positive from innovative women and female students—they expressed feeling finally recognized.

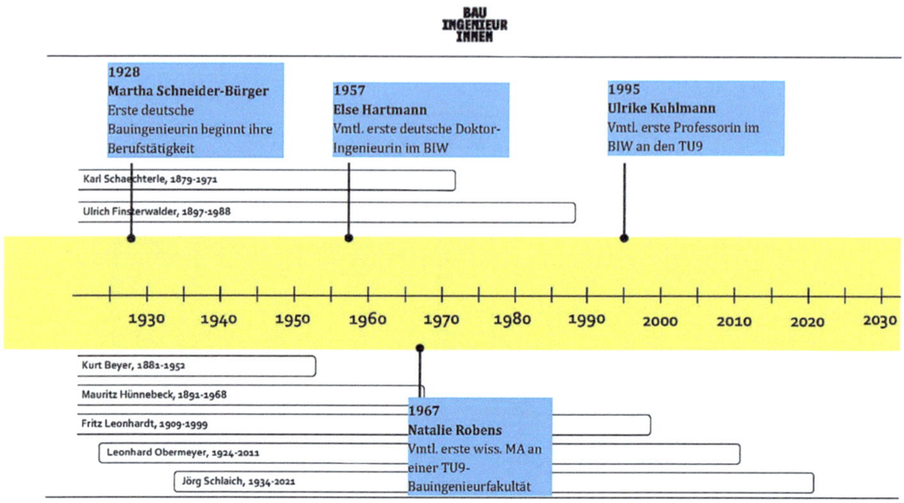

Fig. 10.8 Selected academic milestones of female civil engineers at the TU9 universities

The portrayal of role models plays a crucial role in bringing especially young people, particularly young women, closer to the diversity of civil engineering and to inspire them for this profession. In view of the future challenges in civil engineering, which can only be overcome collectively, many bright minds are essential. The motto "*You can only become, what you can see*" underlines the importance of female role models, who motivate and inspire through their visibility.

For the currently pending sustainability transformation of the construction industry, diverse perspectives and competent professionals are necessary. To attract more women to leadership positions, working conditions and corporate cultures need to be adapted. The excellence of women in leadership roles must be made visible.

The establishment of the network is of great importance for this. Many companies have recognized the need to promote women, especially in view of the shortage of skilled workers. The network is deliberately also accessible for men, to provide them with insights on how they can effectively support women as leaders.

A central concern is to provide students with a platform and to network them with companies. Mentoring is a key element to advance professionally. Civil engineering is a highly complex field, so it is advantageous for students to have many points of contact in practice.

It is also crucial to become aware of one's own prejudices and to dismantle them. Everyone has a shaped perception of the world due to their life experiences and socialization. People with the bias of naive realism, for example, believe to view the world objectively and to evaluate applicants objectively. The similarity effect often leads to people unconsciously preferring applicants who are similar to them, even though diverse team members are advantageous. The gender bias, i.e., the gender-related prejudice, can wrongly assume that women are not interested in a career. It is important to reflect on and recognize these distorted perceptions.

The "Female Civil Engineers" network presents innovative women as examples to mitigate stereotypes and open up new perspectives.

References

Ashby W.R. (1956) Introduction to Cybernetics. Chapman & Hall. ISBN 9,781,614,277,651.

Baccarini, D. (1996) The concept of project complexity—a review. International Journal of Project Management, vol. 14, no. 4, pp. 201–204.

Bennett, J. and Fine, B. (1980) Measurement of complexity in construction projects. Department of Construction Management. University of Reading.

Beer, S. (1972) Brain of the firm; a development in management cybernetics. Herder and Herder. ISBN-10: 0,713,902,191.

Beer, S. (1979). The Heart of Enterprise. Wiley. ISBN: 978–0-471–94,837-7.

Beer, S. (1985). Diagnosing the System for organizations. Wiley. ISBN: 978-0-471-95,136-0.

Damayanti, R. W. et al. (2021). Leadership Competency for Megaproject's Complexity Management: A Conceptual Study. International Journal of Innovation, Management and Technology, 12(4), 68–74. https://doi.org/10.18178/ijimt.2021.12.4.911.

Frahm, M., Roll C. (2022). Designing Intelligent Construction Projects. Wiley ISBN-10: 111969082X.

Frahm, M., Rahebi, H. (2021). Management von Groß- und Megaprojekten. Page 8. Springer, ISBN: 978-3-658-30,983-1.

Frahm, M., Pfiffner, M. (2023) Decrypting the DNA of Megaprojects - A Model-based Management Approach using the Viable System Model (VSM).; PM World Journal, Vol. XII, Issue II, February.

He, Q., Luo, L., Hu, Y. & Chan, A. S. C. (2015). Measuring the complexity of mega construction projects in China—A fuzzy analytic network process analysis. International Journal of Project Management, 33(3), 549–563. https://doi.org/10.1016/j.ijproman.2014.07.009.

Hoffmann, W. (2018) Zum Umgang mit der Komplexität von Bauvorhaben (Indikatorbezogenes Modell zur Bewertung von Komplexität in Bauprojekten). Dissertation. TU Kaiserslautern.

Hovestädt, D. et al. (2022). Beyond evidence: the use of archives in transitional justice. Routledge, 1st Edt. https://doi.org/10.4324/9781003260622-1.

Jackson, M. C. (2019). Critical systems thinking and the management of complexity. John Wiley & Sons.

Jia, F., Xiang, P. & Chen, D. (2023). A two-dimensional complexity evaluation model of megaprojects based on structure and attributes. Ain Shams Engineering Journal, 14(2), 101,852.

Morris, P.W.G. (1997) The Management of Projects. Thomas Telford. London.

Nyarirangwe, M. & Babatunde, O. K. (2021). Megaproject complexity attributes and competences: lessons from IT and construction projects. International Journal of Information Systems and Project Management, 7(4), 77–99. https://doi.org/10.12821/ijispm070404.

Pfiffner, M. (2020). Die dritte Dimension des Organisierens. In Springer eBooks. https://doi.org/10.1007/978-3-658-29247-8.

SCiO (2023) https://www.systemspractice.org/de/page/scio-competency-framework (Accessed 17. November 2023).

Schwanninger, M. (2006). Intelligent Organisations - Powerful Models for Systemic Management. Springer.

Study Programs and Training. Education for Future—Lifelong Learning 11

11.1 Numbers, Data, Facts About Apprenticeship, Women's Share and New Work in the Craft Sector

Dr.-Ing. Dipl.-Ing. (FH) Bianca Weber-Lewerenz

And now, dear reader, if you are ready for a quick introduction to numbers, data, facts about apprenticeship and women's share in the craft sector, you will be presented with a picture of "drive" and astonishment. In exchange with the Central Association of German Crafts (ZDH), in December 2023, the book editor asked the head of the "Image Campaign Crafts" for more detailed insights into the figures and the general development of female skilled workers in the craft sector.

Women are high performers in various positions (Fig. 11.1): whether as an independent entrepreneur, master, founder, with a partner at the top or as a successor in the family business. Almost every fifth successful master's examination in 2022 was passed by a woman (17.9%). Every fourth craft business (24.2%) is run by a woman. Over 74.3% of craft businesses are family businesses, run jointly by a (married) couple. The development potential of women is therefore far from exhausted.[1]

What inspires women in the crafts sector to become self-employed? Independence, professional challenge, personal development, succession in the family business, variety in the activities and the combination of family and career (Ludwig-Fröhler-Institut 2023) (Fig. 11.2). A study conducted an online survey of 627 craft businesses in Germany in the period from May to June 2023 and shed light on the situation.

[1] https://www.zdh.de/ueber-uns/fachbereich-soziale-sicherung/frauen-im-handwerk/ (Accessed 21. December 2023).

B. C. Weber-Lewerenz (ed.), *Diversity in Construction—The Gamechangers*, https://doi.org/10.1007/978-3-658-45308-4_11

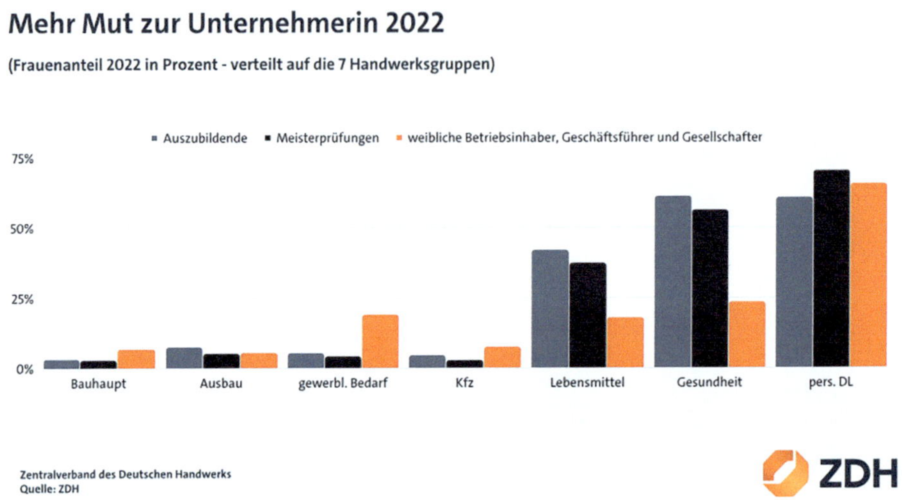

Fig. 11.1 More courage to become an entrepreneur 2022. *Source* cf. https://www.zdh.de/ueber-uns/fachbereich-soziale-sicherung/frauen-im-handwerk/#c4680

Fig. 11.2 Reasons for self-employment. *Source* Study by the Ludwig-Fröhler-Institute for Crafts Sciences, September 2023

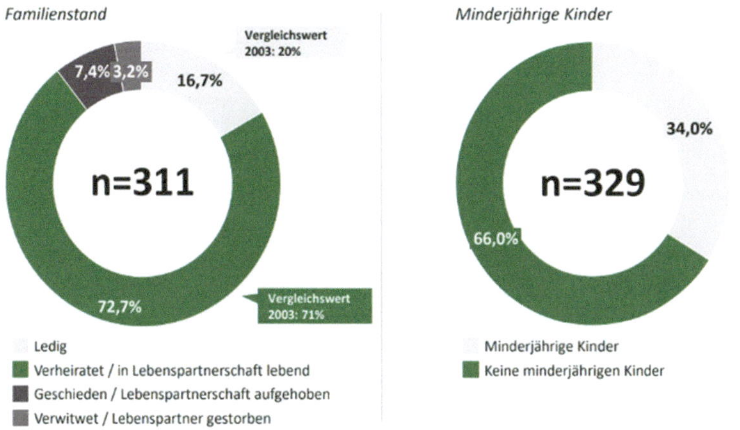

Fig. 11.3 Family situation of women in management. *Source* Study by the Ludwig-Fröhler-Institute for Crafts Sciences, September 2023

A central finding is that 22% of the surveyed craft businesses still operate without female employees. And there is no change since the preliminary survey in 2003. About 26% of all trainees in the surveyed businesses are women. Especially large companies show a significantly higher willingness to strengthen women's quotas.

About 35% of all CEO positions in the surveyed businesses are occupied by women. The proportion of women in the crafts sector is also significantly higher than the women's quota in the management of the 100 largest German family businesses (8.3%), medium-sized businesses (12.7%) and companies in the DAX, M-DAX and S-DAX (14.3%). Younger businesses with up to 10 years since their foundation show women's quotas in the management of about 30%. And about 34% of the female CEOs have minor children to care for in addition to their position in the craft businesses (Fig. 11.3). 53% of all employed women in the surveyed craft businesses are part-time active in the company—significant increase compared to the quota in 2003 (36%). The high acceptance of part-time employment shows the often great flexibility of employment in the crafts sector. The spread of more flexible working models in the form of part-time employment is not only evident in young businesses (Fig. 11.4). Both young and old businesses have a high proportion of part-time employment among women. There is no trend towards greater popularity and acceptance in younger businesses. Part-time and flexible working time models are already offered in the majority of businesses (increase since 2003 evident), a further 6% plan implementation in the near future. The implementation of further health offers as well as the 4-day week is also in many businesses planned. A wide range of further measures to improve the working conditions of the majority of companies due to the requirements of the job profile or company size is seen as not feasible (e.g. sabbatical, home office).

The approach to new concepts in the area of "New Work in Crafts" (ref to chapter) requires the integration of improved, flexible working time and organizational models in companies. This benefits mothers, but also fathers. Re-entry agreements and programs

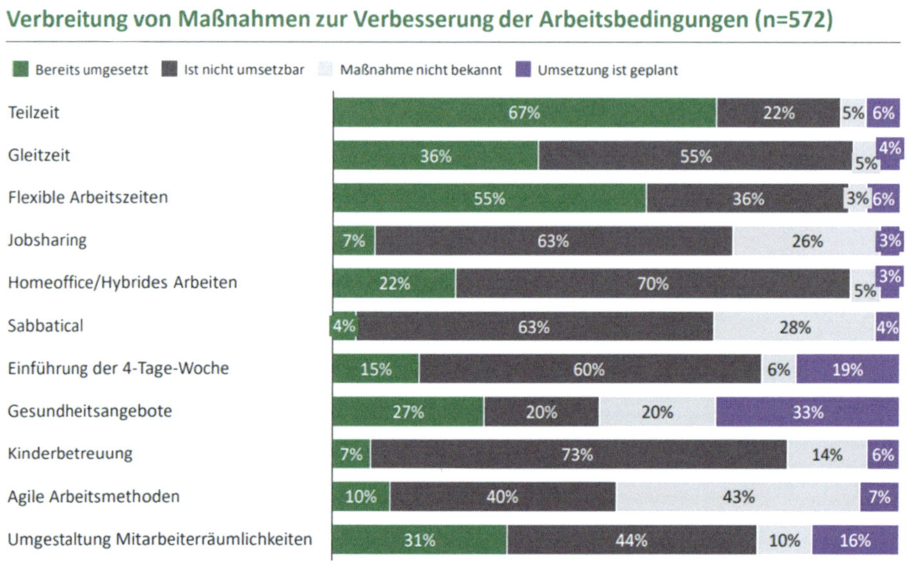

Fig. 11.4 Part-time and flexible working time models in crafts. *Source* Study by the Ludwig-Fröhler-Institute for Crafts Sciences, September 2023

after parental leave, conscious maintenance of contact during parental leave, further training, flexitime, flexible part-time models, trust-based work, individual organizational solutions such as home office, job sharing and leadership in part-time are offered here as an option (Steinbeis Innovation Center, Corporate Development at Pforzheim University 2013). Childcare includes own entrepreneurial childcare offers, purchase of external day-care places, children's office (in an emergency the child can be brought along), takeover/participation in childcare offers, tax-free employer subsidy, children's holiday program and homework supervision in the company, household-related services and support in the care of relatives. Because, where there's a will, there's a way. The success (and happiness) factor family and the reconciliation of work and family promote that core and fun competencies can merge. The need for action is expandable and is attributed to both companies and politics.

The trend in the proportion of women in the professions of bricklayer and concrete worker, installer and heating builder is reflected in the newly concluded contracts (Fig. 11.5). Since 2020, there has been a clear increase every year in new contracts concluded with female forces. In the HLS trade more than in bricklaying. Overall, the percentage of women present is still low, however, the annual growth rate in 2020, 2021 and 2022 shows clear upward trends. The book chapter "Individual profiles in construction—My Story" provides exciting new insights and various reasons and aspects for the increasing enthusiasm of women for professions in crafts and how the environment in crafts and everyday construction is changing. The statements of women coin-

	New Contracts Women			Women's Share		
	2020	2021	2022	2020	2021	2022
Bricklayer and Concrete Builder	63	70	96	1.5%	1.6%	2.6%
Plumber and Heating Installer	217	230	306	1.7%	1.7%	2.2%

Fig. 11.5 New contracts in the craft sector women's share 2020–2022. *Source* ZDH Statistics 2022

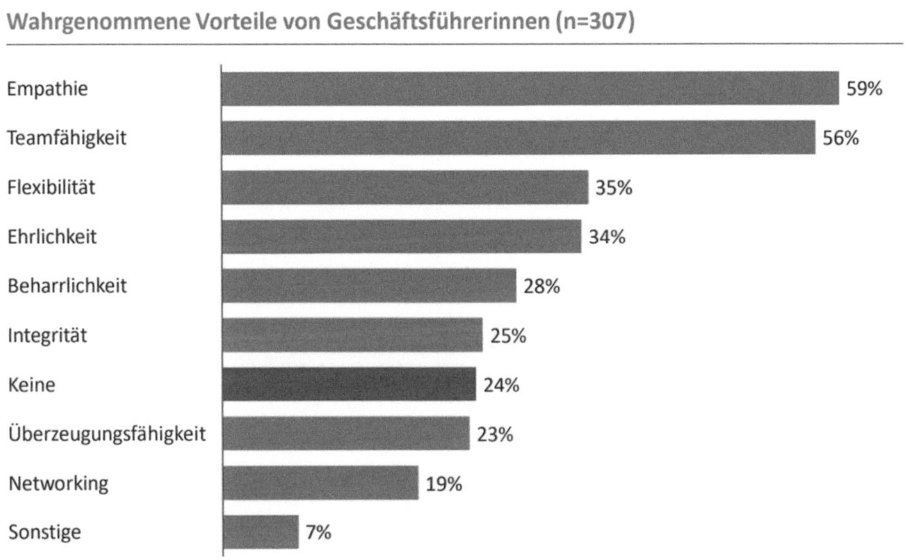

Fig. 11.6 Perceived advantages of female CEOs in craftsmanship. *Source* Study of the Ludwig-Fröhler-Institute for Craftsmanship Sciences, September 2023

cide with the result of the study by the Ludwig-Fröhler-Institute for Crafts Sciences: Empathy, teamwork, flexibility, honesty and persistence are the perceived advantages of female leadership and cooperation in diversely structured work environments in crafts (Fig. 11.6).

Regarding the women's share a) in continuing in the learned profession, or b) in continuing with a master's degree, or c) even switching to studies, ZDH had no data available.

Also, upon request, whether there were any experience values for the reason for taking up the apprenticeship and professional career, e.g. entry into the family business, takeover and generational succession, shortened apprenticeship after high school with

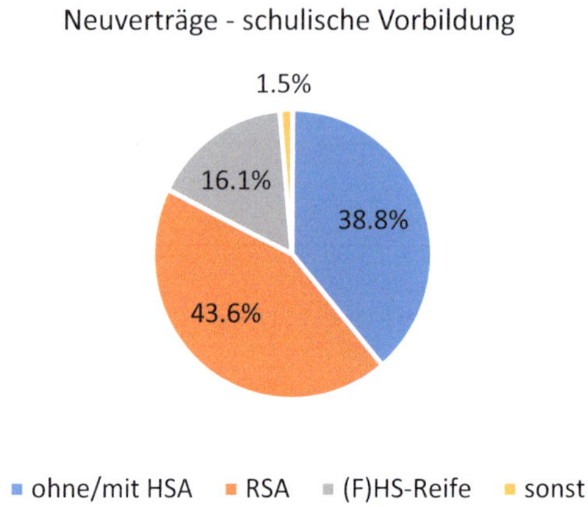

Fig. 11.7 Share of high school graduates starting in craftsmanship apprenticeship. *Source* ZDH Statistics 2023

subsequent studies, aiming for a master's degree and starting a business, ZDH had no numbers and information available.

It would also have been interesting to know whether women start directly after middle school/high school or as career changers after another profession/after studies. Unfortunately, ZDH has no data available on this.

Further insights were provided by ZDH on the share of high school graduates at the start of the craft apprenticeship (Fig. 11.7). High school graduates make a significant impact with a share of 16.1%. Real school graduates start with 43.6%, and with or without a primary school degree, 43.6% start with contract conclusions in the craft sector. The share of "other" with 1.5% is not to be underestimated. This share is increasing and counteracts the shortage of skilled workers in the craft sector, as it is increasingly craft forces with a migration background, from foreigners, refugees and asylum seekers, who successfully integrate in Germany and increasingly rely on craft entrepreneurs. The fact that exactly these forces attend language courses directly after work and achieve school degrees catching up or saddling up, and how many companies support this Herculean task, is unfortunately often overlooked.

The statistics of the Competence Center for Securing Skilled Workers KOFA provides an overview ofthe number of missing skilled workers and training in the crafts sector in 2023.[2] (KOFA 2023) In 2022, a record number of 236,818 open positions in craft professions were reported (Fig. 11.8). At the same time, the number of unemployed craftsmen continued to decline, leaving around 128,891 positions unfilled. Especially in the areas of electrical construction (18,000 skilled workers missing) and sanitary, heating and air conditioning technology (almost 14,000 missing) there is a strong shortage of skilled workers. Although more training places are being offered, many remain vacant because the number of applicants is declining. The strongly increased demand is due

[2] https://www.kofa.de/daten-und-fakten/studien/fachkraeftemangel-und-ausbildung-im-handwerk-2023/ (Accessed 21. December 2023).

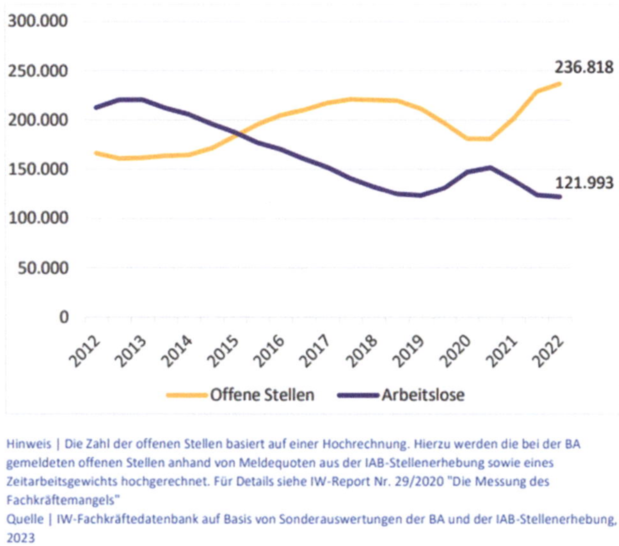

Fig. 11.8 Development of open positions in the crafts sector. *Source* KOFA 2023

to the generally good economic situation, at least for the construction industry, and the political goals in climate protection and housing construction. Dual training proves to be a central way of recruiting skilled workers. Although the number of offered training positions has increased more than the number of applicants, at the same time more training positions remain vacant (Malin and Köppen 2023). Figure 11.9 visualizes the top 5 bottleneck professions divided into professional categories with the skilled worker gap, existing open positions and the job surplus rate. Figure 11.10 presents the development of the training market in craftsmanship over a ten-year period, from 2012 to 2022.

How the supply and demand of training in the predominantly craft-related top bottleneck professions develop is shown in Fig. 11.11. The professions are considered meaningful due to their relevance for the climate change, they currently receive increased media presence, giving young people insights into the professional routine, and they are associated with high job security in times of crisis (KOFA 2022).

Of great interest is also the question: where (in which sectors) is the craft becoming more female? And what has been the trend over the past few years? For this, there are specifically meaningful statistics from the ZDH[3]. The statistics clearly show how underrepresented women are in the industrial-technical professions (Fig. 11.12).

---

[3] Compare Fig. "Die 10 beliebtesten Berufe weiblicher Auszubildender". https://www.zdh.de/ueber-uns/fachbereich-soziale-sicherung/frauen-im-handwerk/ (Accessed 21. December 2023).

	Berufsgattung	Fachkräfte-lücke	Offene Stellen	Stellenüber-hangsquote
Fachkräfte	Bauelektrik	17.846	21.549	82,8
	Sanitär-, Heizungs- und Klimatechnik	13.702	16.787	81,6
	Kraftfahrzeugtechnik	13.680	20.504	66,7
	Holz-, Möbel- und Innenausbau	8.478	14.847	57,1
	Metallbau	6.608	10.337	63,9
Meister:innen	Aufsicht und Führung – Medizin-, Orthopädie- und Rehatechnik	1.935	2.208	87,6
	Aufsicht und Führung – Fahrzeug-, Luft-, Raumfahrt-, Schiffbau	1.885	2.880	65,5
	Aufsicht – Klempnerei, Sanitär-, Heizungs- und Klimatechnik	1.612	1.952	82,5
	Aufsicht – Hochbau	1.600	2.110	75,8
	Aufsicht – Tiefbau	1.199	1.355	88,5
Fortbildungsabsolvent:innen	Aufsicht und Führung – Verkauf	7.443	10.309	72,2
	Facility-Management	1.009	1.369	73,7
	Spanende Metallbearbeitung	809	1.447	55,9
	Wirtschaftsinformatik	323	666	48,5
	Metallbau	264	326	81,0

Quelle | IW-Fachkräftedatenbank auf Basis von Sonderauswertungen der BA und der IAB-Stellenerhebung, 2023

Fig. 11.9 Top5 bottleneck professions in the crafts sector. *Source* KOFA 2023

In the service sector of craft professions, such as hairdressing (Top1), tailoring (85.7%), goldsmithing (73.4%), confectionery (83.9%) and in health crafts such as dental technology (61.9%), optometry (67.3%) and hearing acoustics (53.9%), there is a high proportion of women.

ZDH shared important facts and data around the question: what does the integration of foreign skilled workers (through flight, emigration) look like in the craft sector?

In 2022, the most recent data available from ZDH, 44,953 of the 349,295 apprentices in the craft sector had a foreign passport. These apprentices represented 168 different nationalities. Of all apprentices with a nationality from one of the eight most common asylum access countries (Afghanistan, Eritrea, Iraq, Iran, Nigeria, Pakistan, Somalia, Syria), who are currently undergoing dual training in Germany, about every second one is learning in the craft sector (2022: 47.7%). Migrants work disproportionately often in craft professions. This shows the disproportionately large integration performance of the craft sector compared to other economic sectors. The Institute for Craft and Small Business Economics at the University of Göppingen has researched this (Runst 2016). The study concludes, among other things, that about a decade passes from the time of arrival, until the proportion of an immigration cohort working in the craft sector adjusts to the proportion of Germans in the craft sector. Only through innovative ways can the deficient qualification level of migrants be addressed and thus represents a central politi-

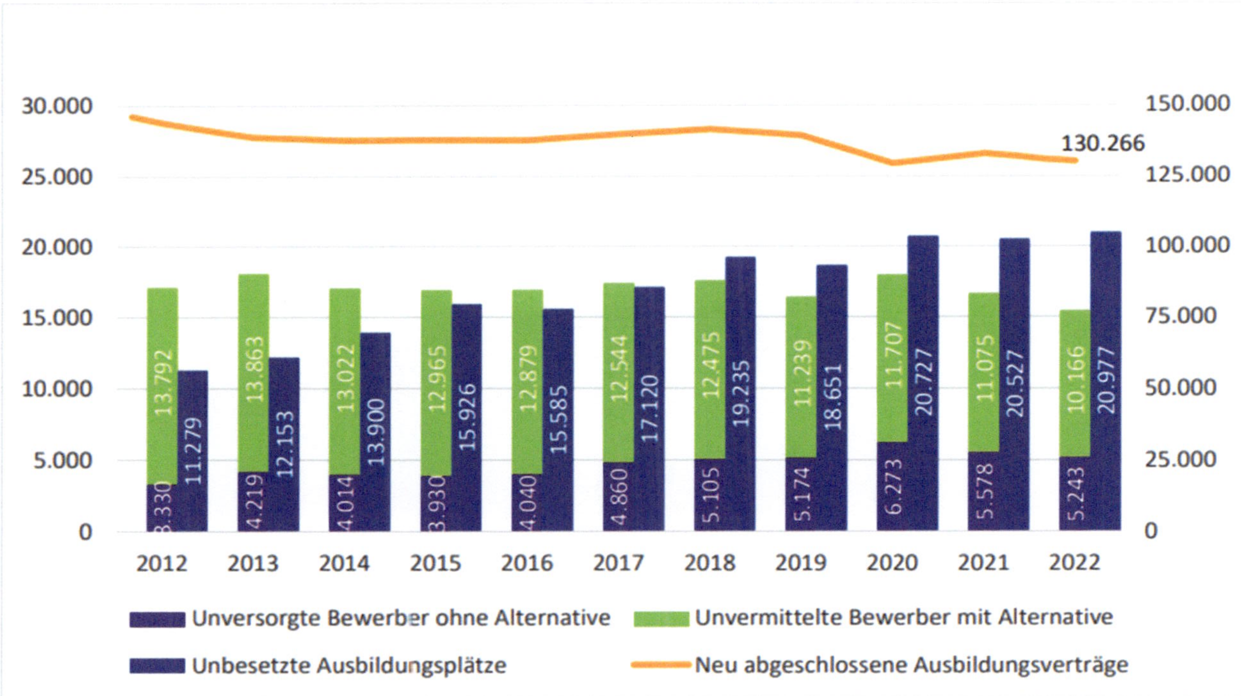

Hinweis | Zu den Bewerbern und Bewerberinnen mit Alternative gehören Jugendliche, die ihren Vermittlungswunsch aufrechterhalten, aber zwischenzeitlich eine Alternative (z. B. weiterer Schulbesuch, berufsvorbereitende Maßnahme) begonnen haben. Nicht enthalten sind die Personen, die sich ohne Angabe eines Verbleibs nicht mehr bei der BA gemeldet haben und damit unbekannt verblieben sind.
Quelle | IW-Fachkräftedatenbank auf Basis von Sonderauswertungen der BA und des BIBB, 2023

Fig. 11.10 Development of the training market in craftsmanship. *Source* KOFA 2023

Berufsgattung	Entwicklung Fachkräftelücke	Fachkräftelücke 2022	Entwicklung Ausbildungs-platzangebot	Entwicklung Ausbildungsplatz-nachfrage
Bauelektrik		17.846		
Sanitär-, Heizungs- und Klimatechnik		13.702		
Kraftfahrzeugtechnik		13.680		
Holz-, Möbel- und Innenausbau		8.478		
Metallbau		6.608		
Verkauf von Back- und Konditoreiwaren		4.873		
Verkauf von Fleischwaren		3.989		
Dachdeckerei		3.619		
Land- und Baumaschinentechnik		3.616		
Maurerhandwerk		2.903		

Quelle | IW-Fachkräftedatenbank auf Basis von Sonderauswertungen der BA, der IAB-Stellenerhebung und des BIBB, 2023

Fig. 11.11 Development of training supply and demand in to the predominantly craft-based top bottleneck professions. *Source* KOFA

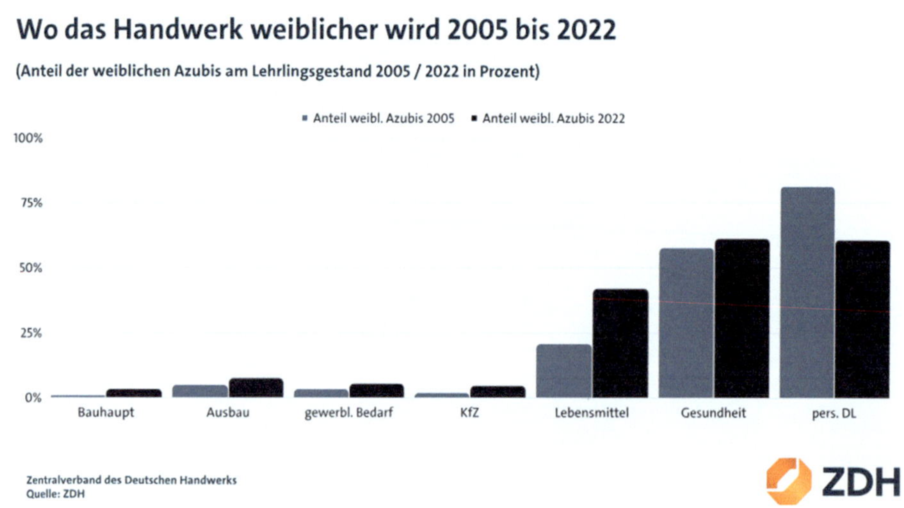

Fig. 11.12 The 10 most popular professions of female trainees. Where the craft is becoming more female, 2005–2022. *Source* ZDH

cal field of action to ensure successful societal and professional integration. The shortage of skilled workers can only be met to a limited extent through migration.

In connection with how diverse and inclusive the craft sector is set up, of particular interest is the question of what facts and data are available on inclusion. In which professional groups and branches of the construction industry is inclusion lived and where are physically impaired people, disabled people promoted and their participation in the profession secured. This also involves existing and increasingly adapted framework conditions that make this possible: a workplace with special work equipment, new technologies at the workplace to facilitate execution, home office regulations, mentoring and coaching. However, ZDH does not have any statistical data on this.

This provides the perfect transition to new knowledge and data and facts about the topic "New Work in Crafts". It would be interesting to know: what information is available regarding the changed work and corporate culture in the craft sector? This is already being researched more deeply in other sectors. Not yet in the craft sector. ZDH cannot (yet) provide any data here.

In the context of digital transformation, completely new job profiles and fields of activity are emerging, and new Training branches or specializations in future-oriented professions in the craft sector are being added. The physical strains of craft professions are decreasing and the need for creative, communicative and design potential is growing. In the executive floors, topics such as e.g. working climate, leadership culture, communication, design of work organization and working hours, work-life balance and personnel development are moving into the focus. Craftsmanship thrives on diversity and there are exciting challenges to discover.[4] This requires a complete rethink. In this context, best practice companies, startups, initiatives, networks and other commitments play important role models. They additionally encourage the courage for more innovation, provide orientation, are often signposts, stimulate value creation through lived diversity at all levels. Diversity and inclusion also form pillars of success here.

The professional qualification backgrounds of those working in the craft sector are shown in Fig. 11.13. The high number of technical specialists and journeymen is noticeable. The statistics are a mirror of diversity itself. The good mix ensures the fulfillment of a wide variety of tasks and fields of application in construction. From the worker, allrounder, expert to company management.

All in all, there is room for improvement and the potentials need to be exploited. ZDH summarizes: around 1,000,000 companies with around 5.7 million people belong to the craft sector. About 350,000 young people receive a qualified training in the craft sector (as of 2023). Thus, 12.3% of all employed persons and 28.7% of all apprentices in Germany work in the craft sector. In 2022, the turnover in the craft sector reached around 735 billion euros (excluding VAT). The craft sector forms the core of the German economy, more appreciation and implementation of digital work processes and innovations would be beneficial to actually generate more value creation and to do justice to the pep and pride that the craft sector embodies.

[4] https://www.zdh.de/ueber-uns/fachbereich-soziale-sicherung/frauen-im-handwerk/ (Accessed 21. December 2023).

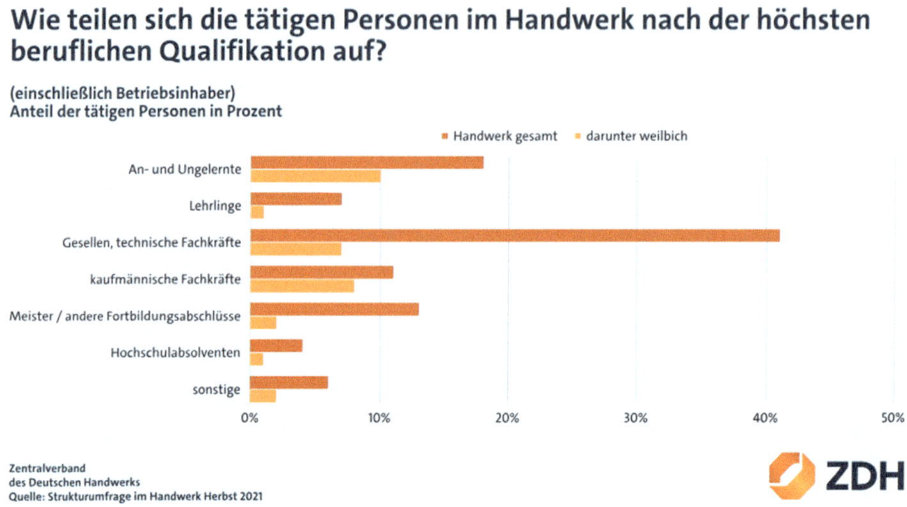

Fig. 11.13 Professional qualification in the craft sector. *Source* ZDH

Further Reading Tips

- https://ifh.wiwi.uni-goettingen.de/veroeffentlichungen/frauen-im-handwerk-status-quo-und-herausforderungen-kurzfassung/
- KOFA analyses on developments in the craft sector

11.2 ... Because My Sweetheart is a Master Painter ... Education and Labor Market Entry of Women and Men in Construction

Dr. Gabriele Wydra-Somaggio

11.2.1 Background

Demographic changes and the current structural change, characterized by digital and ecological transformation, pose major challenges to the labor market. Especially in connection with the ecological transformation and the goals set for housing construction, craftsmanship is given great importance. Trades, such as the main construction trade, the finishing trade (Wiemann 2023) and the craft for commercial needs, which supplies industrial production, play a significant role here. But other areas of craftsmanship, such as the food craft, the hairdressing trade or craft professions in the health sector, provide important and in-demand services. At the same time, the craft sector has been struggling for years with the recruitment of skilled workers and trainees as well as with business succession. In the next few years, the strong birth cohorts of the baby boomers will retire from the labor market due to age. Younger birth cohorts, who are entering the labor market, are demographically smaller and cannot numerically replace the older workers. In addition, younger people are increasingly less interested in dual vocational training in the craft sector. As a result, skilled labor shortages have increased in recent years. In addition to this, the craft sector has been struggling with the increasingly severe supply shortages and the sharply rising material and energy costs (HBS 2022; Koneberg et al. 2022; ZDH 2023b, c). Particularly affected by these developments is the construction industry, which is now under increased pressure, as the tasks in the implementation of the energy and climate turnaround remain (Schmidt 2023).

To accomplish these tasks, it is necessary to recruit skilled workers in various ways (Runst 2022). First, it is important to attract more already trained skilled workers for the craft businesses. In principle, women represent a special potential (Greilinger 2023), although they make up only a small part of the employees in the craft sector. Second, dual vocational training is still an important strategy for recruiting skilled workers (Greilinger 2023). With a share of 28% of all trainees, it makes an above-average contribution to training. In comparison, almost 13% of all employees in the overall economy are employed in the craft sector (Böhme et al. 2023). Thus, the craft sector has a significant share of the overall economy. However, the craft sector is more than other training areas strongly dominated by men. Only just under 17% of trainees are women (ZDH 2023a). In addition, the choice of profession by women, as in other labor market areas, differs significantly from that of men, making it also a strongly gender-specifically segmented labor market area: 29% of men complete their training in a machine and vehicle technology profession, 37% of women in a non-medical health profession (especially hairdressing). This is due to the fact that career choices are made at a young age under the aspect of personal preferences, interests and social norms (Fuchs et al. 2021), which can unconsciously solidify into a gender-specific occupational separation (Polachek 1981). Because ultimately, the career choice decision is based on a value-expectation theory. Young people choose those professions that they assign a high value to and in which they can implement or achieve their goals with relatively high certainty, such as the meaning of the activity, compatibility of work and private life (Beckmann et al. 2006). Especially women still usually choose professions, in where part-time employment is possible, to reconcile care tasks with professional life. This may be at the expense of career advancement opportunities or higher salary prospects. As a result, women often work in lower-paid jobs than men (England 1992).

Although salary prospects in some areas are still lower than in other economic sectors, particularly industry, due to high demand, earnings in craftsmanship as well as employment prospects are likely to continue to improve, also for women. At the same time, studies show that less part-time work is done in male-dominated professions, but they also offer better career prospects (Althaber and Leuze 2020; Haverkamp et al. 2015). Therefore, it is noteworthy that in the craft sector, the proportion of women in areas that have so far been predominantly learned by men, such as automotive mechatronics, electronics or optometry, is increasing. With this development, some of the increased demand caused by the climate change (Malin and Köppen 2023), which is particularly present in these professions, could be covered. After all, male-dominated professions are paid higher than female-dominated professions (Zucco 2019; Kleinert et al. 2023). For example, the average salary in masonry professions is significantly above that in hairdressing professions. Furthermore, the positive economic developments of recent years facilitate the implementation of another important strategy to meet the demand for skilled workers: This strategy states that graduates should be more strongly and longer tied to the training company. Thus, the retention rates of graduates in the craft sector have increased in recent years but are still slightly below those of other training sectors (Hell and Wydra-Somaggio 2023). Therefore, the focus of this contribution is on the training and labour market situation of graduates in the craft and construction sectors. It will be shown who the women in the craft and construction industries are, how their share has developed and what the situation of women and men is at the transition from training to the labour market.

11.2.2 Methodological Approach

11.2.2.1 The Apprenticeship Panel Craftsmanship

The evaluations are based on the Apprenticeship Panel Craftsmanship. This unique dataset combines information from 14 chambers of crafts[5] mainly in West Germany with information from the Integrated Employment Biographies (IEB) of the IAB on all young adults who were in training between 2004 and 2020. The dataset represents a broad spectrum of chamber districts in terms of population density, economic structures and thus also different employment opportunities. An advantage is that with this dataset, trainees in the craft sector can be identified for the first time and their employment histories can be traced. In addition, we have detailed information about the training itself and the course of training. Thus, the contractual start of training, the contractual as well as the actual end of training, which corresponds to the examination day, but also to the date of an early contract termination, the training profession, the exam success, as well as the number of exam attempts and the school leaving certificate are recorded in the chamber data. Finally, the type of training is also included in the chamber data (funded versus non-funded training or retraining). From the IEB, this information is supplemented by daily details on further employment history, such as employment, unemployment and

[5] The participating chambers are: Hamburg, Hanover, Oldenburg, Münster, Düsseldorf, Koblenz, Lower Franconia, Upper Franconia, Middle Franconia, Lower Bavaria/Upper Palatinate, Munich/Upper Bavaria, Swabia, Berlin and Dresden. Since at the time of the evaluations the training data of the HWK Hamburg up to the 2020 graduating class was available due to problems with data transfer, the evaluations are only carried out with the 13 other chambers.

measure times, as well as industry, place of residence and work and the salary. Thus, with the Apprenticeship Panel Craftsmanship, we can investigate to what extent training characteristics or courses have an impact on the transition to the labour market and on further professional life. The two data sources are supplemented by company characteristics, such as the size of the training company and the current employment company, from the Company History Panel. The present study focuses on the graduates of a training program at a craft business in one of the participating chamber districts. The evaluations refer to graduates who have completed their training without interruptions, i.e., without changing professions during their training. However, company changes are allowed, as especially in larger craft companies multiple businesses can be reported. In these companies, the change between individual areas is part of the training. The graduates have all successfully completed the training with at least the grade "sufficient". Only graduates of unsubsidized training are considered.

11.2.2.2 Empirical Approach

The focus of the present contribution is the training situation of women and men in the construction sector as well as the entry into the labor market. The present data set contains all graduates, who have obtained their degree in a craft business, which means that degrees can also be obtained as office and secretarial professionals and in sales professions, which, however, are not craft professions according to the Crafts Code (HWO). In the following, we only consider apprenticeship professions according to the HWO, which can be assigned to trade groups. The trade groups are assigned training professions that belong to activity and partly also economic sectors thematically.[6] By excluding non-craft professions, the proportion of women decreases, because even if they undergo training in crafts, they often do this in office and secretarial or sales professions (Table 11.1).

As a demarcation of the construction sector, we use the classification according to Haverkamp et al. (2020), whose division is based on the Crafts Code Annex A. For the evaluations, we refer to the construction sector as a whole, which consists of professions in the main construction trade and the finishing trade. We combine these areas, firstly because professions in the construction industry are less often learned by trainees than those in the finishing trade: Eight percent of the graduates in crafts are in construction professions, 30% in finishing professions. Secondly, very few women complete their training in construction professions (about 1% are women), in finishing professions it's slightly more (just under 7%). Aggregated analyses of remaining in the training company are exemplarily carried out on the 2015 graduation cohort This is the most recent year in

[6] We differentiate between the trade groups: construction, finishing trades, trades for commercial needs, motor vehicle trades, food trades, healthcare trades, trades for private needs and other trades.

Table 11.1 Career choice of women and men in crafts overall, Top10, 2021. *Source* Vocational Training Panel Craft; own calculations. © IAB

Training profession	Men	Proportion of women	Training profession	Women	Proportion of women
Vehicle-Air-Spacecraft, Shipbuilding tech	5392	4.2	Personal Care	983	75.6
Energy Technology	2805	2.8	Medical, Orthopedic and Rehabilitation Technology	939	64.6
Plumbing, Sanitary, Heating, Air Conditioning	2662	2.0	Office and Secretarial	815	72.4
Wood Processing and Finishing	1885	13.7	Sale of Food	728	75.1
Building Construction	1422	2.4	Food and Beverage Manufacturing	475	43.9
Drywall, Insulation, Carpentry, Glass, Roller Construction	1176	3.8	Wood Processing and Finishing	298	13.7
Metal Construction and Welding Technology	1101	2.8	Vehicle, Aerospace, Shipbuilding Technology	235	4.2
Painting, Plastering, Building Sealing, Building Protection	929	16.8	Painting, Plastering, Building Sealing, Building Protection	187	16.8
Food and Beverage Manufacturing	608	43.9	Sale of Clothing, Electronics, Vehicles, Hardware	143	38.2
Medical, Orthopedic and Rehabilitation Technology	515	64.6	Energy Technology	82	2.8

which the first years of the employment history after completion of the training can be considered for the evaluations, the graduates who remain in their training company after their training were identified through a company number comparison. Thus, with the stay in the training company, the stay in the craft or in the construction sector can be traced. However, no statement can be made about whether graduates who leave their training company also continue to work in a craft business.

11.2.3 Characteristics of Graduates in Construction

The following section provides an overview of structural characteristics of the graduates of a craft training in the entire panel and in the construction sector for the 2015 graduation year.[7] This involves both an insight into the distribution of the graduates on the manifestations of the various characteristics, as well as whether there are gender-specific differences in construction professions compared to the craft professions. First, the distribution by nationality, education, the final grade of the training and the size of the training company in relation to the size of the graduation years is discussed.

In the 2015 graduating cohort, there were approximately 32,170 graduates, of which about 23% were women. In the construction industry, approximately 12,800 graduates completed their training in 2015, accounting for 40% of this graduating cohort. Thus, the construction industry makes a significant contribution to the recruitment of skilled workers in the trades. However, the gender imbalance in construction is even more pronounced than in the trades as a whole. Only just under 6% of this graduating cohort were women. Approximately 94% of the graduates in the trades had German citizenship in the 2015 graduating class. There are no gender-specific differences here. The situation is different in the construction industry. Here too, 94% had German citizenship, but among women in the construction industry it was 98%. This was slightly above the comparative values. Nevertheless, the proportion of graduates with foreign citizenship had risen to almost 12% in the trades as a whole by 2021, starting from a low level. In particular, the proportion of graduates from the TOP8 asylum countries of origin increased, among whom mainly men could be integrated into the labor market through vocational training in the trades (Table 11.1). In the construction industry, the developments correspond to those of the trades as a whole. This increase reflects the fact that in recent years the proportion of foreigners in the population in Germany and thus also in the participating chamber districts and their proportion of trainees in the trades has increased significantly. This means that the trades make an important contribution to the integration of foreign citizens (Runst 2022).

With regard to the educational background of the graduates at the start of their training, those with a secondary school leaving certificate dominated with a share of 45%. Almost 40% of the graduating cohort had intermediate maturity and 11% had a high school diploma. In the construction sector, the level of school leaving qualifications was somewhat below that of the trades as a whole. 49% of the graduates had a secondary school leaving certificate, 38% had intermediate maturity. The differences were even more pronounced in the construction industry. More than one in five women in the construction industry had a university entrance qualification, compared to less than one in ten men. Over time, it can be seen that more and more training graduates in the craft sector had significantly higher school leaving qualifications. The proportion of those with a secondary school leaving certificate fell from over 50% in 2008 to just under 34%. During the same period, the proportion of graduates with a high school diploma rose from six to 17%. The main reasons for this are certainly that for years the federal government's education offensive has been reducing the proportion of school leavers from secondary schools and thus

[7] An overview of the developments in the socio-demographic and educational characteristics of graduates in 2008, 2015 and 2021 can be found in Table 11.2.

the potential for trainees has become significantly smaller. Secondly, more and more people in a school leaving cohort have a university entrance qualification, which increases their chances of learning a trade (in the trades). In addition, young adults who cannot find their way at a university often start training in the trades.

The largest proportion of the graduates' final grades, 36.1%, was "satisfactory" and the second largest, about 33.1%, was "adequate". 12% completed the training with a "good". 36% of the men achieved a "satisfactory" in the final certificate of the training. This grade was the most common. Women completed their training with better grades than men. Although the most frequently achieved final grades were also a "satisfactory" (almost 38%) among female graduates, however significantly more women than men achieved a "good", but also significantly fewer achieved a "satisfactory" on their final certificate. In the construction industry, the final grades were slightly worse than in the craft sector as a whole. As with school education, the gender-specific differences in the achieved final grades are significantly larger. Unlike the men, women in the construction sector completed their training on average with a "satisfactory". At the same time, the proportion of those with the final grade "good" was significantly higher than that of the men. On the other hand, compared to men, fewer women completed their training with a "satisfactory".

Traditionally smaller businesses play an important role in training in the craft sector. Therefore, it is not surprising that 47% of graduates in 2015 completed their training in a company with fewer than 20 employees. Still, just under a quarter of the graduates completed their training in a company between 20 and 49 employees. About 7% completed their training in companies with more than 250 employees. It is noticeable that women were more likely to successfully complete a craft training in smaller companies. The proportion of women who completed their training in companies with fewer than 20 employees was 48%. But women also disproportionately often completed their training in companies with 250 or more employees. The construction industry is even more characterized by smaller training companies. As with educational characteristics, gender-specific differences in the company size in the construction industry compared to the craft sector as a whole were significantly more pronounced. At 64%, almost two thirds of women had a training company with fewer than 20 employees. Already during the training, women sort themselves into smaller companies. Reasons can be diverse, including career choice-related, for example, training in hairdressing and sales professions, which are often dominated by women, relatively often take place in smaller companies. These usually also pay slightly lower wages than larger companies (Card et al. 2016).

11.2.4 Apprenticeship Status Quo in Crafts and Construction

This chapter takes a closer look at the construction sector. For this purpose, we will first look at how between 2008 and 2021 the numbers of female and male graduates in the craft sector as a whole and in the construction sector have developed. Afterwards, the choice of profession in the construction sector will be examined in more detail and possible gender-specific differences will be highlighted.

Looking at the development of graduate numbers, it is noticeable that the number of male and female graduates in the craft sector as a whole fell from 35,150 people to 28,450 people between 2008 and 20,221. This shows a significantly stronger decline among women than among men (Table 11.2). While in 2021, 14% fewer men achieved a vocational training degree in the craft sector than in 2008, the decline in the same period

Table 11.2 Structural characteristics of women and men in craft training, graduation cohorts 2018, 2015, 2021 in %. *Source* Training Panel Crafts; own calculations. © IAB

	2008			2015			2021		
Characteristics	**Men**	**Women**	**All**	**Men**	**Women**	**All**	**Men**	**Women**	**All**
Total	25,941	9208	35,149	24,674	7494	32,168	22,739	5715	28,454
In %	73.8	26.2		76.7	23.3		79.9	20.1	
Nationality									
Other	3.3	3.3	3.3	3.6	3.0	3.4	5.1	6.2	5.3
Turkey	1.1	1.8	1.3	1.8	2.6	1.9	1	1.2	1.1
TOP8-Asylum countries of origin	0	0.1	0.1	0.3	0.2	0.3	5.8	1.3	4.9
Germany	95.5	94.8	95.3	94.4	94.2	94.4	88.1	91.3	88.7
Education									
Without school leaving certificate	3.8	2.4	3.5	1.9	1.4	1.8	2.6	1.2	2.3
Secondary school certificate	51.8	48.6	50.9	45.8	40.7	44.6	34.6	29.6	33.6
Intermediate school certificate	31.8	35.8	32.8	40.2	39.1	39.9	43	41.2	42.7
High school diploma	4.6	9.2	5.8	9.7	17.3	11.5	15.3	24.7	17.2
Other	8.0	4.0	6.9	2.4	1.6	2.2	4.5	3.2	4.2
Final grades									
Very good	2.9	2.5	2.8	1.9	2.2	2.0	2.6	5.4	3.1
Good	9.8	13	10.6	11.6	13.7	12.1	13.5	19	14.6

(continued)

Table 11.2 (continued)

Satisfactory	35.8	41.5	37.3	35.6	37.7	36.1	33.6	36.3	34.1
Adequate	36.2	31	34.9	33.8	30.9	33.1	34	25.4	32.3
No information	15.3	12.1	14.4	17.0	15.6	16.7	16.3	13.8	15.8
Company size									
No Info	1.6	1.3	1.5	1.1	0.9	1.1	0.6	0.6	0.6
Up to 20 employees	49.3	57.9	51.6	46.4	48.8	46.9	43.2	48.5	44.3
20 to under 50 employees	21.8	17.3	20.6	24.2	20.5	23.3	26.2	22.8	25.5
50 to under 100 employees	10.6	8	9.9	12.2	11.4	12.0	13	11.2	12.7
100 to under 250 employees	8.9	7.3	8.5	9.2	10.2	9.5	9.9	9.9	9.9
250 and more employees	7.8	8.2	7.9	6.9	8.2	7.2	7.1	7.1	7.1
Business group									
Others	5.9		41.7	15.2	7.3	40.4	15	7.4	32.4
Construction industry	11	0.2	8.2	11.6	0.5	9	12	1.3	9.8
Finishing trades	36.6	5.7	28.5	37.5	8.8	30.8	39.1	13.2	33.9
Crafts for commercial needs	12.9	1.2	9.8	14.4	1.9	11.5	12.6	2.4	10.6
Automotive industry	22	0.9	16.4	21.5	1.7	16.9	20.5	3.8	17.2
Food industry	7.3	6	7	3.5	7.2	4.3	2.9	8.4	4
Health industry	1.8	8.3	3.5	2	12.5	4.4	2.4	17.1	5.4
Crafts for private needs	2.3	35.3	10.9	2	26.5	7.7	2.8	21	6.5
Crafts	0.2	0.7	0.3	0.3	0.5	0.3	0.2	0.3	0.2

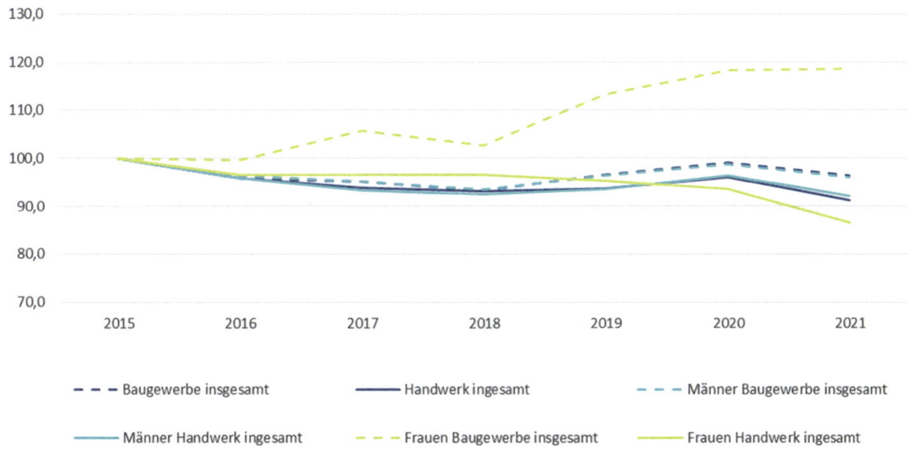

Fig. 11.14 Development of women and men in the craft and construction industry as a whole; Index = = 2008. *Source* Training Panel Crafts; own calculations. © IAB

among women was 28%. Looking at the construction industry as a whole, i.e. expansion and construction, the number of graduates has also declined, although not as much as in the craft sector as a whole. However, it is noticeable that the number of female graduates in the construction industry between 2008 and 2021 increased significantly more than in the other sectors (Fig. 11.14), although starting from a low level.

Table 11.3 shows which of the construction professions as a whole have gained importance and how between 2015 and 2021 the proportion of women in these has developed. The top-7 professions in the construction sector have, with Except for energy technology and drywall construction, there has been a decline. The highest decrease was recorded in the professions in the field of painting, plastering, building waterproofing, and building protection. Here, the number of graduates fell by 18%. In structural engineering, there were still 10% fewer graduates in 2021 compared to 2015. The number of graduates in energy technology, on the other hand, increased by 4.3%, and in drywall construction by 4.7%.

In all top construction professions, the number of women increased between 2015 and 2021, in some cases by over 100%, such as in the fields of energy technology and drywall construction, the only growth areas. But also in the other areas, which overall recorded a declining number of training completions, the proportion of women increased significantly. However, it is noticeable that in the field of painting and plastering work, which has the highest proportion of women in the construction sector, the number of women also declined. However, the decline was not as strong as that among the graduates, so that women in this area continue to gain importance.

Overall, however, the increasing interest of women in training in the construction industry does not compensate for its general decline. This is because the growth rates are

Table 11.3 Top professions in the construction industry, development between the graduation cohorts 2015 and 2021, total and the proportion of women. *Source* Training Panel Craftsmanship; own calculations. © IAB

	2015		2021		Change from 2015 to 2021	
Training profession	Total	Proportion of women	Total	Proportion of women	Total	Women
Energy technology	2695	1.4	2810	2.7	4.3	102.6
Plumbing, Sanitary, Heating, Air Conditioning	2473	1.2	2406	2.0	−2.7	62.1
Wood processing and treatment	2249	10.4	2181	13.7	−3.0	27.4
Structural engineering	1622	1.3	1457	2.4	−10.2	66.7
Drywall, Insulation, Room, Glass, Roller Construction	1167	1.7	1222	3.8	4.7	130.0
Painting, Plastering, Building Sealing, Building Protection	1319	15.7	1070	17.2	−18.9	−11.1
Paint and Varnish Technology	437	9.6	412	13.3	−5.7	31.0

so high because women are still represented in small numbers in the construction professions.

11.2.5 Transitions from Apprenticeship to the Labor Market

11.2.5.1 Employment Situation in the First Professional years

Once apprentices successfully complete an apprenticeship in the crafts, the question arises about the course of employment. The following analyzes the whereabouts by employment status for the 2015 class of craft graduates. The class of 2015 is used as an example because a longer observation period is available after the end of training. To investigate how many people succeed in transitioning to employment after completing their training, we look at the employment status of the graduates in the first five professional years after the end of training. We distinguish between a job subject to social security contributions, further training, and without employment, i.e., unemployment or

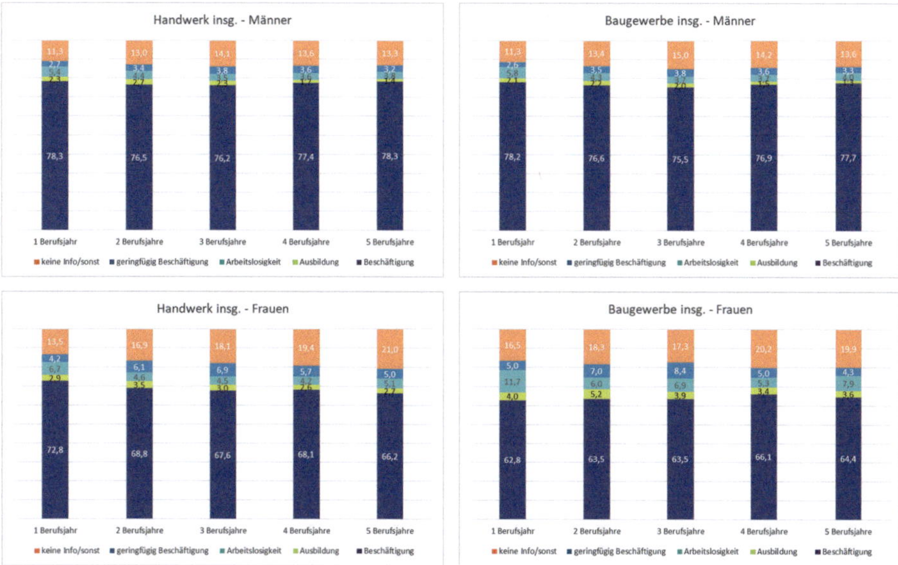

Fig. 11.15 Employment status of women and men by economic group (Crafts, Construction). *Source* Training Panel Craftsmanship; own calculations. © IAB

temporary withdrawal from the labor market. In this case, our dataset does not contain any information about the people and their status outside the labor market.

Taking up employment subject to social security contributions was the most common entry into the labor market. 78% of graduates In the trades, one year after the end of training, they were employed subject to social security contributions (Fig. 11.15 top left). Even five years after the end of training, just as many graduates were employed subject to social security contributions as after one year. About six percent were unemployed after one year. This value decreased to just under four percent during the first years after the end of training. 11% of the graduates left the labor market after the end of their training. This proportion was 13% after five years. Similar patterns are shown in the construction industry for labor market entries. 78% of construction industry graduates entered the labor market with employment subject to social security contributions. Five years after the end of training, it was still the same number of people. Just under six percent started their working life with unemployment, but this reduced to four percent five years after training. The development of the proportion of those without labor market information corresponded to that of the trades overall. Women with training in the trades also started their working life in most cases in employment subject to social security contributions. The proportion was slightly lower at just under 73% one year after the end of training than that of men. Five years after the end of training, only two thirds of the female graduates were still employed subject to social security contributions. This proportion was significantly lower than that of men with the same labor market experience. Accordingly, women were more likely to be unemployed, more likely to pursue marginal employment or to leave the labor market (temporarily). While the proportion of unemployed female graduates decreased in the first years of their career, the proportion of those without labor market information increased by eight percentage points during this period. Looking at women in the construction industry, it is clear that the proportion of employment subject to social security contributions was significantly lower at about 62% than that of women in the trades overall. At the same time, significantly more women were unemployed than in the trades overall (10%). Likewise, more women than in the trades overall left the labor market. It is noticeable that this proportion increased more strongly in the first five years after training than among men, but less strongly among female graduates in the construction industry than in the trades overall.

People who are not in the labor market in the first five years after apprenticeship may have started studying or further training, which could also include master training. After a master's training, when a business is founded or taken over, they are then no longer registered in the dataset as self-employed. For women, leaving the labor market could also be associated with parental leave.

11.2.5.2 Development of Retention Rates in the Training Company

Another strategy for securing skilled workers from a company's perspective, in addition to training its own skilled workers, is to retain them and benefit from the performance of the journeymen and journeywomen. This can cover at least part of the company's own skilled worker needs. If the graduates leave the training company, the company loses skilled workers and the training may only have paid off to a limited extent or, from a monetary point of view, possibly not at all. However, it should always be borne in mind when evaluating retention in the training company that companies may train more than they need and thus plan for the graduates to leave after completing their training. The extent of training beyond demand cannot be derived from the data. In earlier years, it was often assumed that the trades trained more than other sectors (Haverkamp and Gelzer 2016). However, Haverkamp et al. (2020) found that even in the years 2013 to 2015, the retention rate in the trades was only about 4% points lower than in other sectors. In addition, training companies that make a retention offer give their graduates the opportunity to gain professional experience in their own company. But it is also advantageous for the graduates themselves to accept a retention offer. Staying in the company also makes them attractive to other potential employers attractive. It shows that they have proven themselves as apprentices and have also gained professional experience beyond the training period (Hell and Wydra-Somaggio 2023; Mohrenweiser et al. 2020). The following figure shows how the retention in the company has developed over the different graduating cohorts (Fig. 11.16).

The retention rate in both the craft trades and the construction industry has been steadily increasing since 2015. For men, it was 55% for the craft trades in 2015 overall, in the construction industry 60%. In 2020, 60% of craft trade graduates remained in the training company, in the construction industry two out of three graduates. Of the women,

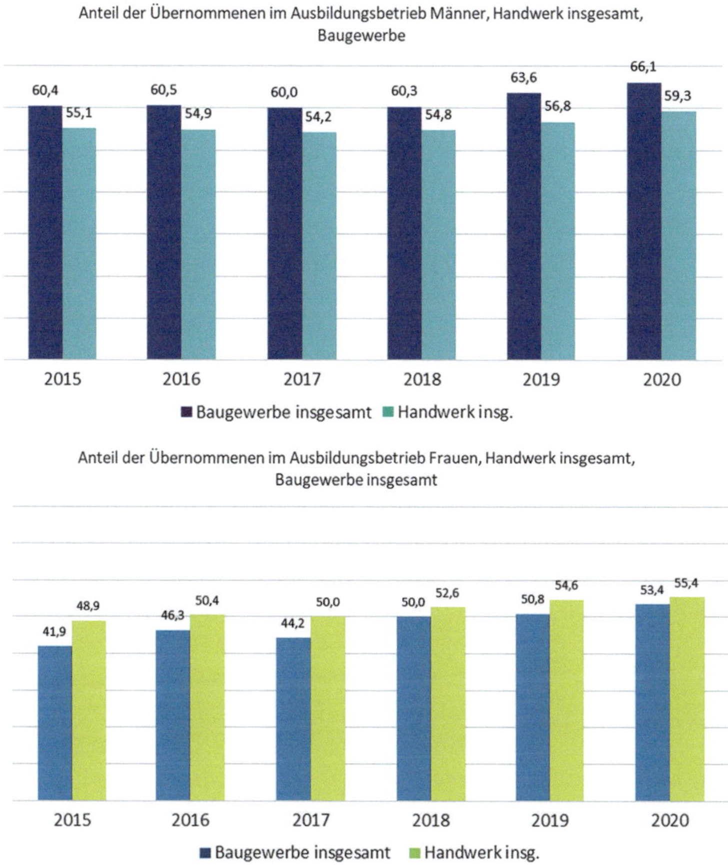

Fig. 11.16 Proportion of those retained in the training company by gender, craft trades overall, construction industry overall, graduating cohorts 2018 to 2021. *Source* Craft Training Panel; own calculations. © IAB

just under 49% remained in the craft trades in 2015 after the end of their training, in the construction industry it was 42%. This proportion increased to 55% in the craft trades overall by 2020, in the construction industry to 53%. It is also noticeable that the retention rates for men in the construction industry were higher than in the entire craft trades. This pattern has not changed over the years. On the other hand, there are slightly lower retention rates for women in the construction industry compared to the craft trades. The proportion of female graduates in the construction industry who began their working life in the training company has increased more than in the other craft sectors. Thus, the gap has almost closed over the years. This means women who have completed an apprenticeship in a construction profession now remain in that company almost as often as their female colleagues in the other craft sectors.

The figures show that the chances of being taken on and staying for women in the craft trades as well as in the construction industry were significantly lower than those of men. In the craft trades overall, this could be due to the choice of profession. For example, training in the hairdressing trade, which is still chosen by women, often involves a change of company after training. It should also be considered that women, when they decide to train in the craft trades, usually had contacts in the family or circle of friends beforehand (Mischler and Ulrich 2018). So the relatively more frequent changes could also be due to the fact that they switch to the family business after their training in order to continue it later.

The positive trend of recent years in the increase in retention rates in the company is partly due to the simultaneous decline in the number of training graduates (Leber et al. 2023). The greater change in retention rates in the construction industry among women is therefore probably due to the stronger increase in the number of female graduates in the construction industry. However, this cannot compensate for the declining numbers of male apprenticeship graduates and thus the demand for skilled workers, although the retention rates have continued to rise there as well. On the other hand, the construction industry benefited greatly from the positive economic development until 2021, so that more skilled workers were also in demand. This development was accompanied by improved employment prospects, which is indicated by the increased retention rates.

The next figure shows how long training graduates remain in the training company, who started their working life in it immediately after their training.

Figure 11.17 shows that nearly 70% of the men from the 2015 graduation cohort who remain in the apprenticeship company immediately after training are still employed there one year after the end of training, and after five years it was just over 30%. The retention rates did not differ between graduates from construction and those from the craft trades overall. For the women, the retention rates of those who entered the labor market in the training company were similarly high as for their male colleagues. However, female graduates from construction were less likely to remain in the apprentice company at the beginning of their professional lives than their colleagues in other craft trades and male colleagues. Nearly 60% of the female graduates taken on were still employed in the training company after one year. However, women in the craft trades change companies more frequently after two years of professional life than female graduates in construction, so that the retention rates have leveled off five years after the end of training. Overall, however, there is a high fluctuation in the first years of professional life across all areas and for both genders.

11.2.6 Retention of Graduates Outside the Construction Industry

The previous chapters revealed that a large part of the graduates change their company in the first years of their professional life. This raises the question of whether changing companies is associated with a change to another area. Figure 11.18 shows how many of

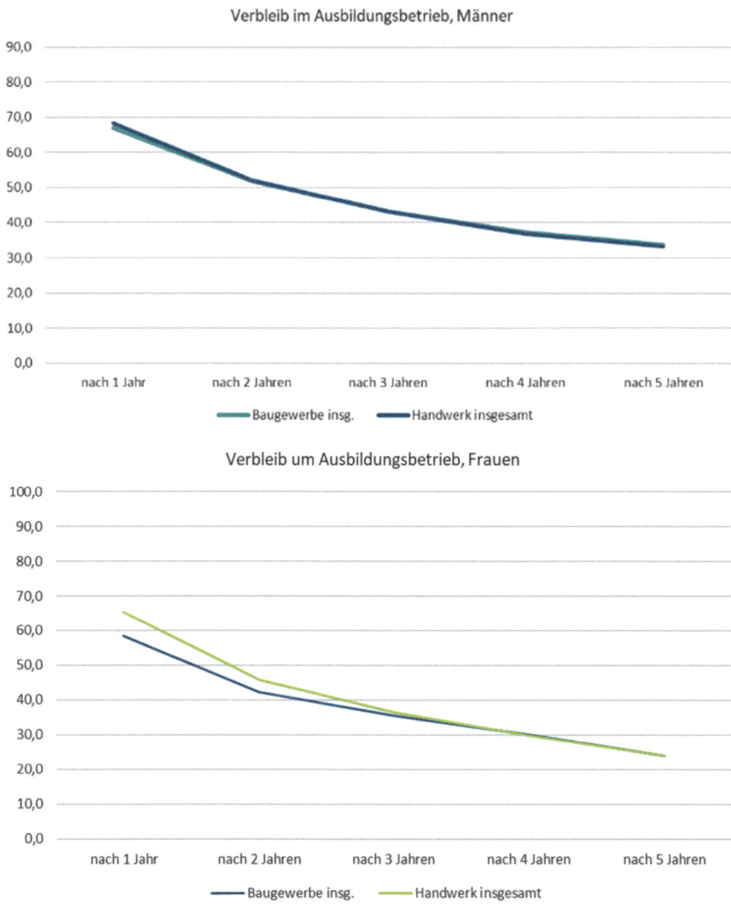

Fig. 11.17 Retention in the training company of those taken on after training, by gender, overall craft trades, construction and non-construction trades up to 5 years after the end of training. *Source* Training Panel Crafts; own calculations. © IAB

the male and female graduates who change companies in the first years of their professional life after the apprenticeship also change from the construction industry or continue to work in this area.

The figures for the 2015 graduation cohort show that women not only leave the apprentice company more frequently than men, but also the construction sector. The proportion increased significantly in the first four years of professional life for both genders and weakened slightly in the fifth year of professional life. While in the first year of professional life, nearly 37% of men changed from the apprenticeship company to the construction sector, by five years after the end of training it had been over half. For over half of the women, leaving the company was associated with a change from the construction sector already one year after the end of apprenticeship, five years after training the pro-

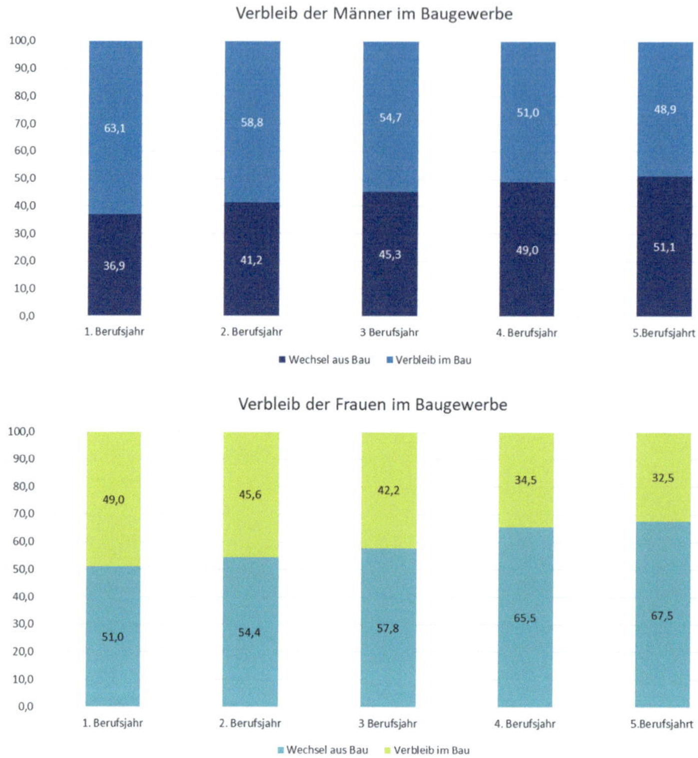

Fig. 11.18 Retention of male and female graduates in the construction industry who change companies, 2015 graduation cohort, in %. *Source* Training Panel Crafts; own calculations. © IAB

portion rose to over two thirds. The relatively more frequent sector changes by women from construction professions could result from the fact that they complete training in these areas, but then leave them because they used the training time as a screening of working conditions (Mohrenweiser et al. 2020). Women and to some extent men find that their expectations do not match the conditions in the training sector. This can also create a so-called "revolving door effect" (Jacobs 1989), according to which women, who initially learn male-dominated trades in the crafts, return to gender-typical areas after some time. change or exit the labor market.

Do certain types of women or men with specific educational characteristics tend to change the construction sector more often? The following Table 11.4 shows the retention in or change from the construction sector by gender and educational characteristics.

The training characteristics of women and men still differ significantly. However, there was little difference between those who left the construction sector and those who stayed in the construction sector. This applies to both genders. Women with a secondary school diploma and men with a middle school diploma tended to stay in the construction sector. Both men and women had a higher retention rate with better final exams. Gradu-

Table 11.4 Sociodemographic characteristics of graduates who remain in the construction industry, graduation cohort 2015 in %. *Source* Training Panel Crafts; own calculations. © IAB

	Men		**Women**	
	Change from construction sector	Remain in construction sector	Change from construction sector	Remain in construction sector
School education				
Secondary school certificate	53.3	51.9	37.6	38.0
Intermediate maturity	33.6	37.8	42.4	42.7
High school diploma	7.4	6.0	18.6	17.1
Final grades				
Very good	2.5	2.1	1.9	3.3
Good	7.2	10.2	13.8	16.9
Satisfactory	29.2	33.7	39.5	37.1
Adequate	44.2	35.4	26.1	24.2
No information	17.0	18.6	18.8	18.6
Company size				
No info	1.4	0.8	1.1	2.3
Up to 20 employees	54.9	56.5	63.1	65.4
20 to under 50 employees	21.3	22.6	12.5	14.1
50 to under 100 employees	9.6	9.4	8.8	7.2
100 to under 250 employees	6.9	6.3	6.2	6.0
250 and more employees	6.0	4.3	8.3	5.0

ates who completed their training in micro and small businesses also tended to stay in the construction sector.

When they leave the construction sector, men usually take up a profession in the field of crafts for commercial demand. The businesses of crafts for commercial demand (including trades in metal construction, precision mechanics and information technology) are primarily dependent on the needs of the industry as a supplier. Women, on the other hand, often go into the field of crafts for private demand, which mainly includes hairdressing professions. However, female graduates mainly work in professions of natural stone, mineral and building material production, as well as in textile manufacturing professions in this trade group. This craft sector has a very high proportion of women (Table 11.1). This suggests that women who leave the construction sector then also switch to gender-typical areas within the crafts, because there, their expectations of working conditions may be more likely to be met than in gender-atypical professions.

11.2.7 Concluding Remarks

In view of demographic change, the lack of affordable housing, digital and ecological transformation, the craft sector faces major challenges. The existing skills shortages threaten to intensify. This endangers the implementation of the energy transition, where the craft sector is a central actor, especially the construction industry, which is largely responsible for the implementation of political measures. Therefore, various strategies for meeting the demand for skilled workers are considered. Dual training, but also women as a potential source of skilled workers, as well as retaining skilled workers to meet the demand for skilled workers, are of great importance. This article has shown that the graduating classes in the craft sector, but also in the construction industry, are becoming increasingly smaller. Reasons for this are, on the one hand, cohorts of school leavers with low birth rates and, on the other hand, the increasing tendency to take up studies after school. Nevertheless, in recent years the number of women who have obtained a degree in the construction industry has increased. However, the increase in women does not offset the decline in men, so that the construction industry is training fewer skilled workers overall. Nonetheless, this development is to be evaluated positively, as women are increasingly interested in the construction industry and are completing training in gender-atypical professions. This positive trend could encourage more women to learn gender-atypical (construction) professions and thus provide more skilled workers in this area.

In addition, women have more difficult transitions in the craft sector as a whole and in the construction industry. The retention rate in the training company is significantly lower for women than for men after completing training in the construction industry. In addition, job changes for women are much more often associated with a change out of the construction sector than for men. The higher probability of women changing jobs could also have personal reasons, because they prefer to work in mixed teams, for example, or see the compatibility of private life and work in male-dominated areas less often than in female-dominated areas. Men could also better assess their career opportunities in gender-typical areas. A study by the German Crafts Institute also found that typical role models in craft professions are even more pronounced than in other areas and than the respondents would wish. Therefore, there is a starting point for the craft sector to support the reduction of clichés (Greilinger 2023), so that young people, and especially young women, choose the craft sector. The existing networks that specifically address women in the craft sector and are available to them must therefore be further strengthened and expanded. They can be contact points for women in career questions, but also in questions about the compatibility of work and private life. Because the craft sector faces enormous challenges. For some time now, many craft businesses have been complaining about staff shortages, lack of young talent and difficulties with business succession—problems that are likely to intensify considerably in view of demographic change (Bennewitz et al. 2022). Further training offers for young people and a (increased) use of digital technologies, which could relieve physical activities, could make the craft sector more attractive for both women and men (Wydra-Somaggio 2023). Through various channels, such as school visits, trade fairs or offers of internships and job shadowing, as well as appearances in social media, young people can be reached. This increases transparency regarding work and career opportunities in the craft sector, which could make it more attractive for young people. Because young people still attribute a more negative image to the craft sector in terms of working conditions, content and career paths than experts do, according to a study by Mischler and Ulrich (2018).

References

Althaber; A. & Leuze, K. (2020). Der Einfluss der beruflichen Geschlechtersegregation und beruflicher Arbeitszeitarrangements auf Teilzeitarbeit. Gleiche Übergangsbedingungen für Frauen und Männer? In: Kölner Zeitschrift für Soziologie und Sozialpsychologie, (Suppl. 1) Vol. 72, p. 317–349.

Beckmann, J. et al. (2006). Motivation durch Erwartung und Anreiz. In: Heckhausen, Jutta; Heckhausen, Heinz (edt.): Motivation und Handeln. Heidelberg 2006, p. 105–142.

Bennewitz, E. et al. (2022). Zwei Jahre Corona-Pandemie: Die deutsche Wirtschaft zwischen Krisenstimmung und Erholung – Ein Vergleich der Jahre 2019 und 2021 – Ergebnisse des IAB-Betriebspanels. IAB-Report No. 20.

Böhme, S. et al. (2023). Die Bedeutung des Handwerks für Beschäftigung und Ausbildung ist regional sehr unterschiedlich. In: IAB-Forum 14. June 2023, https://www.iab-forum.de/die-bedeutung-des-handwerks-fuer-beschaeftigung-und-ausbildung-ist-regional-sehr-unterschiedlich/ (Accessed 3. November 2023).

Card, David, C. et al. (2016). Bargaining, sorting, and the gender wage gap: Quantifying the impact of firms on the relative pay of women. The Quarterly Journal of Economics, Vol. 131, p. 633–686.

England, P. (1992). Comparable Worth: Theories and Evidence. New York.

Fuchs, J. et al. (2021). Projektion des Erwerbspersonenpotenzials bis 2060: Demografische Entwicklung lässt das Arbeitskräfteangebot stark schrumpfen, in: IAB-Kurzbericht 25 (2021), https://doku.iab.de/kurzber/2021/kb2021-25.pdf.

Greilinger, A. (2023). Attraktivität einer Ausbildung. Empirische Analyse zu Erwartungen von Schulabsolventen an einen Arbeitgeber und an Berufsorientierungsmaßnahmen. Ludwig-Fröhler-Institut Forschungsinstitut im Deutschen Handwerksinstitut im Handwerk.

Haverkamp, K. & Gelzer, A. (2016). Verbleib und Abwanderung aus dem Handwerk: Die Arbeitsmarktmobilität von handwerklichen Nachwuchskräften. Göttinger Beiträge zur Handwerksforschung, No. 10, Göttingen.

Haverkamp, K. et al. (2015). Frauen im Handwerk – Status Quo und Herausforderungen. Göttinger Handwerkswissenschaftliche Studien, No. 97, Duderstadt.

Haverkamp, K. et al. (2020). Sektorale und berufsspezifische Ausbildungsstrategien und Übergangsmuster nach der dualen Ausbildung. In: Zeitschrift für Wirtschaftspolitik, Vol. 69., Iss. 1, p. 44–68.

HBS (Hans-Böckler-Stiftung) (2022). Handwerk. Zu wenig Hände für zu viel Arbeit. In: Mitbestimmung, H. 2, online.

Hell, S. & Wydra-Somaggio, G. (2023). Duale Ausbildung im Handwerk: Der Anteil der jungen Menschen, die nach der Ausbildung im Ausbildungsbetrieb bleiben, steigt. In: IAB-Forum 14. September 2023, https://www.iab-forum.de/duale-ausbildung-im-handwerk-der-anteil-der-jungen-menschen-die-nach-der-ausbildung-im-ausbildungsbetrieb-bleiben-steigt/ (Accessed 3. November 2023).

Jacobs, J. (1989). Revolving Doors. Sex Segregation and Women's Careers, Stanford: Stanford University Press.

Kleinert, C. et al. (2023). Occupational sex segregation and its consequences for the re-production of gender inequalities in the German labour market. In: S. Weinert; G. Blossfeld; H.-P. Blossfeld (Eds.), Education, Competence Development and Career Trajectories. Analysing Data of the National Educational Panel Study (NEPS), Cham: Springer, S. 295–317. https://doi.org/10.1007/978-3-031-27007-9_13.

KOFA Study KOFA-Kompakt 74/2022.

KOFA Kompakt 05/2023 – Fachkräftemangel und Ausbildung im Handwerk 2023.

Koneberg, F. et al. (2022). Energie aus Wind und Sonne: Welche Fachkräfte brauchen wir? KOFA-Studie, No. 3/2022, Institut der deutschen Wirtschaft (IW), Kompetenzzentrum Fachkräftesicherung (KOFA), Köln.

Leber, U. et al. (2023). Die betriebliche Ausbildung vor und während der Corona-Krise: Besetzungsprobleme nehmen zu, Anteil der Betriebe mit Ausbildungsberechtigung sinkt. (IAB-Report 3/2023).

Ludwig-Fröhler-Institut (2023). Die Beschäftigungssituation von Frauen im deutschen Handwerk. Studie des Ludwig-Fröhler-Instituts für Handwerkswissenschaften, September 2023.

Malin, L. & Köppen, R. (2023). KOFA Kompakt-Fachkräftemangel und Ausbildung im Handwerk. Studie Stand Juni 2023. www.kofa.de/studien. (Accessed 21. December 2023).

Mischler, T. & Ulrich, J. G. (2018). Was eine Ausbildung im Handwerk attraktiv macht. Ergebnisse einer Befragung von Jugendlichen. In BiBB Report. Nr. 5, Bonn.

Mohrenweiser, J. et al. (2020). Information advantages of training employers despite credible training certificates. In: Oxford Economic Papers, 72(3), p. 651–671.

Polachek, S. W. (1981). Occupational self-selection: A human capital approach to sex differences in occupational structures. The Review of Economics and Statistics, 63, p. 60–69.

Runst, P. (2016). Bedeutung von Migranten als Auszubildende, Fachkräfte, meister und Selbständige im Handwerk. Veröffentlichung des Volkswirtschaftlichen Instituts für Mittelstand und Handwerk an der Universität Göttingen Forschungsinstitut im Deutschen Handwerksinstitut e. V. https://doi.org/10.3249/2364-3897-gbh-12.

Runst, P. (2022). Migration und Handwerk: Fachkräftemangel und integratives Potenzial. Bundeszentrale für politische Bildung, https://www.bpb.de/themen/migration-integration/kurzdossiers/515507/migration-und-handwerk-fachkraeftemangel-und-integratives-potenzial/ (Accessed 20. October 2023).

Schmidt, C. (2023). Deutlicher Rückgang ja, Absturz nein. In Ifo-Schnelldienst: Zur Diskussion gestellt: Bauwirtschaft: Droht ein Absturz in der Baubranche? No. 1, p. 3–6.

Steinbeis Innovation Center, Corporate Development at Pforzheim University (2013). Women in leadership positions in Baden-Württemberg Top women. An empirical investigation of the status quo, personnel policy measures and areas of action in politics and the private sector. Study of the top women in Baden-Württemberg March 2013.

Wiemann, B. (2023). Quo vadis Bau? Die Bauwirtschaft in herausfordernden Zeiten. In Ifo-Schnelldienst: Zur Diskussion gestellt: Bauwirtschaft: Droht ein Absturz in der Baubranche? Nr. 1, S. 7–9.

Wydra-Somaggio, G. (2023). Was braucht das Handwerk, um für junge Leute attraktiv zu sein? In: STARK – Kreishandwerkszeitschrift Dortmund-Lünen, No. 3.

ZDH (Zentral des deutschen Handwerks) (2023a). Daten und Fakten: Junge Frauen im Handwerk, https://www.zdh.de/ueber-uns/fachbereich-soziale-sicherung/frauen-im-handwerk/#c20773 (Accessed 28. September 2023).

ZDH (Zentralverband des Deutschen Handwerks) (2023b). Unsicherheit dämpft die Stimmung. Konjunkturbericht 2/2023.

ZDH (Zentralverband des Deutschen Handwerks) (2023c). ZDH-Betriebsbefragung zu den aktuellen wirtschaftlichen Herausforderungen für die Handwerksbetriebe. Berlin 08.09.2023.

Zucco, A. (2019). Große Gender Pay Gaps in einzelnen Berufen hängen stark mit der überproportionalen Entlohnung von langen Arbeitszeiten zusammen. DIW-Wochenbericht 86(10), p. 127–126.

Part V
COMMUNICATION

What Makes Good Communication in Architecture and Construction? 12

12.1 Architecture is Built Communication

Andreas Grosz, Master of Business Administration (Dipl.-Betriebswirt), in an interview with Bianca Weber-Lewerenz in January 2024.

What Was Your Very Personal Approach to Architecture?

My interest in architecture developed intuitively. Growing up in the Ruhr area in the 1950s/60s, I was impressed as a student by the massive industrial architecture, the mine towers, blast furnaces, factory halls, all brightly lit at night. But it also sensitized me to the sometimes brutal interventions in nature and landscape. I studied economics and political science, always combined with a penchant for architecture, fine arts, and design. At the Technical University TU in Braunschweig, Germany, I made my first contacts with architecture students and aspiring artists from the local School of Fine Arts. A good balance for the often so theoretical and practice-distant study of economics, which helped to broaden my view. Architecture is built communication because the houses can tell a story about the life of the city's people. The Ruhr area was rather history-less, the cities almost interchangeable—everything was shaped by coal and steel and industrialization. The upheaval and the end of coal mining in the Pott, the associated sometimes radical restructuring of the Ruhr area, has accompanied me since my youth. But also the destruction caused by the devastating 2nd World War shaped my perception of the built environment. Over a quarter of a century later, I lived in the 2000-year-old cathedral city of Cologne. Cologne, for example, has not yet recovered from the severe damages caused by the war.

Architecture and urban planning are like an open book to me, an orientation that guides through life. This perception and this interest led early on to my involvement, especially in local politics, in debates about the future of the city (Fig. 12.1*).*

B. C. Weber-Lewerenz (ed.), *Diversity in Construction—The Gamechangers*, https://doi.org/10.1007/978-3-658-45308-4_12

Fig. 12.1 Andreas Grosz in Rheinau Harbor Cologne. *Source* Albrecht Fuchs

In the 1980s, I opened a design gallery in Braunschweig that dealt with the emerging architecture and design movements of the time. The post-modern attacked the Bauhaus dogma, which threatened to stifle the development of architecture and design. The square was blown up, colorfulness as in pop art, dynamic forms, as the zeitgeist expressed, were reflected in architecture and design language.

During a visit to Milan, I met Ettore Sottsass. This stroke of luck later enabled me to have an exhibition with the legendary Memphis Group. I am still in friendly contact with the architects Michele de Lucchi and Matteo Thun (co-founders of Memphis).

During this time, I first published the culture magazine LIVING in Braunschweig, later nationwide (Fig. 12.2 *and* 12.3*). For me, the magazine was a kind of catalyst because I could address all the topics and people that interested me with the magazine. True to the motto of Francis Picabia: "The head is round so that thinking can change direction". Architecture and urban development played a major role in LIVING, but not with a purely professional perspective, but as a social or cultural topic that fundamentally affects people's coexistence. One of the issues in the early 90s dealt explicitly with the architecture of the new capital Berlin. Here, Kristin Feireiss, the founder of the AEDES architecture gallery, advised us, many photos of the issue come from her partner Hans-Jürgen Commerell. AEDES is one of the first architecture galleries worldwide, then still in its beginnings, today an institution that shapes the debate about architecture and city not only in our country but also internationally.*

My cultural commitment, now also advising companies, brought me to EXPO 2000 in Hanover as founding managing director in the 1990s. Here I was responsible for the development of the architecture and the theme park of the world exhibition. A job that

Fig. 12.2 Covers from 1993–1994 of the culture magazine LIVING *Source* Andreas Grosz

Fig. 12.3 Covers from 1993–1994 of the culture magazine LIVING. *Source* Andreas Grosz

now also professionally connected me with the world of planning and building and with the heads of the international architectural profession. A formative experience was, for example, the collaboration with Albert Speer and his architectural office, which was commissioned with the development of the master plan of the EXPO site. Even after leaving the management, I remained project-oriented connected to EXPO 2000. Thus, in cooperation with the Austrian architect and university teacher Zamp Kelp, founding member of the avant-garde group Haus-Rucker-Co, and in cooperation with Ute Bernhardt, chairwoman of the Schaumburger Landschaft, the "Millennium View" was created as a "Decentralized Project of EXPO 2000" (https://www.zamp-kelp.com/jahrtausendblick/).

What Does Architectural Communication Mean to You?

Architecture was still largely a specialist topic until the end of the 1990s, an internal event that was only slightly accessible to a broader public. Building took place in narrow circles: architectural competitions are not public, who ultimately and why gets the chance, these questions were usually only understandable to a professional audience. Architects were little known to the public. The professional organizations aimed more at equalization, the star architect widespread in the Anglo-Saxon context was frowned upon at least in Germany. German architects built in Germany, everything else was an exception. The abstinence has something to do with our recent history, which made it difficult for German architects after 1945 managed to gain a foothold abroad. For this, the German market was open to the international architecture community: With the glamour of international star architects, resistance against unpopular projects could be cracked in Germany. Investors refined their construction projects with resonant names, municipalities thus upgraded their city marketing. In the course of growing globalization, the German architecture community also intensified its international marketing. As an example, the Architecture Export Network (NAX) is mentioned, which has been pursuing the goal of strengthening the brand "Architecture Made in Germany" abroad since around 2010 (www.nax.bak.de/).

The construction boom in Asia and the Emirates (especially Dubai) benefited this development. German architecture firms have now successfully established themselves internationally. Pioneers here include the offices of GMP, Ingenhoven, HPP, AS + P and KSP Jürgen Engel Architects.

It remains to be noted that the globalization of markets for architects and construction in the last quarter of a century has far-reaching effects beyond the scene and has significantly brought the topics of architecture, urban development and construction into the focus of public perception. This is noticeable, among other things, in the enormous increase in city trips, where the hip metropolises, always associated with extraordinary architecture, are in the foreground. (Cities like Barcelona, Venice or Amsterdam already have to protect themselves against the travel boom because the visitor rush is hardly manageable anymore.) But also domestically, the changes within the city structure are already accompanied by a sensitized public. The necessary transformation of cities due to climate change, energy crisis, mobility transition, housing shortage, demography, digitization, change of the city has catapulted questions of architecture and urban planning into the center of social disputes.

Let's note: Globalization and the associated rapid change of cities from an industrial to a service and knowledge society have significantly increased interest in vibrant urban design, which can be clearly seen since the beginning of the 2000s in the founding of a large number of initiatives and forums for urban design.

The restructuring of cities—architecture, planning and construction—became a central urban societal issue. If architecture was previously more a topic of the feuilleton or the trade press, architecture and urban planning moved more and more into the field of view of the general public and the popular press. Anyone who wanted to build had to explain themselves and often learn the still unfamiliar dealing with the public, the citizen. Architecture communication was the order of the hour, was a new profession in architecture offices and administrations.

How Did the KAP Forum Come into Being or How Can Good Communication of Architecture (in Construction) Succeed?

As already explained, professional architecture communication is a still young discipline. It is based on the growing interest of the public in the development of our built environment. People increasingly want to know in advance of a construction measure why, who, how and for whom is being built.

The founding of the KAP Forum for Architecture & Urban Development may serve as an example of architecture communication at the beginning of the 2000s.

The background was the redesign of the Cologne Rheinauhafen (Cologne Harbor), one of the largest construction measures in Cologne after 1945 and at times also one of the largest construction sites in Europe. The project was controversial and much discussed in the city, not least because of its size and because it would significantly influence the silhouette of the cathedral city. At the very south of the almost two km long project, the KAP am Südkai was created as a prelude, developed and built by KSP Engel + Zimmermann and Primus Developments. Also against the background of poor communication and the poor public response to the Rheinauhafen project, Michael Zimmermann (KSP) and Achim Nagel (Primus) were invited at the beginning of 2004 to develop a concept for an architecture forum. The forum was to be located on the ground floor of the KAP on Südkai, which is directly on the Rhine, and thus provide a location for the overarching architecture and urban planning discussion in the Cologne Rhine area. The project was to be privately financed, explicitly without public funds, but like an entrepreneurial initiative with partners from architecture-related and construction-related companies. To cut a long story short: By autumn 2004, the KAP FORUM project was ready, the rooms were ready for occupancy, and the partners had been found.

Tobias Groß, www.studio.cologne, was able to win the competition for the corporate design. At that time still working for frogdesign in Silicon Valley, California, this was his return ticket to self-employment in Cologne. His design concept from the logo to the web design to the invitation cards and the KAP magazine were decisive for the growing popularity and have shaped the KAP Forum into a recognizable brand. The KAP postcards, like the multiple award-winning KAP magazine, have become collector's items.

The founding partners in 2004 included: Alape, Carpet Concept, BASF, Dornbracht, Gira, Kvadrat, Wilkhahn, and Zumtobel. There were changes over the years. Kusch + Co, Cor, Godelmann, and Zehnder joined as partners, others left. The companies saw the KAP Forum as an interface for communication between architects and building suppliers, as well as a communication and knowledge platform. That's how it was intended from the start:

The preamble to the KAP Forum states: „Building is culture and symbolizes the quality of life of a society. Building is a complex process that combines technical, design, social, economic, and political aspects. The KAP Forum promotes dialogue between all those involved in the building process with its program: architects, designers, urban planners, project developers, real estate and construction industry. The companies are united by a high demand for quality and the constant search for innovative ways of planning, building, and furnishing. Building bridges!"

What was intended to be temporary for five years is now entering its 20th year in 2024. There was intensive cooperation with the KAP partners, the climate was characterized by mutual respect and interest in the event topics. The internal communication of the KAP partners conveys the topic of architecture and urban development into the participating companies. Especially Thomas Trenkamp from Carpet Concept, a founder type through and through, and Andreas Dornbracht were among the drivers of the KAP Forum, were always approachable and have been real KAP fans for almost 20 years.

Architects shy away from deceptions, i.e., events and invitations that turn out to be simple acquisitions. The KAP Forum has been free of this for all these years. The focus was always on the content and the commitment to good architecture. And long-term relationships also developed between many architects and our partners through encounters and exchanges. We understand architecture as a holistic topic, with a view to the entire value chain of planning and building. From this understanding, the KAP Forum was one of the first platforms to equally address and integrate architects, investors, project developers, and the construction industry into the program. This was anything but self-evident until the late 1990s. People kept strictly to themselves and nurtured their prejudices. The architect, as an artist type, with a black turtleneck sweater and Saab driver, had little in common with the real estate investor. According to the prejudice, the investor was only return-driven, which ultimately diluted the architect's noble design, while the investor saw the architect merely as a cost factor who knows nothing about building. Little constructive was to be expected on this basis and so people kept officially better out of the way. The practice has certainly always looked different from what the advantage wants to make us believe. But there is a kernel of truth in it. Remnants of it have survived to this day, although the image of togetherness has clearly changed for the better, that is to say, towards togetherness.

For the KAP Forum, addressing all trades of planning and construction was constitutive. This was also the new and different and certainly a essential building block of success. Over 400 events in the 20 years, exhibitions, discussion rounds, lectures, workshops, seminars, excursions, the Rheindesign festival, celebrating parties—all this belonged to the KAP Forum. The actual caesura was experienced by the KAP Forum through Corona, when the program was cancelled for two years. As a real located communication platform we had to reinvent ourselves virtually in the net. We have since given up our premises, work predominantly on the net. How the journey continues is open. End well, everything in good flow …

How Has the Program of the KAP Forum Developed?

The topic and program development of the KAP Forum were among my very own work focuses, which I could shape independently as a curator. Since I like to proceed conceptually, I have from the beginning shaped event series that were equipped with a long-term perspective. With a series "Profiles and Positions" we have for many years invited clever and interesting heads from the field of architecture for lectures or work reports.

The aim of the KAP Forum was not to achieve a unique position, but rather to enable networks, partnerships and cooperations. The long-standing event series "On the Future of the European City" may stand as an example. Here there was a comprehensive cooperation with the Department for Planning and Building of the City of Cologne, with the Kölner Stadt-Anzeiger and the IHK zu Köln. From the event planning to the execution we did it together. Especially the Cologne building directors Bernd Streitberger and later Franz-Josef Höing have put all their energy into this format. City planners and architects from many European countries were guests at the KAP and presented their future plans for urban development for discussion. Guests included cities such as London, Liverpool, Vienna, Barcelona, Lisbon, Lille, Paris, Turin, Rotterdam, Copenhagen, Stockholm, Zurich, Frankfurt, Munich, Hamburg, Leipzig, Berlin.

A special highlight was the hosting of RheinDesign, a festival that made Cologne a mecca for architecture & design in the summer of 2007. Curated in cooperation with the culture journalist Inken Herzig, organized by the KAP Forum, which was also the central venue, the festival attracted about 50,000 visitors. Over 50 initiatives from different areas participated in the program design. Partner cities were Paris and Brussels. The TGV as project partner, brought designers and their guests from Paris and Brussels to the cathedral city for free. With extraordinary poster actions, a strongly designed program folder, with media partnerships in print, radio and online, the festival was present at all levels at that time.

Exciting and worth mentioning is also the cooperation with the futurologist Klaus Burmeister from forsightlab.de in Berlin, with whom we among other things in 2015 launched the city conference "City in Transition—Rethinking the City". He was from the beginning one of the most important network partners and content advisors. His co-initiated initiative d2030 is a communication platform that wants to advance the future dialogue between economy, politics, sciences and civil society. An open think tank, which fits well with the intentions of the KAP Forum for the future development of architecture and society.

Outlook—or Where is the Journey Going?

As described, the public perception of architecture and building activity has fundamentally changed in the last 25 years. The transformation of our cities, in architecture and construction, are part of our Discussion on how we want to live, work, and reside in the future. PR and public relations are now a matter of course for architects, real estate, and construction industry. In almost all larger communities, there are architecture and design forums, such as the HdA House of Architecture in Cologne, the aed in Stuttgart, the Hamburg Foundation for Building Culture, the Building Culture NRW, AEDES and the Architecture Gallery in Berlin, the Architecture Museum in Frankfurt. But above all, on the net, the information offer has increased enormously. From german.architects, over the BauNetz, from the Chambers of Architects to the BDA. The work of the Federal Foundation for Building Culture, which is based in Potsdam, is particularly noteworthy. The team around Reiner Nagel now creates a comprehensive mediation activity on architecture and urban development nationwide and integrates very different partners from the world of planning and building.

When I started my career, the construction sector was largely male-dominated. This has fortunately changed fundamentally. Even if gender balance has not yet been fully achieved is, the path to it is paved and women are present at all levels and in all areas of architecture and construction. This is not a pure question of statistics, but a question of values. In addition, women bring new, usually less hierarchically oriented ways of thinking into the work one. Building can only benefit from diversity in general, because building significantly influences our lives. For this the views and ways of thinking of different genders and people from different cultural circles can make indispensable and valuable contributions. This can be experienced quite tangibly by anyone who, for example, visits an internationally operating architecture office today.

In a master's degree program in Architecture Media Management at the Bochum University of Applied Sciences, young architects can learn communication concepts, presentation techniques and marketing strategies. Architecture mediation, public relations, cultural management, product and service marketing are the future fields of activity for the aspiring media managers.

Although the dialogue in the construction process has become significantly more transparent and open, the bankruptcy of the major investor Benko (Signa Group) shows that building can still also be an object for speculators and quick money. Here, once again, control and communication have equally failed at the expense of the general public. Almost every decade still has such a scandal about real estate and money laundering.

The spread and compliance with the ESG guidelines (Environment, Social and Governance), which are based on the UN Agenda for Sustainable Development and the Sustainable Development Goals (SDGs), are supposed to ensure that this changes fundamentally. Architecture should be in harmony with people and nature. The ICG Institute for Corporate Governance in the German real estate industry e. V." also stands for these values. According to its statutes, the association represents the general ideal interests of the German real estate industry. It stands for value-oriented corporate governance, professionalism, transparency, integrity, and sustainability in the German real estate industry.

High goals that require comprehensive communication and an open exchange with all participants. Goals for which every effort is worthwhile, considering which and how many resources building consumes worldwide and in what threatening state our globe is already in today.

Finally, as a closing word, a quote from the magnificent work by Werner Sobek "non nobis—about the building of the future", Volume 1: One must start from what is. Stuttgart 2022, p. 16.

"As future civil engineers, we see our task in designing the world sustainably and well. How we implement this in our industry, we are not taught. We have to find out for ourselves." *Four young students of civil engineering at the Koblenz University of Applied Sciences in a letter to the author, dated December 19, 2017.*

12.2 From Walkie-Talkies to Messaging Apps: Exploring the challenges and opportunities in Communication and Data Collection in Construction Projects

M.Sc. Diego Cisterna

The construction sector is a fundamental and evolving component of the global economy. Prior to the outbreak of the COVID-19 pandemic, the AEC industry (Architecture, Engineering, and Construction) reached an investment milestone of nearly 12 billion US$, accounting for about 15% of global GDP (Mazhar and Arain 2015). Despite emerging fears of an economic downturn due to geopolitical tensions and disruptions in the global supply chain after the pandemic, the industry is on track to achieve a turnover of 391.3 billion euros in Germany by the year 2025 (Statista 2022). And this with an expected average annual growth rate of 6.05% (Technavio. 2023).

However, the enormous size of the industry also poses challenges, particularly in terms of productivity, which has stagnated over the last thirty years. This trend is documented in various studies both globally (Fig. 12.4) and within Germany (Fig. 12.5).

The stagnation of productivity can be attributed to several factors. A crucial aspect underlying these challenges is communication. Ineffective communication has been identified as the main cause of low productivity in the industry. The fragmentation of construction projects, in which numerous stakeholders are involved, exacerbates this problem and leads to delays, misunderstandings, and errors. Coupled with the low degree

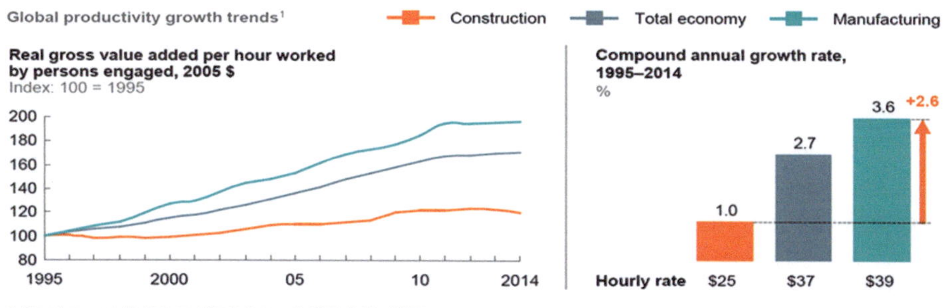

Fig. 12.4 Global Productivity Growth Trends. *Source* OECD; WIOD; GGCD-10, World Bank; BEA; BLS; national statistical agencies of Turkey, Malaysia, and Singapore; Rosstat; McKinsey Global Institute, 2017

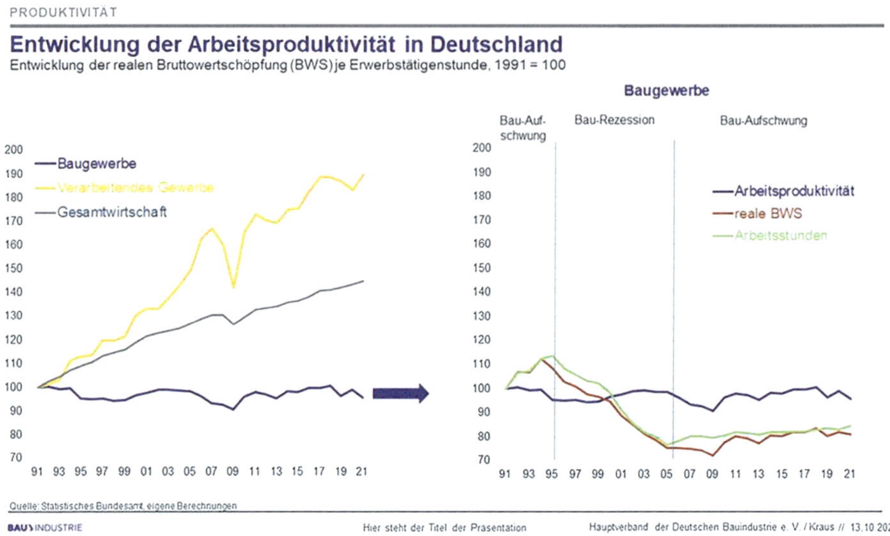

Fig. 12.5 Development of Labor Productivity in Germany. *Source* Federal Statistical Office, Main Association of the German Construction Industry e. V./Kraus, 2022[1]

of digitization of the industry, which is just ahead of agriculture and hunting (Fig. 12.6), the construction sector struggles to introduce more efficient methods of information exchange and project management. This leads to a heterogeneous IT ecosystem of construction-related software products as well as fragmented and disconnected knowledge silos with numerous media and information gaps (Puolitaival et al. 2018).

If a critical problem in construction projects lies in suboptimal communication, it is imperative to thoroughly investigate this problem to find viable solutions. The first step towards improvement is the evaluation of communication. Quantifying communication within a team poses a challenge due to its intangible nature. Nevertheless, numerous tools have been developed over time to facilitate this evaluation. Among these, Social Network Analysis (SNA) stands out as a particularly effective tool.

SNA has evolved into a significant method for analyzing social structures and has gained importance in various sectors since its targeted development in the 1950s. It uses mathematical frameworks from graph theory and matrix algebra to scientifically model social relationships. Thus, it provides a tangible bridge between abstract social theories and empirical data. The concept of a "sociogram" introduced by Moreno in 1934 marked the first visualization of a social network, mapping the relationships between students within a classroom.

[1] (Hauptverband der Deutschen Bauindustrie e. V./Kraus 2022).

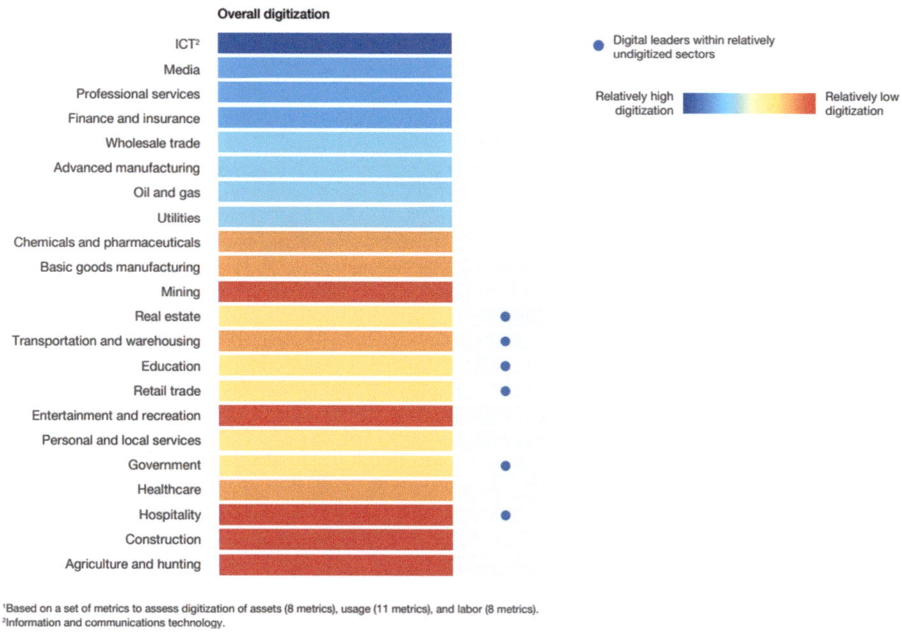

Fig. 12.6 Industry Digitization Index. *Source* McKinsey Global Institute 2017

To grasp the potential of SNA, please consider Fig. 12.7, which illustrates the response of students of various grade levels to the question "Who wants to sit next to whom?" Here, the orange circles (nodes) represent girls and the white ones represent boys. The lines (edges) between these shapes indicate a relationship between two children, while the size of the circle indicates the number of connections between them. Subsequently, the significance of the individual within the network is determined by its size and centrality in the network, which are determined by the algorithm used in this method.

Considering the above, the analysis of Fig. 12.7 concludes that children in lower grade levels prefer to sit with individuals of their own gender, and that this tendency disappears in higher grade levels, where the network shows more mixed color patterns. It is also easy to identify which individuals are most popular and that there are several isolated groups in the fifth grade.

By translating social relationships into diagrams, SNA provides a powerful framework for understanding how individuals or units within a network are connected. This graphic representation, in conjunction with a mathematical analysis, provides profound insights into the nature and dynamics of social structures, making it a valuable tool for tackling communication challenges in construction projects or other collaborative endeavors. Importantly, SNA promotes diversity by encouraging the participation and inclusion of a broader spectrum of voices in project decisions and ensuring that information from various sources is considered.

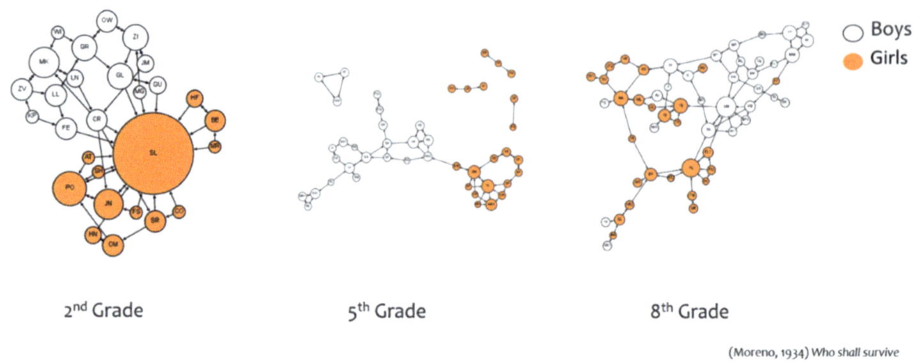

Fig. 12.7 Sociograms representing the seating preferences of children in the classroom in response to the question "Who wants to sit next to whom?" *Source* Moreno (1934)

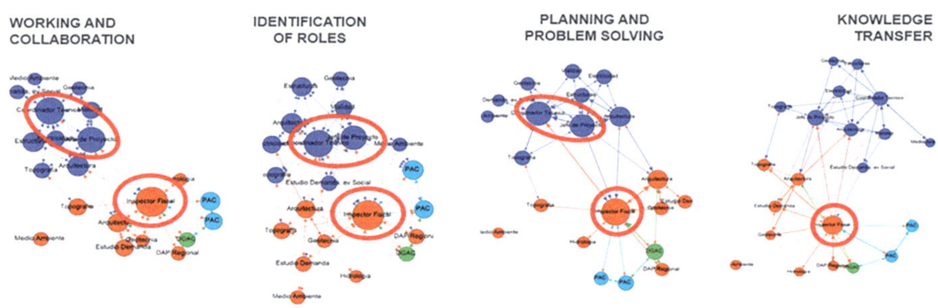

Fig. 12.8 Social networks in different dimensions in an infrastructure project in Chile. *Source* Cisterna et al., 2018

Similarly, the communication network of a construction project could be visualized. Each team member can be represented as a node and the communication or flow of information between them can be represented by edges connecting the nodes. The size of the nodes could illustrate the volume of information that each team member processes. Figure 12.8, for example, shows the results of an SNA that evaluated the communication and information exchange across various dimensions in an infrastructure project conducted in Chile (Cisterna et al. 2018). These networks show how the various teams within the project interact with varying intensity depending on the measured dimension. Clusters and isolated groups, bottlenecks in Information flow, knowledge receivers and providers, as well as the most connected nodes were identified, which were expectedly the project managers.

In this project, SNA helped to understand how information flowed throughout the project team—or not—and enabled the leaders to make informed decisions about where to intervene to improve communication. If only a few team members had key information (represented by a group of connected nodes), measures were taken to spread the information and knowledge further. Conversely, if certain team members had no or very few connections (isolated nodes), strategies were formulated to better integrate them into the communication network. Through such measures, SNA not only identifies communication gaps, but also promotes a more inclusive environment in which diverse contributions are valued and integrated into the decision-making process.

Another central aspect of communication is the channel through which information flows. In construction projects, information typically circulates in three main ways, namely analog, verbal, and digital:

- **Analog communication** involves the exchange of information in physical, written formats. This channel includes printed materials such as blueprints, photos, reports, graphics and planning diagrams, as well as handwritten notes on these documents. It is a practical way of information exchange that is based on physical objects that can be held, seen, and discussed. However, this type of information exchange primarily generates data that must be processed manually.
- **Verbal communication** offers immediacy and flexibility, whether it's personal conversations, group meetings or discussions facilitated by telecommunications devices such as radios and phones. This direct form of communication allows for quick exchange and adaptation to the fluidity of construction projects. However, as they are only short-lived, there are often no permanent records for future references or reviews.
- **Digital communication**, especially via emails and messaging apps like WhatsApp, is leading in modern communication strategies in the construction sector. This approach not only ensures a quick and unrestricted exchange over distances but also lays a robust framework for data organization, analysis, and visual representation. The main strengths of the digital realm—easy distribution of updates, quick exchange of visual aids for consultations, and the joint modification of documents—underline its ability to improve project management by increasing efficiency and improving decision-making processes. A significant disadvantage of using this communication channel is that the parties involved must have knowledge of digital technologies for data transmission. If the participants are not familiar with these technologies, training and change management initiatives are required to culturally acclimate them to the use of these tools.

Among the three discussed communication channels, digital communication is the most promising way to counteract the stagnating productivity in the industry. By digitizing the transmitted information, data is generated that can be effectively used. This data can be organized, compiled, and are statistically analyzed to identify patterns, as well as used for training Artificial Intelligence (AI) models to support and automate processes.

In industries such as manufacturing, where productivity growth rates are high, digitization has been identified as a key factor for increasing productivity (Kupriyanova et al. 2023). However, the AEC sector faces major challenges in its efforts towards digital transformation. Building Information Modeling (BIM) is the digital backbone of con-

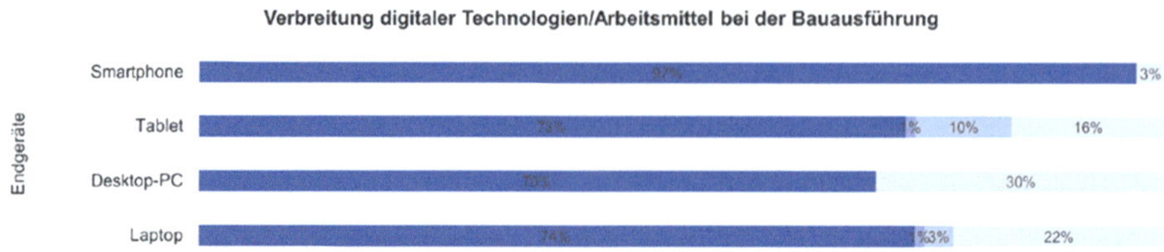

Fig. 12.9 Spread of digital technologies/tools in construction execution. *Source* Erdogan, 2023

struction, but the adoption of this technology has been extremely slow. Various studies highlight resistance to change, the cost of technology, and a lack of experts as the main obstacles for the BIM implementation (Ahmed 2018; Lourenço et al. 2024; Ma et al. 2023) Unfortunately, this resistance extends to all levels of implementation of digital tools in construction. Many digital initiatives fail not because of software defects, but because they do not adequately capture construction site information, often, because users do not familiarize themselves with new software or find it too complex and therefore do not invest time in populating these systems with data.

Here lies the potential of communication via emails and messaging apps. While people working on construction projects may not be exactly thrilled about the introduction of new software, they can usually use these familiar tools without any problems for everyday communication. This tendency is driven by the widespread availability and user-friendliness of these technologies on devices that are already widely used on construction sites. A recent survey from Germany found that smartphones and laptops are the most commonly used devices on these sites, with an astonishing usage rate of 97% for smartphones Erdogan (2023) (Fig. 12.9). (Sample: 1150 contacts, 681 companies from the construction industry. Participation: 154 people).

Smartphones are indispensable on construction sites and serve far beyond making phone calls. According to a study by BauInfoConsult (BauInfoConsult 2020), 72% of construction workers in Germany use WhatsApp for internal coordination and 55% for documentation of construction site activities. The widespread adoption of messaging apps effectively overcomes the hurdle of low acceptance of new technologies (Ahmad Pozin et al. 2019). WhatsApp, for example, as shown in Fig. 12.10, is found on 92% of German smartphones (Brandt 2021), with usage evenly distributed across age groups (Fig. 12.11), showing a similar share among people aged 35 to 44 and those aged 45–64 (Morning Consult 2022). This broad age penetration is an achievement that few software platforms can boast, making it an ideal tool for capturing information directly on construction sites. Its use alone has been proven to improve communication between employees at the workplace and thereby stimulate work performance in various companies (Terkan and Celebi 2020). The use of such apps not only facilitates communication, but also promotes a diverse flow of insights and ideas, enriches project decisions, and fosters a more inclusive workplace culture.

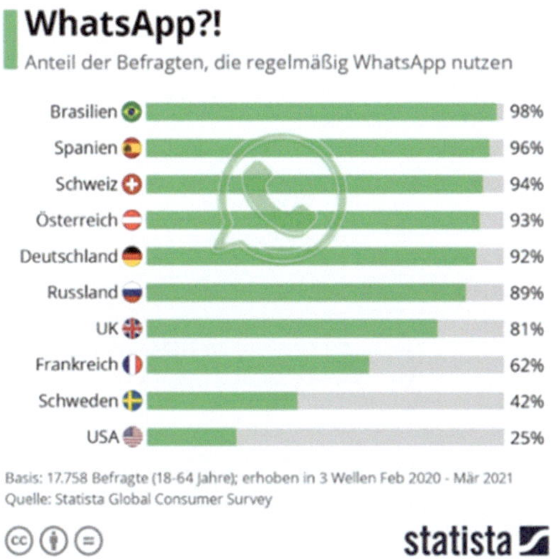

Fig. 12.10 Percentage of respondents who regularly use WhatsApp. *Source* Statista, 2021

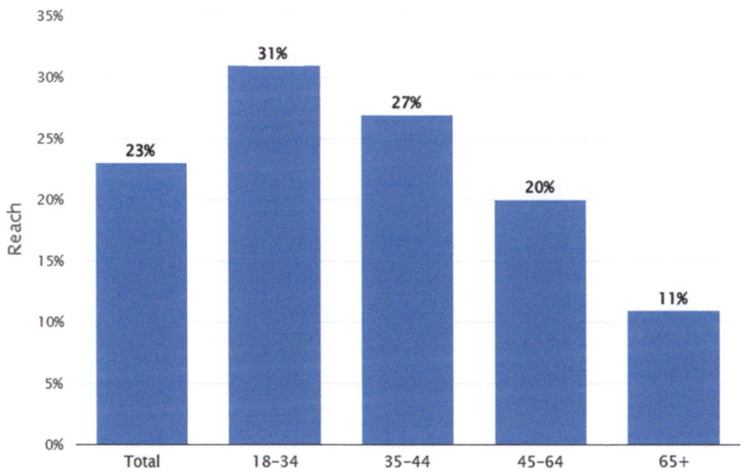

Fig. 12.11 Share of adults in the United States who have a WhatsApp account in May 2022, by age group. *Source* Statista, 2022

In this environment, the potential is clear to use communication tools like WhatsApp, MS Teams, Telegram, and others to improve construction project management. By integrating their user-friendliness and broad acceptance with specialized project management software, communication, documentation, and coordination can be significantly optimized on a large scale. This strategy not only builds on the existing digital habits of construction workers but also offers a seamless approach to capturing and managing project-related data (Valoon GmbH 2024).

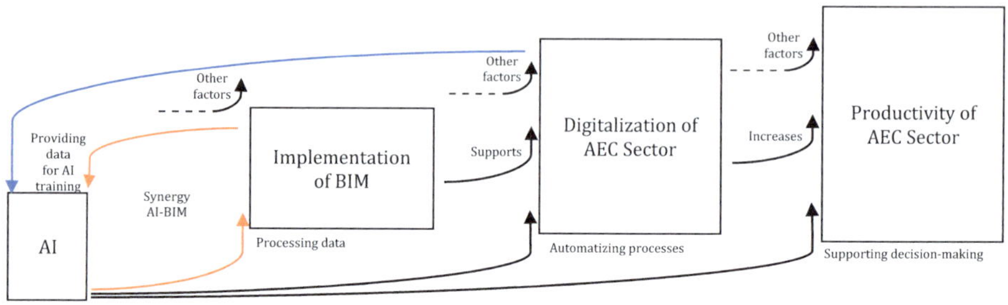

Fig. 12.12 The impacts of AI on BIM Application, digital transformation and the growth of productivoity in AEC Sector. *Source* Cisterna, 2024

The use of solutions like Valoon or similar platforms paves the way for a significant improvement in data collection and structuring directly on the construction site. As already mentioned, the conversion of raw data into a structured format is a crucial step towards the broader adoption of AI in construction projects (see the blue arrow in Fig. 12.12), a development that is attracting great interest in the AEC sector. More and more companies are using AI to recognize patterns in unstructured data sets, streamline decision-making processes, and automate routine tasks, thus facilitating the digital transformation of the industry (Regona et al. 2022). At its core, this technology is based on the analysis of extensive data sets, enabling machines to learn and adapt. This process aims to extend human capabilities in daily tasks and provide support and insights that streamline operations and increase efficiency.

Building Information Modeling (BIM) plays a crucial role as a facilitator for the application of AI, as it effectively digitizes building information and creates a rich data ecosystem from which AI can draw (Kyivska and Tsiutsiura 2021). In return, AI can enhance BIM capabilities by improving data collection, manipulation, and use within its framework (Zhang et al. 2022). The symbiotic relationship between AI and BIM illustrates a positive cycle (see the green arrows in Fig. 12.12), in which the integration of AI not only increases the efficiency of BIM but also extends its functionality beyond its original scope.

Therefore, it is possible to sketch a causal chain that details the impacts of AI use in the AEC industry (Fig. 12.12). By using platforms that improve communication and project data collection, the construction industry can create a solid foundation for AI implementation, significantly advance its digital development, and improve the overall efficiency and effectiveness of the project.

For decades, the construction industry has faced the challenge of countering stagnant productivity. Digitization offers a potential solution, but its implementation is slow. A closer look reveals that communication problems and inefficiencies in data collection are the main causes. However, the tide is turning: the increasing acceptance of digital communication tools is beginning to change the landscape. The integration of these tools into project management software can have significant impacts by enabling structured data collection. This change not only aims to boost productivity but also opens the door to the use of advanced technologies like AI.

The acceptance of innovations and the commitment to a Cultural changes hold enormous potential for the construction industry. By transitioning from traditional communication methods to modern digital platforms, the sector opens new doors to address long-standing productivity issues. This move towards digitalization, coupled with a strong commitment to diversity and inclusion, not only promotes collaboration between different teams, but also deepens the decision-making process through the integration of various viewpoints and data sources. If the industry embraces this digital and cultural change, it will not only improve productivity, but also create a more inclusive, innovative, and progressive atmosphere.

12.3 iMHD: The Interactive Module Manual for Universities to Promote Procedural, Digitized and Collaborative Work in the Construction and Real Estate Industry

Prof. Dr.-Ing.-Habil. Anica Meins-Becker, M.Sc. Agnes Kelm and M.Sc. Nils Koch to Krax

12.3.1 Background and Objectives

Analyses and findings by the authors of this article regarding obstacles and hurdles to the implementation of digital transformation in the construction and real estate industry led to the following result: "little understanding for other planners/"silo" thinking", "separation of planning and execution", "no or insufficient digital equipment of the authorities", "no consideration of the overall system, relationships are severed; the building is thought of as autonomous and not as part of the entire energy system". The causes are seen, among other things, in a lack of transparency, collaboration and communication in the construction process from project development, planning, construction, operation and demolition. The topics mentioned are strongly interconnected and the question arises as to how these problems can be solved.

From the authors' point of view, the key lies in a uniform understanding of the construction process. The question to be clarified is whether there can also be reference process models for the construction industry that contribute to clearly defining responsibilities, but at the same time also creating an understanding for each other's activities. But what could such a model look like and to what extent can it specifically contribute to improving transparency, collaboration and communication along the entire construction process and provide the basis for digitization and then also for digital transformation? Through years of research work at the BIM Institute of the Bergische Universität Wuppertal, an extensive process model for the construction and real estate industry was developed together with numerous experts.

In order to prepare future civil engineers, architects, and architects during their studies for collaborative work through mutual process understanding and the digitization or digital transformation of the industry, the research project "iMHD—Interactive Module

Handbook", funded by the Foundation for Innovation in Higher Education under the auspices of the Toeper Foundation GmbH, was conceived and implemented at the Bergische Universität Wuppertal over the past two years.

The aim is not to change the teaching content, but to assign existing teaching content to the construction process along the entire life cycle of the buildings in construction, as a so-called backbone of teaching, for better orientation and to promote process understanding among students. The motivation also lies in networking the complex and very versatile knowledge taught in these disciplines transparently using processes to form an overall picture.

As at any other university/higher education institution, the various modules are covered by the professors/chairs/institutes. All content is coordinated via the module handbook. However, the module schedule is not directly related to the construction process from project development, planning, construction, operation and demolition at any university. Especially at the beginning of their studies, students cannot assign the individual lectures to the future practical construction process and there is often a lack of content orientation. In addition, there is a lack of an overview of which tasks fall within the area of the other construction-related study programs (here architecture and civil engineering). This bridge is to be built by the iMHD. The implementation of the iMHD research project at the Bergische Universität Wuppertal is predestined, as it is to some belongs to the few universities where architecture and civil engineering were combined into one faculty.

The research project investigates how the use of the construction process model and the assignment of a multitude of existing teaching content can contribute to an improvement of teaching and a consolidated understanding of the graduates of the interrelationships of the teaching content in construction practice. The challenge is to find an appropriate process level that is deep enough to capture the knowledge to be conveyed and shallow enough not to counteract the desired understanding.

To put the processes and the content of the teaching into context and make them accessible to the students, a Learning Management System (LMS), here Moodle, was used. This is already known to the students from their everyday university life and is accessible to all. Not only can the construction process and the teaching content be presented via this, but additional incentives can also be created to use the information provided. Gamified elements were integrated to increase the motivation of the students and to make studying easier, especially at the beginning.

12.3.2 Methodology

12.3.2.1 Component 1: Process Model Over the Life Cycle of High-Rise Construction Projects

At the center of the iMHD is an already existing process model of the German construction process, which has been created since mid-2015 in collaboration with the researchers of the BIM Institute and experts from various areas of the construction and real estate industry and is continuously being developed. Its aim is to depict the processes and the flow of information over the entire life cycle of a property in a universally valid way in order to derive concrete digitization approaches for the industry from this.

To achieve the goal of the research project presented here, a suitable level of process depth had to be found. It must not be too superficial and must provide enough anchor points for teaching content of the modules of the courses. If it is too detailed, there is a risk that especially students of the first semesters will lose the connection and transparency is not given. The process model adapted to the specific needs of the research project then forms the red thread to assign the teaching content to this within the framework of the iMHD.

12.3.2.2 Component 2: Teaching Content of the Modules of the Undergraduate Courses in Architecture and Civil Engineering of the Faculty

The second essential component of the iMHD is the teaching content of the modules of the undergraduate courses of the faculty. These are the Bachelor's and Master's degree programs in architecture and civil engineering. For this purpose, the relevant module manuals and examination regulations as well as the existing timetables were evaluated. The timetables are not binding, but serve as a template for a coordinated compilation of the modules. This scheme was transferred to the iMHD and the modules were arranged to correspond to the timetables. In order to relate the modules to the construction process, the contents described in the module manuals were recorded. As preparation for the provision, a matrix was created. Here, the contents were noted in bullet points. For further information, the person responsible for the module, the chair, and the intended semester were assigned to each module. Dependencies both within the respective modules and between the examination performances were recorded from the examination regulations. In this way, it is now apparent to the students which module must be completed so that a certain further module can be taken. In addition, the type of examination and the examination duration were noted.

12.3.2.3 Module 3: Presentation of the Chairs and the Associated Laboratories

The third module of the iMHD is the presentation of the chairs and the associated laboratories. To provide students with the easiest possible access to relevant information about the teachers and their facilities, videos of the chairs and their facilities were created, in which the responsible persons introduce themselves and their research and teaching focuses briefly. For each chair, a site plan was also created. Here, the locations of the respective facilities on the campus are shown in a mapped representation.

Basically, this information was already publicly available. The essential added value that the iMHD provides is in the interlinking and accessibility of this information. Their combination and visualization ensure the transparency and accessibility of the information. were.

12.3.2.4 Integration of Modules 1–3 into the iMHD

The construction process was chosen as the framework for the integration and networking of all three modules. To accomplish this, the contents of the teaching contents of the modules were evaluated and a content-related assignment to each construction process was made. This means that each process description previously created on the information basis of the entire model was linked with the worked out teaching contents of the modules. As a result, a collection of modules or teaching contents could be attributed to each construction process, in which the students are taught necessary knowledge and skills to carry out the activity described in the process. Conversely, this assignment was used to assign processes to each module for which they are relevant. Only in the case of the basic subjects was an assignment omitted (in the civil engineering bachelor's program, these are, for example, mathematics, statistics and construction materials science). These subjects are not irrelevant, but rather significant for almost all processes. Therefore, an assignment was omitted for these.

12.3.3 Results

The result of the data processing are four separate Moodle courses: Bachelor Architecture, Master Architecture, Bachelor Civil Engineering and Master Civil Engineering. Each degree program has its own independent iMHD. For the landing page of the courses another graphic was designed, which at a glance shows all functions available to the students. The following graphic (Fig. 12.13) exemplarily shows the interactive mod-

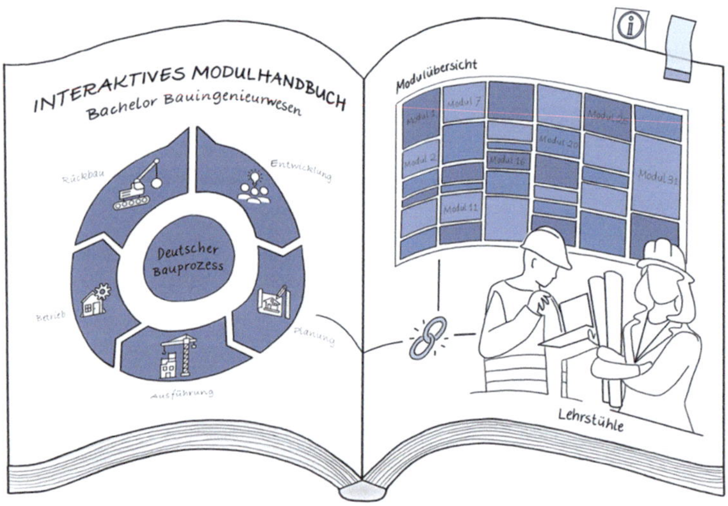

Fig. 12.13 Landing page iMHD Bachelor Civil Engineering. *Source* Institute for the Management of Digital Processes in Construction and Real Estate

ule handbook for the Bachelor Civil Engineering program. The three developed modules are represented in the graphic as *German Construction Process*, *Module Overview* and *Chairs* and are provided with a link. When these are clicked, the students are taken to the respective subpages of the iMHD. Since this use exceeds the usual functions of the Moodle platform, a workaround is used with a link via an image map, which can turn each graphic into a clickable button. The "Info" bookmark at the top right displays a welcome text and a brief explanation of the functions of the course. The graphic is deliberately simple and clearly designed to make all information accessible to students at a glance. The usual presentation with tiles and text inserts is omitted in order to make the entry into the use as barrier-free as possible.

The subpage for the German Construction Process, Module 1, provides a more detailed overview of the processes assigned to the lifecycle phases of Level 1 (Fig. 12.2). Here, students get a first impression of the interrelationships of the processes and can proceed to the detailed views of the sub-process models. This expanded control loop serves as a navigation interface to all processes. The figure also serves as a control element and is clickable.

Excursion Process Model:
The reference process model, on which the process representation in iMHD is based, is structured as follows. At the center are the lifecycle phases of a building (the inner circle in Fig. 12.14*), also called process level 0: development, planning, execution, operation, and demolition. Each of these phases is assigned further processes, which go into more detail with each additional level. Figure* 12.14 *shows another process level, level 1. Here, a first structuring of the sequences takes place. Process level two, visible in Fig.* 12.15*, is represented as Business Process Modeling Notation (BPMN) modeling. Here, roles are already assigned to the workflows, making the interaction of the individual participants*

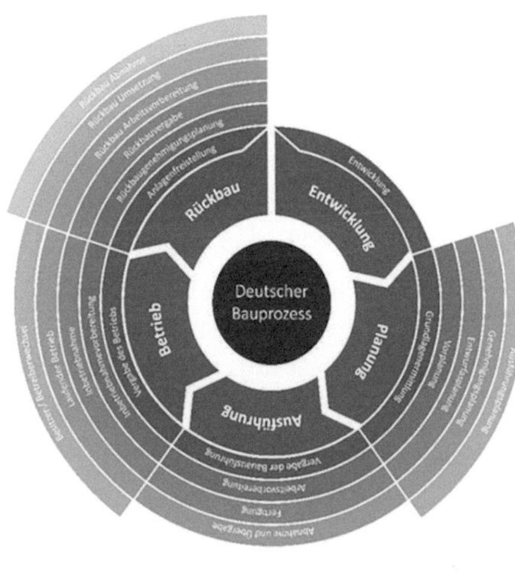

Fig. 12.14 The German Construction Process—Navigation Interface in iMHD. *Source* Institute for the Management of Digital Processes in Construction and Real Estate

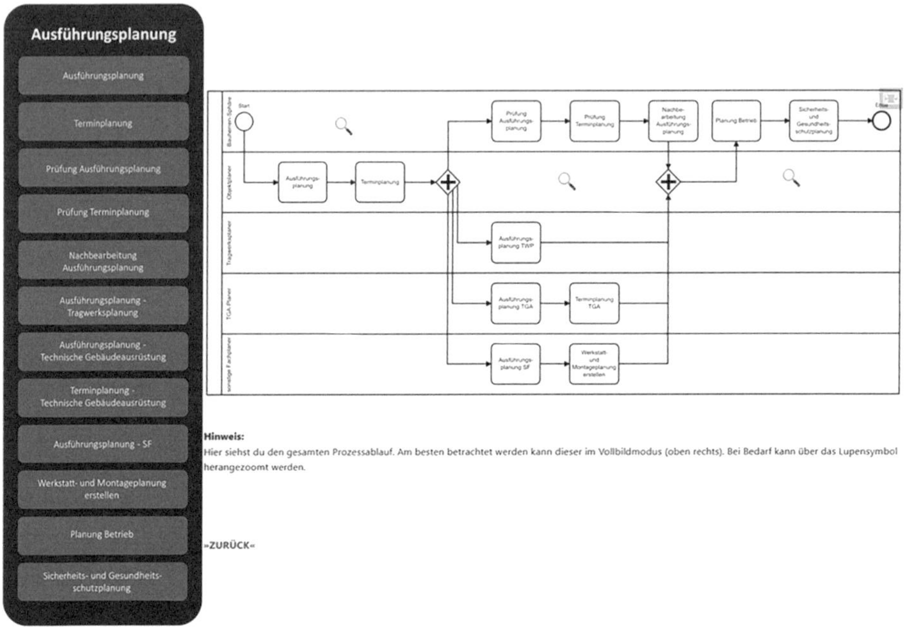

Fig. 12.15 Detailed view of the sub-process model. *Source* Institute for the Management of Digital Processes in Construction and Real Estate

visible. There are further detailed process levels (level 3–5), which contain further information. In iMHD, these are not displayed to ensure clarity. The information, however, has been incorporated into the description of the processes of level 2.

Figure 12.15 shows an example of the overview of all processes of the second level for execution planning. For the appealing presentation of the information and the use of the process model, the individual processes were modeled in BPMN at the desired level of detail. To ensure clarity, the sub-process models were supplemented by a graphic that serves as a button. The BPMN diagram shows the interrelationships of the work steps and their process-responsible roles. Via the button at the top right, the diagram can be viewed in full screen and additionally zoomed in using the magnifying glass symbols. The button on the left side allows a redirection to the process-specific page. Each subpage has a "Back" button to further facilitate navigation.

This level of process representation not only offers students the opportunity to understand the necessary work steps, but also to understand the interaction of the various actors involved in construction. Who needs to provide which information to whom, which work steps are necessary before another can be tackled, are questions that are answered here. If students navigate further to the individual process view, as seen in Fig. 12.16, they receive more detailed information about the process content and the linked modules of their degree program. Under "What happens in this process?" the pro-

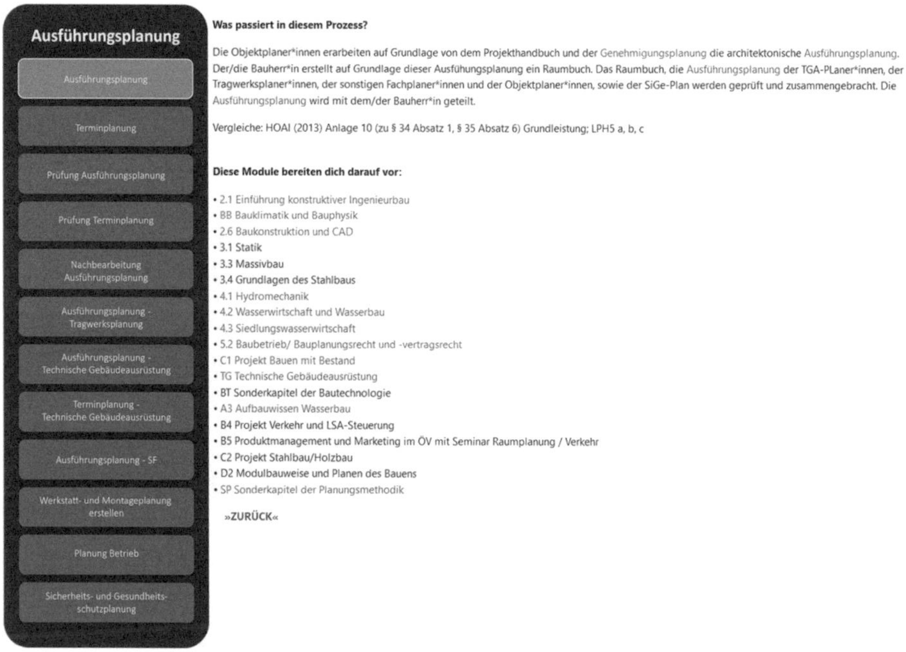

Fig. 12.16 Detailed view of the process. *Source* Institute for the Management of Digital Processes in Construction and Real Estate

cess description, which is based on all the information available in the model, is provided. was generated, displayed. "These modules prepare you for" lists all modules previously associated with the process in the data collection. This not only increases the transparency of the process model, but also provides significant added value for the students. This representation is intended to enable students to classify what they can use the knowledge they are learning for in their later profession. The aim of this is to increase the motivation of the students to continue using the iMHD and to complete their studies with a specific goal.

The list of modules associated with the process is also linked. When a module is selected, students are taken directly to the module detail page, part of building block 2. Here all information about the modules is displayed. Each of the building blocks can also be reached via the graphic of the landing page. The contents are structured as follows:

- Chair: *Link to the institution responsible for the module (students can reach the third pillar of the iMHD, the overview and presentation of the chairs,* via *the link)*
- Credits: *Weighting of the examination performance*
- Content: *Bullet point list of the teaching contents*

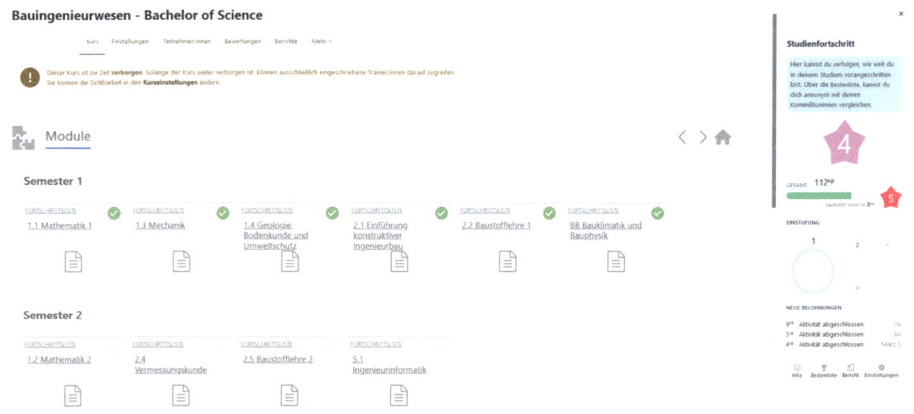

Fig. 12.17 Excerpt from the module overview. *Source* Institute for the Management of Digital Processes in Construction and Real Estate

- Required as a prerequisite for: *List of dependencies among the modules*
- What you learn here prepares you for the following processes: *Linked list of all relevant processes from which the detailed view of the processes can be called up*
- Examination performance: *List of all module-internal examination (pre)performances necessary for a module completion (Students can independently track their progress by ticking off the individual performances. The completion of a module leads to a point award. In an overall ranking, students can track their study progress and also compare it anonymously with other students. The point number of a module corresponds to the respective credit points and the sum of all points to be achieved corresponds to the total credits required for the completion of the studies.)*

Above the individual module pages is the module overview, building block 2, arranged. Here the modules are listed semester by semester according to the proposed timetables. Modules that have already been marked as completed by the students are marked with a green tick. The points in the overall ranking can also be displayed via a navigation bar from Moodle. Figure 12.17 shows a section from the module overview including the navigation bar to the study progress:

The third building block is presented as an overview of the chairs depicted (see Fig. 12.18). Here, a conscious effort is made to ensure long-lasting information content. This area is intended to make it easier for students to start their studies. At a glance, they can see who is responsible for the chair and where each facility of the chair can be found. The latter is done via a specially created map, on which these facilities are located. At this point, the respective video portraits also find their place, which have been created by the majority of the chair holders. They are embedded above the maps and

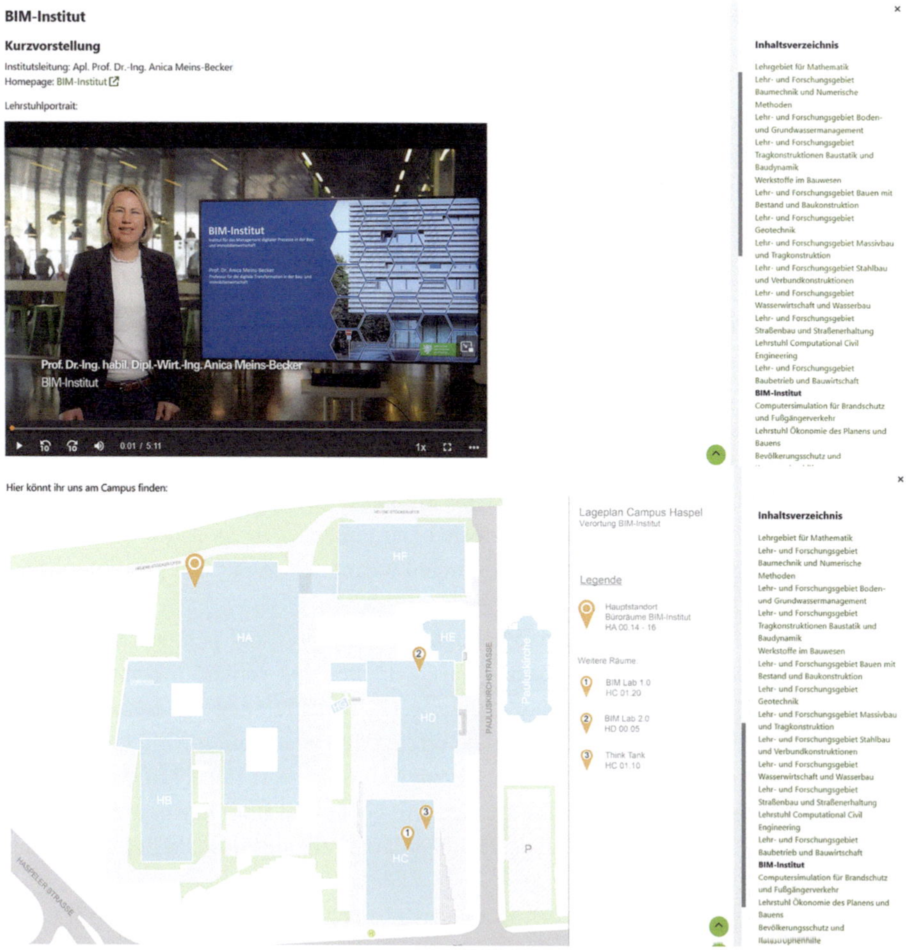

Fig. 12.18 Presentation of the chairs – Video portrait and Location plan. *Source* Institute for the Management of Digital Processes in Construction and Real Estate

simultaneously form a visual anchor. Here, the responsible persons and their facilities briefly introduce themselves and give an overview of the teaching and research topics. Care is taken to ensure that not individual projects or one-time events are advertised, so that the video retains its relevance for as long as possible. To obtain more detailed information about current courses and specific research projects, students have the opportunity to access the chair's own website directly via an embedded link.

These compact pieces of information serve to provide students with a as simple and barrier-free a start to their studies as possible.

12.3.4 Outlook

The focus of the research project is the conception of the iMHD and the networking of process data with the teaching content of the Faculty of Architecture and Civil Engineering at the Bergische Universität Wuppertal. Special attention is paid to a user-friendly implementation in the Learning Management System Moodle, which makes it easier for students to start their studies and offers continuous added value.

The next step is the planned validation of the application with the students of the faculty. This will identify and correct any incorrect links and erroneous information.

Furthermore, there is an intention to add further functions to the iMHD. The linking of the processes in the life cycle of a building with the teaching content of the construction-related degree programs in Architecture and Civil Engineering is intended to improve understanding of individual professions and collaborative work in practice. Here, individual professions could be assigned to the roles in the process model in the iMHD, for the completion of the work steps. Similar to the video portraits of the chairs, profiles for these professions could be created and gradually integrated into the iMHD. A possible feature could also be a bulletin board where companies can post internships, student jobs and job offers.

To investigate the effects of the iMHD on students, pre- and post-use surveys will be conducted. Planned are topic-specific questionnaires that capture the assessment of cross-role process understanding, individual career orientation and the course of study. In addition to this, discussion rounds will be offered. In collaboration with the individual chairs, the behavior of the students will be analyzed. A possible impact of the iMHD on the dropout rate and the general grade average will be derived from key figures. It will be investigated whether an improved understanding of the relationships is reflected in easier contact initiation by the students.

A future focus will be the updating of the information as processes change, chair leaderships change and examination regulations can be renewed. This requires continuous maintenance of the iMHD, which also needs to be planned and carried out.

Once it has been clarified who will take care of the maintenance of the existing information, further information on changing Courses such as elective subjects and project work can be integrated.

12.3.5 Conclusion

The research project has taken a significant step towards the further development of the construction industry and its connection with academic education. A successful bridge has been built between the complex processes of the German construction process and the curriculum of construction-related courses. The iMHD presents itself as an experiential construct, in which the process model serves as a guide to make the multifaceted construction industry understandable and tangible for students.

By integrating information about course content, examination modalities, and the ability to document and compare individual progress in studies, the study experience is enriched. This creates a direct connection to the students' world of life and enables them to better understand and internalize the complex process knowledge.

This holistic understanding of the processes in the construction industry gives students insights into various professional groups and workflows that they would not otherwise receive. It forms a solid foundation for a successful career in the construction and real estate industry.

However, it is particularly important that this gained understanding and knowledge is an indispensable prerequisite for the successful digitalization and digital transformation of the industry. Often, digital solutions are isolated and only address the processes of individual professional groups or companies. This leads to difficulties in communication with the various process participants outside of these solutions. The knowledge about the interrelationships of the processes in the life cycle of a building, promoted within the framework of the iMHD project, contributes to developing a better understanding of the concerns of other professional groups and roles in later professional practice.

Overall, this project not only helps to strengthen the connection between theory and practice, but also prepares students for the challenges and opportunities of a constantly changing industry. It lays the foundation for a future-oriented education and a successful career in the construction and real estate industry.

References

Ahmad Pozin, I. et al. (2019). An Ability of WhatsApp Usage in Industrialised Building System (IBS) Construction Project. International Journal of Interactive Mobile Technologies (iJIM), 13(04), 188. https://doi.org/10.3991/ijim.v13i04.10548.

Ahmed, S. (2018). Barriers to Implementation of Building Information Modeling (BIM) to the Construction Industry: A Review. Journal of Civil Engineering and Construction, 7(2), 107. https://doi.org/10.32732/jcec.2018.7.2.107.

BauInfoConsult. (2020). Social Media am Bau: Stärkerer Einsatz von WhatsApp im Berufsalltag. https://bauinfoconsult.de/social-media-am-bau-staerkerer-einsatz-von-whatsapp-im-berufsalltag/ (Accessed 15. January 2024).

Brandt, M. (2021). WhatsApp?! Statista.

Cisterna, D. et al. (2018). Application of Social Network Analysis in Lean and Infrastructure Projects. 412–421. https://doi.org/10.24928/2018/0483.

Erdogan, M. (2023). [Master Thesis]. Karlsruhe Institute of Technology.

Hauptverband der Deutschen Bauindustrie e. V./Kraus. (2022). Produktivität im Bau(haupt-)gewerbe – ein statistischer Befund. https://www.bauindustrie.de/zahlen-fakten/auf-den-punkt-gebracht/produktivitaet-im-bauhauptgewerbe

Kupriyanova, M. et al. (2023). The impact of digitalization on the manufacturing industry. E3S Web of Conferences, 431, 05030. https://doi.org/10.1051/e3sconf/202343105030.

Kyivska, K., & Tsiutsiura, S. (2021). Implementation of artificial intelligence in the construction industry and analysis of existing technologies. Technology Audit and Production Reserves, 2(2(58)), 12–15. https://doi.org/10.15587/2706-5448.2021.229532.

Lourenço, M. P. et al. (2024). Barriers to the Implementation of Building Information Modeling (BIM) in Late-Adopting Countries in the European Union: The Case of Portugal. https://doi.org/10.20944/preprints202403.1301.v1.

Ma, L. et al. (2023). Barriers and strategies for building information modelling implementation: A comparative study between New Zealand and China. International Journal of Construction Management, 23(12), 2067–2076. https://doi.org/10.1080/15623599.2022.2040076.

Mazhar, N., & Arain, F. (2015). Leveraging on Work Integrated Learning to Enhance Sustainable Design Practices in the Construction Industry. Procedia Engineering, 118, 434–441. https://doi.org/10.1016/j.proeng.2015.08.444.

McKinsey Global Institute. (2017). Reinventing construction: A route to higher productivity. https://www.mckinsey.com/~/media/mckinsey/business%20functions/operations/our%20insights/reinventing%20construction%20through%20a%20productivity%20revolution/mgi-reinventing-construction-a-route-to-higher-productivity-full-report.pdf (Accessed 15. January 2024).

Moreno, J. L. (1934). Who shall survive?: A new approach to the problem of human interrelations. Nervous and Mental Disease Publishing Co. https://doi.org/10.1037/10648-000.

Morning Consult. (2022). National Tracking Poll #2204197 April 30—May 03, 2022. https://assets.morningconsult.com/wp-uploads/2022/05/09135830/2204197_crosstabs_MC_TECH_TWITTER_Adults_v1_CC.pdf (Accessed 15. January 2024).

Puolitaival, T. et al. (2018). What's the real story around digital technologies in construction management? The 42nd Australasian Universities Building Education Association Conference in Singapore.

Regona, M. et al. (2022). Opportunities and Adoption Challenges of AI in the Construction Industry: A PRISMA Review. Journal of Open Innovation: Technology, Market, and Complexity, 8(1), 45. https://doi.org/10.3390/joitmc8010045.

Statista. (2022). Baugewerbe in Deutschland: Industrien & Märkte. https://de.statista.com/statistik/studie/id/7432/dokument/baugewerbe-in-deutschland/ (Accessed 15. January 2024).

technavio. (2023). Construction Market in Germany by Type, and End-user—Forecast and Analysis 2023–2027. https://www.technavio.com/report/construction-market-industry-in-germany-analysis (Accessed 15. January 2024).

Terkan, R., & Celebi, S. I. (2020). How WhatsApp Changes the Way Business Work? International Review of Management and Marketing, 10(5), 179–184. https://doi.org/10.32479/irmm.10769.

Valoon GmbH. (2024). Valoon—Software für auf'm Bau Website [Computer software]. https://www.valoon.chat/.

Zhang, F. et al. (2022). Integrated applications of building information modeling and artificial intelligence techniques in the AEC/FM industry. Automation in Construction, 139, 104289. https://doi.org/10.1016/j.autcon.2022.104289.

13 AI Meets Art

Dr.-Ing. Dipl.-Ing. (FH) Bianca Weber-Lewerenz

> *"The poet writes something, the architect builds something,*
> *the teacher educates and creates education*
> *and all of them are creators!*
> *Creative activity delights the human soul extraordinarily,*
> *Therefore god has shuffled the cards for us*
> *And given us his creative spirit."*
> **P. Henri Boulad S. J., Egyptian Jesuit**

This brief excursions stimulates, creates islands of tranquility and inspires. The interpretation and definition of the term "diversity" can be extensively discussed. This book at hand deals, among other things, with the diversity of innovations, technologies, in sum the diversity of technical feasibility—beyond construction: In the Arts.

Have you, dear readers, ever dealt with questions around AI and art? Then now at the latest.

Can Artificial Intelligence (AI) do art? Is that art? And who is the artist behind, and who is the inventor? Because AI only works when the programmer or a person has fed their data to the algorithm (Fig. 13.1). The result is, for example, a painting generated by AI. There is already AI software, including the software DALL-E developed by OpenAI[1] (Fig. 13.2), which analyzes and reproduces the works of renowned painters. DALL-E can create realistic images and artworks from a text description in natural language. Neuroscientist Matthias Bethge has conducted various research on this. In an interview with the federal government in 2020, he said, "*This form of AI collects experiences, ana-*

[1] https://openai.com/dall-E-2 (Accessed 17. December 2023).

B. C. Weber-Lewerenz (eds.), *Diversity in Construction—The Gamechangers*,
https://doi.org/10.1007/978-3-658-45308-4_13

Fig. 13.1 Hierarchical text-conditional image generation with CLIP latents, April 2022. *Source* https://openai.com/dall-e-2

Fig. 13.2 DALL·E Illustration: Justin Jay Wang, September 2022. *Source* https://openai.com/blog/dall-e-now-available-without-waitlist

lyzes structures, detaches itself from the past and creates something new, surprising on this basis. A creative human does not do it any differently." [2]

Can humans keep pace? Some artists themselves are already euphorically talking about a "real partnership between a painter and artificial intelligence". The procedure looks roughly like this: These newly developed artificial assistants analyze the artist's images, break them down into their components and reassemble them in a new way. This machine assistance serves the painter as an extended arm and expanded memory in one.

[2] https://www.ruprecht.de/2022/06/14/art-or-artificial/ (Accessed 17. December 2023).

Fig. 13.3 Conducting Robot EveR-6. *Source* Kim Sunjoo

Goodbye to creative crises.

Not only in painting, but also in the art of music, composers rely on virtual tools. Machine Learning (ML) makes it possible. This human–machine music even causes listener ecstasies and goosebump effects.[3] In 2023, the robot EveR-6 had its premiere as a conductor at the National Theater in Seoul.[4] Through programming, it nods to the beat of the music and conducts (Fig. 13.3).

A Berlin start-up has invented the AI-based app "Enote".[5] It is a portable, digital music library to make all sheet music from the twenty-first century accessible at any time.

Nevertheless, art is HANDCRAFT. Art is creativity, love, desire, passion, feeling for artistic creative design. AI can only be a supporter here, as it relies on instructions, information, and data from humans. Felix Rebitschek, Managing Director and Scientific Director of the Harding Center for Risk Competence at the University of Potsdam, researches and communicates results that help people make decisions. He is firmly convinced, "*In doubt, I would trust a person with experience*."[6]

---

[3] https://www.kunstloft.de/magazin/ki-kunst/ (Accessed 17. December 2023).

[4] https://www.youtube.com/watch?v=aFRB8Uk5-ro (Accessed 17. December 2023).

[5] https://enote.com/ (Accessed 17. December 2023).

[6] https://www.ruprecht.de/2024/02/01/in-doubt-for-the-human/ (Accessed 17. December 2023).

So the artist Hans Furer and the computer scientist Bernhard Egger, Professor for Cognitive Computer Vision, at the FAU Erlangen-Nuremberg, asked the question: "Can AI understand an artist and his work?"

In answering where AI meets art, there was a vernissage of an AI & Art experiment at the FAU Erlangen-Nuremberg on International Women's Day on March 8, 2024. The art spectacle was underscored by four short and exciting lectures: "*A look back: The history of digital art*" by Sabine Lang—PostDoc Digital Art History. "*Generative Artificial Intelligence*" by Bernhard Egger—Junior Professor for Cognitive Computer Vision. "*Feminist AI for a fairer distribution of power*" by Eva Gengler – PhD student Women's Rights in the Age of AI & Co-Founder of enableYou and feminist AI. "*The AI experiment has changed my work*" by Hans Furer—Artist and Lawyer.

The panel discussion rounded off the celebration of AI in art. Here, Annabelle Hornung, Director of the Museum for Communication Nuremberg, Museum for Communication Nuremberg, Vladimir Alexeev, writer, researcher and artist, Hans Furer, artist and lawyer, Daniel Bucher, master's student in art history, and Eva Gengler, PhD student Women's Rights in the Age of AI, provided deeper insights. Frank Bauer, Academic Council, Chair of Graphic Data Processing at the FAU, moderated and let his professional expertise flow in.

Although software programs are increasingly approaching human craftsmanship, Artists approach, they lack any awareness of social, emotional or societal factors. That is, there is a lack of a very central drive of creative creation. One of the great advantages of machines is that they can evaluate and process huge amounts of data. However, their imagination only goes as far as the data set that the human prescribes to them. Thinking along and making cross-connections to other areas of life and experience are things machines cannot do. In fact, heated discussions have erupted about the clarification of the authorship of the artwork. It remains open, does it belong to the artist, the person who has fed the machine that creates art with data, the software programmers or the coders? A legal field that is crying out for the exploration of new insights to secure artistic creation and intellectual property, orientation and the sensible sustainable use of AI.

The visible increasing interest in AI in art is reflected in societal debate and various dialogue formats. Artistic exhibitions in cooperation with research institutions such as at FAU Erlangen-Nuremberg[7] are proof that we are critically engaging with the newly opened up field of topics. Perhaps once with a laughing eye and more humor when experimenting. Where early testing, testing, refining, throwing away again, starting anew is done, there is actual interest and a engaging with the subject matter.

To achieve progress and innovation, people do well to test themselves and overcome boundaries again and again, to redefine, to critically examine trends, to gather new knowledge and experience. Knowing that human creativity and artistic creation remain with the human.

---

[7] https://www.fau.de/2024/03/news/experiment-at-the-fau-what-happens-when-an-ai-should-create-paintings/ (Accessed 17. December 2023).

Initiatives and Networks—Social Media and Public Perception 14

14.1 Speech by Federal President Dr. Frank-Walter Steinmeier at the Annual Celebration of the German Academy of Engineering Sciences on October 10, 2023 in Berlin (Steinmeier 2023)

"I was looking forward to this evening, to speeches, to discussions, to learning a lot about innovations and technology, and I was also looking forward to an exciting musical program. It is being designed by Tamar Halperin and Guy Sternberg. They will accompany us tonight, and both of them are from Israel. At the moment they are still waiting behind the stage for their performance. And I think we should say tonight how horrified and shocked we are in the face of the terrible news and images that reach us from Israel, and I would like to tell you, dear Mrs. Halperin, dear Mr. Sternberg, on behalf of the entire Israeli people: We mourn with you and with our Israeli friends for the dead; we share in the fate of the many injured; we fear for the families, whose relatives have been kidnapped. We stand firmly by your side. You are not alone.

"*Time does not change us, it only unfolds us.*" This is written by Max Frisch in his diary. Even though the sentence is a bit older, I think it fits wonderfully into this time, into this event tonight. Because much is changing, and we are in a time when some are worried about being transformed too much by this change, by new technology. I sometimes even get the impression that many people are currently less curious about change,

B. C. Weber-Lewerenz (ed.), *Diversity in Construction—The Gamechangers*,
https://doi.org/10.1007/978-3-658-45308-4_14

but very worried about being transformed against their will. Some are worried about becoming redundant in these processes, they see their jobs, sometimes entire industries, at risk. The word transformation sounds technically abstract to some, to others it always sounds a bit like loss of control. As if a completely different, new life was being imposed on them. I think we should align our perspective more with Max Frisch and trust ourselves as individuals and society alike to be able to shape the future. And that means: developing perspectives, expanding the horizon, and yes, also daring to try new things again and again. We need all of this in the phase of upheavals we are in.

Holding on to the past, ignoring change, refusing change is not an option—especially not in an open society like ours. But we must give change—that is the task of politics, economy and science—a direction!

Here with you, I know, I don't have to convince anyone. On the contrary: The German Academy of Engineering Sciences (Acatech) is an association where courage to change is almost a prerequisite. Technology and innovations are guarantors for our prosperity, and Acatech is an important political advisor, enabler and networker, so that this prosperity is maintained in times of change.

But that's not all you do: With all technically feasible, you also think about the people, whose lives, whose work is naturally changed by technology, and who sometimes look with skepticism and concern at these changes. Dear members of Acatech, I fondly remember the conversations with Joachim Milberg before and during the founding of Acatech. And today, 20 years later, I am gladly your patron, because you are doing important work here for the future viability of our entire country. And for that tonight my heartfelt thanks.

All of them are on their way to this celebration just passed a construction site. The Gendarmenmarkt is being made fit for climate change: The rainwater, I don't know if you were able to read the signs and information boards, will be collected underground in the future, pre-cleaned and then flows back into the groundwater. This relieves the sewer system and prevents flooding during heavy rain, and it helps to make the stone Berlin more to make a so-called sponge city. At the same time, this concert hall is being renovated to be energy efficient. These construction sites around us, right in the capital, are quite representative of what needs to happen all over the country and what is already happening in many places in the fight against climate change.

We have clear goals in mind: We want to operate and live climate-neutrally by the middle of the century—and at the same time, this is the challenge, remain a strong industrial nation with reliable, sustainable and also affordable energy supply. We want, especially with regard to demographic change, to have enough qualified specialists in the country. And we want to shape digital progress, whenever possible, for the benefit of all. We want to reduce one-sided dependencies on other countries, and the urgency of this task is made clear to us every day by the consequences of the war of aggression against Ukraine.

Achieving these goals is primarily a matter for politics, but not solely a matter for politics, of course not solely a matter for the economy, not solely a matter for science.

It requires a concerted effort by the whole of society, by each individual. And above all: It requires confidence and trust in ourselves that we can actually master this change together and successfully.

When I'm on the road, some of you may follow this on the internet or on social media, I often visit places where the transformation is already succeeding or at least has begun. In Freiberg in Saxony, for example, I visited a company that emerged from a former state-owned enterprise, where gallium arsenide crystals are grown, which are needed for high-speed communication, and which are installed in 70% of all mobile phones worldwide. In Zwickau, just a few weeks ago, I visited the only factory of a German car manufacturer that only produces vehicles with electric drive. In Dresden, a few months ago, I visited the new semiconductor production facility. In Völklingen and Duisburg, steel mill workers told me about their hopes associated with green, climate-friendly steel production.

I do all this as part of a series of visits that I call "Workshop of Change". And with this "Workshop of Change" I will continue to be on the road in the coming months and years. Of course, on the one hand, to look around myself and to stay up to date on new initiatives, new ideas and commitment. Above all, I want to show the population, with the possibilities I have, where exemplary transformation is already succeeding. Examples that hopefully help to reduce fear and create self-confidence that we are indeed up to the challenges of this time here in Germany. We know where we need to catch up; we know where we need to speed up. But we must not constantly belittle our abilities and give even more space to the spreading faint-heartedness! This is also an attitude that we must show!

The transformation of our economy, which affects our everyday life most intensely, will only succeed if we make the bridge to the future so wide that as many as possible, ideally all, can cross it. Transformation succeeds, as I have said elsewhere, when in principle everyone has something to gain. We may not always be aware of it, but we are already in a permanent learning process. Because what mobility will look like tomorrow, how the perfect circular economy works, how microchips become more powerful, how pharmaceuticals can be used more precisely, how steel really becomes green—this is not only the subject of debates, but is already the subject of ongoing change. But this process of change will continue to be accompanied by research and development, and will also need to continue to be accompanied by it.

In order for as many people as possible to participate in and keep up with this process, we need one thing above all: better education and better schools. The situation today is, I don't want to exaggerate, but in view of recent studies, somewhat It's dramatic when every fourth fourth-grader in Germany can't read properly. And that the shortage of teachers is intensifying, we don't need new studies for that at the moment, all 16 federal states are noticing this right now at the beginning of the new school year.

After all, it's clear: Only with good education will we succeed in restructuring our economy. Only with good education will new perspectives emerge for people in our country in the face of change. No top-level research without basic education! But also no innovative power without multiplication tables!

It is all the more important that we first address the shortage of skilled workers where the future is being made: in our schools. If teachers are missing today, then we will also be missing skilled workers tomorrow!

And we have to tell parents: Apprenticeship education is not a disadvantage for the son, the daughter, and it is certainly not the end of a career. On the contrary: Never before have young people in industrial-technical training professions been in such high demand as today! We certainly still need good technical colleges and universities, but I believe we also need a renaissance, a new attractiveness for dual education, especially in industrial-technical professions.

We also need education to awaken curiosity and openness for change. We can only remain a high-tech country if technological progress is understood as a gain, if it finds broad acceptance. We can only develop and apply the most modern technology if we do so with enthusiasm. If the hope for improvements in our quality of life through technical innovations is greater than the discomfort about it. If the hope is greater than the fear of possibly not penetrating the technology to the last fiber. This is especially true for a country like ours, which is poor in resources, rich in ideas.

And that's exactly why we also need a public debate about how we can actually secure human self-determination in our everyday life, which is increasingly influenced by technology. About how we can manage AI-driven processes, decisions and responsibility for them, how we can ensure that ultimately the human being remains the determining factor. I am grateful to you here at Acatech for not only being an important part of this debate with numerous impulses, but also for participating in it with strong presence.

Especially in the field of AI, we see the light and shadow of technological developments. Yes, of course, AI can and will make our lives easier in many areas. When I think of the nurse and the doctor who might have more time for the patient with the use of more AI in hospitals, in care. Who would be spared administrative work that robs them of time today. Or when I think of paraplegics who might be able to get up from their wheelchairs again with AI-controlled prostheses. These are undoubtedly fascinating examples, and we can all assume that these examples are just the beginning.

But also a look at the downsides: For the liberal democracies, which live from the debate and exchange of controversial opinions, the dangers are obvious. AI-generated photo fakes, which are causing the greatest outrage at the moment, are, I believe, only the smaller part of the problem. Above all, it is the governing, regulating algorithms in social media that mainly offer people in social networks news, true, false, such news that constantly reinforce their own worldview, and in which controversy is increasingly rare. If one, and this is the important thing, is only confronted with one opinion, then the plurality of opinions and the substance of democratic debate are passé.

I admit, this observation is not entirely new, and yes, it is also true that politics and platform providers are reacting, perhaps not enough, but they are reacting to this danger, in Germany with the Media State Treaty, in Europe, in the European Union with the Digital Services Act. But we also need resistance from a mature, self-confident citizenry that does not let itself be led by the nose ring of a few large tech giants through the opinion arena. A civil society that understands the boundaries between what can be said and what is unspeakable. A civil society that recognizes the dangers of radicalization and polarization. A civil society that knows that a functioning democracy depends on citizens being fully informed and exchanging ideas, and being able to trust the authenticity of news. This is a crucial constituent of democracy. As helpful and useful as AI can be in many cases—I am convinced: technology will never, not even in the form of AI, ultimately lead to better, more common good-oriented results than we humans can achieve. For justice, for compromises, for forward-looking decisions—for this, and this is not technophobic, but my political conviction, it requires democracy, it requires people.

Therefore, we need smart regulation. A regulation that leaves a lot of room for innovations, for new possibilities, but also takes into account dangers like the ones just described for liberal democracies.

We must not lose touch in Germany and Europe, especially with a key technology like artificial intelligence. We need the ambition to catch up again, to shape technologies and the rules for their application, and only when we are up to date, only then can we co-design! Otherwise, we will not be taken seriously and not be needed.

This imperative must also accompany us in another technological field: space travel. A few weeks ago I was in Belgium and was able to speak with astronauts from the European Space Agency. Among them were our Alexander Gerst and Matthias Maurer, and Samantha Cristoforetti, because she has been at the German Center for Air and Space Travel for so long, I can almost count her as one of ours. They told me about enormous potentials, which come from research in space—and especially in zero gravity. These are very practical projects that are being researched there, projects that could improve all our lives. For example, in the development of drugs against cancer, which can be researched in space in a completely different way under the conditions of weightlessness, because the growth of tumor cells there is not burdened by the conditions on earth, the analysis capabilities are better. It's hard to believe, but far away in the International Space Station, even very earthly questions can be researched, such as building materials that are more climate-friendly for house construction.

The enthusiasm and ambition with which the astronauts told me about these examples, we need as a common basic attitude towards technology and innovation. We should—together in Europe—confidently be guided by the question: How can we be among the best? How can we set standards for technological progress? And we should occasionally remember how difficult the start of the European project Galileo was. I know that we have struggled every year in the federal government about the budget, and everyone was of the opinion: far too much money for the project. In the end, the decisions were always contentious. But today, Galileo is probably the system that generates the best geodata worldwide and offers it worldwide.

We do not live on an island when it comes to regulation and innovation. In our world, where many things are changing faster and faster, of course—one thing is connected with the other—the half-life of scientific knowledge is decreasing. The technological waves of knowledge follow each other more and more quickly. All the more we depend on exchange, on international networking!

This "we", when I speak of "we", by no means only means our country. We can only succeed globally if we think and act European. The negotiation processes with the tech and data giants, as well as the competition with global players, for example, we can only try to shape in our sense as Team Europe. But even as Team Europe, we are of course not alone on the field. Because global challenges such as the fight against climate change, given the starting position, we can only solve in constructive cooperation with the international community.

Isolation, we should have learned, creates stagnation. Networking, on the other hand, creates ideas—ideally a competition of the best ideas. Therefore, my request to you: Raise your voice against all those who advocate isolation! Raise your voice for a free, open, internationally networked society!

I am particularly grateful to Acatech for working in international networks and thinking European. You do this under the guiding principle of strategic and technological sovereignty. I am aware: The path to this varies in different industries and technologies. The claim is great, but the goal is completely correct. We must not become driven by a technical or geopolitical development, but must influence it, shape it or—as I said at the very beginning—give this development a direction!

Ladies and Gentlemen, to give direction, we need you, we need more trust in our technological innovative power and more self-confidence of a society that should have learned and can remember in 75 years of Federal German history, in 33 years of unity, that we have repeatedly overcome crises and setbacks and in this way have also reinvented ourselves again and again. And to some in our country we have to say that democracy does not stand in our way, but that democracy is a prerequisite for this future to succeed! Let's nurture this democracy!"

14.2 Speech, Lecture and Discussion at the Brandenburg State Center for Political Education, Potsdam, 2023 by Klara Geywitz: "City, Country, Woman? Urban and Regional development by and for Women. Building and Women: The Cultural Change in Construction must Continue." (Geywitz 2023)

The spoken word applies.

"Ladies and Gentlemen,

"Women and Building" is my topic. I start anecdotally and then come to the hard facts: In my 15 months as Minister of Construction I have gotten to know the construction and real estate industry quite well. And there I have met many women—from the chambermaid to the real estate agent. I have seen: These women do great work in their field. That was very encouraging.

Of course, I have also experienced how male-dominated the construction industry still is. Two weeks ago, for example, I opened a large trade fair in the heating and sanitation sector. The industry is highly innovative—when it comes to technical questions. At the opening, there were seven important men and exactly one woman on stage. A picture like from the 1960s. At least I was allowed to stand in the middle.

We need to talk about both when it comes to "Women and Building": About women who "stand their ground". And about men who unfortunately sometimes stand in the way of women. Many thanks therefore to the State Center for the invitation to this evening!

And now to the hard facts. Let's start at the top, in the boardrooms of some of the largest publicly traded German construction companies—total turnover: 30 billion euros:

Bauer AG: 100% men on the board
Strabag: 100% men on the board
Hochtief: 80% men on the board (4 men/1 woman)

So at the top, the situation is bad to very bad in terms of equality. But it doesn't get much better below that: In no other industry are so few women employed as in construction—it's 11%. Construction is still largely a woman-free zone.

This certainly has historical reasons: In West Germany, women have only been allowed to work in construction since 1994. It's hard to believe, but before that it was forbidden. (At least in this area, the GDR was already further along.)

The situation is somewhat better among architects and urban planners: Nationwide, "only" 63% of architects and urban planners are male and 37% are female—which is still below their population share.

Several problems are hidden behind these numbers. Quite fundamentally: The idea of equality has only partially arrived in construction. We cannot ignore this if we take equality as a value and as a constitutional norm seriously.

This is not a natural state of affairs—it has something to do with the structures of the construction industry and the related professional fields. And ultimately with the structures of our society.

This is a problem for women who work as roofers, drywall installers or architects or would like to, but are deterred by the circumstances. And it is also a very tangible problem for the industry itself. Keyword shortage of skilled workers. Female roofers, glaziers, carpenters and civil engineers are urgently needed. But they don't exist.

This is not going to get better quickly. Because the next generation is missing: Currently, only 2.8% of apprentices in the main construction trade are female. The proportion of women among students of civil engineering is on average 30% nationwide (in Brandenburg slightly less with 27.5%). Among the students of architecture and real estate economics the proportion of women is quite decent at 58% and 45% respectively.

The fact that women are overall underrepresented in the construction industry has another consequence: Women make up more than half of the population. However, houses, cities, and communities are often planned from a traditionally male perspective. Women only appear on the sidelines in this perspective.

The Argentine architect and urban planner Zaida Martinez summarized this global phenomenon as follows: "*Cities were predominantly planned by men and for the needs of white upper-class men in their prime and physical condition.*"

The post-war idea of the car-friendly city, for example, was—at least in West Germany—oriented towards the traditional ideal of the bourgeois nuclear family: The male breadwinner moved to work by car. The "housewife" performed unpaid reproductive work and stayed—just: at home. But the East German cities were not designed as bicycle- or family-friendly oases either.

The 50s are long gone and luckily they are not coming back. But even today, millions of people live in car-friendly and tendentially gender-unfair cities. Men benefit significantly more from this than women. Even today, 62% of cars in Germany are registered to men and only 38% to women.

The relationship between women & construction is therefore not entirely unproblematic—although construction seems to have more problems with women than vice versa. In short, a few ideas on how it can get better:

Women need promotion, representation, networks, and mentors.

These women's networks already exist in the construction sector and they are getting stronger. There are the very active "Immo-Women", who also have a regional group in Berlin-Brandenburg and award an annual prize for graduates. The Federal Chamber of Architects has a project group on the subject of equal opportunities and the Brandenburg Chamber of Architects has the AG Equal Opportunities—whose chairwoman Katja Melan is also on the panel.

The construction industry has also recognized the signs of the times. On this year's 8th of March, it launched a "Women's Network Construction", for which I have taken over the "patronage"—that's what it's actually called. Mentoring structures are supposed to make it easier for women to enter the construction industry and to advance there.

Of course, the construction industry did not transform into the spearhead of feminism overnight. Instead, there is massive economic pressure behind it. And that's exactly what makes me optimistic that it won't just stop at declarations of intent.

Certainly, the topic of women & construction also needs a cultural change. This starts with the images of women that we convey: Do women also drive a digger in children's books or are recognizable as site managers? This may sound silly, but books—including children's books—convey worldviews. Why shouldn't they also anticipate the future as we wish it to be?

The cultural change must continue in construction. We all have the common construction worker clichés in mind—which are not always, but unfortunately sometimes true. It is proven, however, that the presence of women has a positive influence on the communication style. Another argument for more women in construction.

This also includes: debunking myths. Women don't need their own changing rooms, just separate usage times. (This seems to be a common argument against women in construction: "We don't have extra changing rooms.")

The objection that women physically couldn't keep up is only partially waterproof. Firstly, not all men are sports cannons and weightlifters. And secondly, women in the physically very demanding care professions prove day after day how strong they are.

With the advancing technologization and digitalization, it will be in The future in construction will increasingly be about smart solutions and expertise. And in this, women are certainly not inferior to men.

For the cultural change, we also need a different narrative about women and we need female role models: Two or three famous architects probably come to mind for most people: Walter Gropius, Norman Foster or here in Potsdam Karl Friedrich Schinkel. But three famous female architects?

Zahah Hadid might still be known. Perhaps also Margarete Schütte-Lihotzky with her "Frankfurt Kitchen". But otherwise? Ms. Melan—you surely have a few tips on which names one should remember.

The cultural change does not come by itself. It needs civil society engagement. That's what the Women's Political Council of Brandenburg has been doing for 30 years with admirable perseverance. Thank you very much at this point for this important work!

Of course, it also needs appropriate political framework conditions. And politicians who advocate for gender equality.

Politics—I explicitly include myself in this—can promote gender aspects in all policy areas. In the Federal Ministry of Construction, we do this, for example, in urban development funding.

The guiding document of our national urban development policy, the "New Leipzig Charter", has a clear goal: the just city—which also includes gender justice. People of all genders should have equal opportunities.

We can also promote lighthouse projects for women friendly, inclusive urban design. There are good examples, such as the non-profit "Women-Work-City" in Vienna.

We can anchor gender budgeting more firmly—as agreed in the coalition agreement. This can be a strong lever.

These are all important steps. But I believe we need to tackle the issue of gender equality more comprehensively and at the root.

And that means, above all, addressing the gender pay gap and the infamous "glass ceiling". Women are affected by this everywhere in professional life. Of course, also in the construction industry, in architecture and real estate.

Women in the construction industry earn 18% less than men. Female architects even earn a quarter less than their male colleagues. This also has to do with the fact that women often work part-time because they still take on a large part of the "care work".

Among architects, for example, only 10% of men work part-time, but 40% of female architects. Certainly not because the job pays so well that you can get by with part-time.

But because women primarily take care of the household and the children and sometimes also care for the grandparents.

They perform unpaid care work, therefore earn less and also have fewer opportunities for advancement. This then leads to them earning less, having less say and decision-making power—and thus also less influence on urban development and architecture.

In the real estate industry courses, we have 42% female students. But only 11% of them make it to the executive floors of the real estate industry. More than half of the graduates in the field of architecture are women (58%). But only 3% of the board members in architectural firms are female.

We need to address the "gender pay gap" and the "glass ceiling". If necessary, also with quotas. So far, we only have these for a very manageable number of large companies. But there we see: The quota works. Since we have the "Leadership Positions Act", the proportion of women on the supervisory boards of the affected companies has risen to 35.6%.

And in the end, this benefits not only women, but also companies. Studies by McKinsey and Boston Consulting show: Where more women make decisions, profits are higher and stock prices are better. Companies led by women also perform better in terms of creditworthiness and solvency, and have a lower risk of insolvency. However, this requires a critical mass of at least 30% of women at the top. In other words, companies without a gender quota are harming themselves.

The public sector must take the lead here. Firstly, in terms of leadership positions. The federal government has set a clear goal with the Leadership Positions Act: by the end of 2025, women should have equal participation in leadership positions. There is also still a lot to do below the management level. Currently, among the salaried architects and urban planners, 75% are men and only 25% are women.

My conclusion at this point: Women & Construction—this not only rhymes (in German: Frau am Bau), it actually fits together quite well. And we need to talk about this "actually". We will have the opportunity to do so shortly.

Thank you!"

14.3 The Crafts Image Campaign. Women in Craftsmanship: They Know What They're Doing

Silke Stark (Journalism and Communication Science)

Since 2010, the craft sector, supported by leading creative agencies, has been making a pointed effort to draw attention to itself, giving a strong unified appearance to a million craft businesses with their 5.7 million employees. The craftsmanship image campaign has set new standards in association communication in its kind and with its success. It was the first time that an economic sector with its entire organization appeared in this form and with this force. "The economic power from next door" has thus become an established brand. The slogan "We know what we're doing." confidently expresses the attitude and claim of the craftsmen and craftswomen. Over the past years, the image campaign has significantly increased the perception and significance of craftsmanship. With over 50 awards, it is one of the most awarded campaigns in Germany.

Strengthening the visibility of women in craftsmanship is also a central concern within the framework of the image campaign and its diverse measures. The craft sector is becoming increasingly female, as the campaign also shows in its external representation. Almost one in five successful master examinations in 2022 was passed by a woman (17.9%). Every fourth craft business (24.2%) is led by a woman. In addition, over 74.3% of craft businesses are family businesses, jointly managed by a (married) couple. However, the potential for women's development is far from exhausted. Therefore, it is always also the aim of the campaign to inspire women for a craft profession or the diverse career opportunities in the craft sector.

First and foremost, this is done through the selection of campaign ambassadors, who can be seen on posters and advertisements as well as in film clips and digital advertising media. Here we offer real craftswomen and craftsmen a stage to show what modern craftsmanship looks like. We not only pay attention to a balanced gender ratio, but also like to present craftswomen from trades that were previously understood as "male professions" in order to break down clichés and build role models. Exemplary are Jule Janson (concrete and reinforced concrete builder), Jule Rombey (carpenter) and Luisa Buck (plumber) (see poster motifs depicted, Fig. 14.1).

These images show modern, young women who are going their own way. In combination with strong headlines, the campaign shows that what we do makes us who we are. In craftsmanship, every woman can find her calling and realize herself—whether as an apprentice, journeywoman, master or entrepreneur.

In addition, reference should be made here to a very current campaign motif (2024). The image shows the electrical engineer Mona Kulka, who is installing a photovoltaic system on a roof (Fig. 14.2). Under the guiding principle "Time to act", the campaign this year puts the will to shape and the drive of craftsmanship in the foreground. Only with craftsmanship can the central challenges of our time be create—from digitization to the energy transition. And of course, craftswomen also play a very central role. With skill and passion, they get involved and shape the future of our country.

An important component of the campaign are the so-called career insider videos (https://www.handwerk.de/infos-zur-ausbildung/berufsinsider), which also include poster, advertisement, and social media motifs. The films portray craftswomen and craftsmen in their everyday work and are primarily aimed at a young target group who are looking for professional perspectives. To date, 39 clips have been published on handwerk.de and on the YouTube channel of the crafts as well as short versions on the social media channels of the campaign. These films also show that the craft sector is becoming increasingly female and women are very successful here. For the construction and expansion sector, the videos of Julia Schäfer (bricklayer), Anne Heidrich (tile, slab and mosaic layer), Nièl Braun (roofer), and Leonie Schnetz (carpenter) are mentioned.

Fig. 14.1 Image Campaign Poster Motives Jule Janson, Jule Rombey und Luisa Buck, 2021. *Source* DHKT/Das Handwerk Copyright

Fig. 14.2 Image Campaign Poster Motive Mona Kulka, 2024. *Source* DHKT/Das Handwerk Copyright

In addition to important topics of self-realization and fulfillment, the videos also address prejudices, stereotypes and clichés that the young women sometimes faced. In response to the question "What, you on the construction site—as a girl?" Anne Heidrich replies quite laconically: "Sure, why not?" At this point, the films provide educational work and show that physically demanding jobs can also be performed by women. And they encourage young women and reassure them, to go their own way—and not to be deterred from it. Because as Doreen Gaumann, brewer (Fig. 14.3), puts it very aptly: "If I had believed every saying I used to hear, then I would not be here where I am today."

The campaign also accompanies theme and world days such as Girls' Day and International Women's Day with communicative measures—especially in the social media area. The Women's Day videos of the past years show various creators in their respective crafts and call to follow their example.

All advertising materials of the campaign are available to the entire crafts organization—i.e. all chambers of crafts, professional associations, district crafts associations and companies. They can use these for their own communication work—both in print, moving image and digital area as well as on site at events, trade fairs, career and information days. In this way, we increase the reach of the campaign and its messages and ensure a uniform visual appearance. Our campaign ambassadors thus become inspiring role models for many young girls and women in Germany.

Fig. 14.3 Image Campaign Poster Motive Doreen Gaumann, 2021. *Source* DHKT/Das Handwerk Copyright

References

Geywitz, K. (2023). Speech and discussion on "City, Country, Woman? Urban and regional development by and for women. Building and Women: The cultural change in construction must continue." at the Brandenburg State Center for Political Education, Potsdam. Speech Bulletin of the Federal Ministry for Housing, Urban Development and Construction. https://www.bmwsb.bund.de/SharedDocs/reden/Webs/BMWSB/DE/2023/frauen-und-bauen.html (Accessed 17. January 2024).

Steinmeier, F.-W. (2023). Speech at the annual celebration of the German Academy of Engineering Sciences on October 10, 2023 in Berlin. Bulletin of the Federal Government No. 109-3 of October 10, 2023. https://www.bundesregierung.de/breg-de/service/newsletter-und-abos/bulletin/rede-von-bundespraesident-dr-frank-walter-steinmeier-2229742 (Accessed 12. February 2024).

15 Conclusion and Outlook

"Let's make Construction sexy again!", says Jan Wolber from Valoon—a research start-up that has rapidly developed into a GmbH and is driving AI in construction full steam ahead.

"*Diversity enables growth*", an innovation manager is firmly convinced of this.

In times of increasingly complex construction processes and the growing longing for coherent, above all livable spaces, it is all the more important to focus on the actual task of building and craftsmanship—using meaningful, innovative technologies. What adds to more growth, more long-term success, than doing this in entrepreneurially responsible environments where common values connect and reliability counts.

"Just" erecting a building or a city or an infrastructure has nothing to do with construction. A livable space must touch people in their innermost being, emotionally challenge and promote and not just provide protection from the weather. We should think about the emotions and needs of people when "building", and create emotionally touching, socially sustainable spaces—in a diverse environment that gives the whole its luminosity of sustainability for future generations and resilience.

Where **strengths**, **skills**, **innovation and technologies**, **knowledge**, **communication,** our **values** and **rights** come together and interlock, great things can happen.

Study results provide facts and underline the successes of diversely positioned teams. They make the critical pillars visible. Only in this way can we jointly create a sustainable future in construction with innovative, meaningful technical aids.

New technologies, new approaches and methods, a new culture of learning and thinking, new knowledge are globally available to us. Mutual inspiration in inter- and transdisciplinary exchange and access to cross-disciplinary treasure of experience—almost as an "excursion" into other industries—strengthen us in going new ways.

With the participating co-authors and thanks to the respective great, voluntarily contributed contribution gift, a firework of personal, unique ways, approaches and per-

B. C. Weber-Lewerenz (ed.), *Diversity in Construction—The Gamechangers*, https://doi.org/10.1007/978-3-658-45308-4_15

spectives on innovations and corporate cultures in construction has been achieved. A celebration of diversity in construction!

In 1924, Le Corbusier explained this as follows:

"One brings stones, wood, cement;
one builds houses, palaces with them,
that is a matter of construction.
The inventive spirit is at work.
But suddenly it touches my heart, it feels good,
I am happy, I say:
This is beautiful. This is architecture. Art is present."

From: Vers une architecture, Le Corbusier, éd. G. Crès, 1924, p. 123

The diversity of nature with the change of seasons, the morning bird concerts, the sprouting, ever more colorful nature, sun and warmth have made their own contribution in the finalization of this book project.

This English book version allows the international readership and the construction industry around the globe to participate.

La Champagne

16

Dr.-Ing. Dipl.-Ing. (FH) Bianca Weber-Lewerenz

Diversity and variety are shown sparklingly effervescent in the glass. No product and no wine region is so strongly shaped by women's power and history as La Champagne (Figs. 16.1 and 16.2). The book editor discovered her affinity at a young age during family study trips to Champagne wineries in France.

The names reflect the Champagne myth, Veuve Barbe Nicole Clicquot Ponsardin, Jeanne-Alexandrine Louise Pommery to Louise Lanson de Nonancourt up to Lily Bollinger.[1]

The women who led and lead the Champagne houses brought about groundbreaking renewals of Champagne culture. The timeline of the "widows" (frenchVeuves) is legendary, whether they lost their husband through illness and war, they took the scepter into their own hands. Veuve Clicquot[2] was from 1805 the first woman to ever run a Champagne estate. The traditional method with the shaking of the bottles on the shaking table, to dissolve the yeasts, goes back to her. Already in the first half of the nineteenth century, she created the vintage champagne, as well as the rosé champagne by adding red wine. The prestige champagne "La Grande Dame" stands to this day for her memory. In 1972, Veuve Clicquot launched the annually internationally awarded "Business Women Award" to appreciate the brave entrepreneur. What no other woman achieved: she increased sales with exports to Russia and the USA by 4500%.

Madame Pommery,[3] who took over the Champagne estate after the death of her husband in 1858, is credited with the idea of Champagne as a food companion, the "Brut". At that time an absolute novelty, as in Champagne was almost exclusively bottled sweet.

---

[1] https://www.champagne-bollinger.com/ (Accessed 27. November 2023).

[2] https://www.veuveclicquot.com/de-de/homepage.html (Accessed 27. November 2023).

[3] https://www.champagnepommery.com/home (Accessed 27. November 2023).

B. C. Weber-Lewerenz (ed.), *Diversity in Construction—The Gamechangers*,
https://doi.org/10.1007/978-3-658-45308-4_16

Fig. 16.1 Champagne, Epernay, September 2022. *Source* Bianca Weber-Lewerenz

Fig. 16.2 Drive through Champagne, April 2017. *Source* Bianca Weber-Lewerenz

Another Femme de Champagne is Mathilde-Émilie Perrier, who took over the management in 1887. Her merger with another producer led to the house Laurent-Perrier.[4] In 1889, she produced the very first "Grand Vin Sans Sucre", so the original Ultra-Brut,—just in the year of the completion of the Eiffel Tower.

[4] https://www.laurent-perrier.com/en/ (Accessed 27. November 2023).

At Taittinger[5] with Vitalie Taittinger, Pommery, with Natalie Vranken, Laurent-Perrier, with Lucie and Alexandra Pereyre de Nonancourt and Stéphanie Meneux de Nonancourt, women are still the creative masterminds today.

The oenologist Séverine Frerson shapes as the first female cellar master, with professional years at Piper-Heidsieck, the style of the house Perrier-Jouët.

Margareth Henriquez leads as president and CEO the top brand Krug.[6] The oenologist Julie Cavil is Cellar master at the Krug Champagne house. Alice Téttienne, on the other hand, moved from Krug to Champagne Henriot.[7] The oenologist Séverine Frerson is in the service of Perrier-Jouet.

Taittinger, with its president Vitalie Taittinger, sells more than 6.5 million bottles of champagne annually, generates 180 million EUR in revenue, and employs 235 people. A study by the University of Reims shows that 40% of all champagne houses in the Champagne region are led by women.

Carol Duval-Leroy leads the brand of the same name. Her prestige cuvée is called "Femme de Champagne", meaning Champagne woman, and with cellar master Sandrine Logette-Jardin, she pursues pioneering strategies in adaptation to climate change with the corresponding acid control.

Les dames have one thing in common: they lead the champagne houses into the next generations. They secure the future of the Champagne with new target groups, change and preserve. They revolutionize not only their estates, but the entire industry, both in terms of taste and marketing. Transmission parfaite![8] Champagne truly has no shortage of female expertise. A group of young representatives from seven independent champagne houses, known as "Les Fa'Bulleuses de Champagne", showcase the feminine elegance of their cuvées and continue to expand it powerfully.[9]

The driving forces responsible for the future of the champagne houses confirm: "*What counts is work, talent, energy—not gender.*" They see themselves as a link in the chain, where the security of permanent employees without layoffs is far more important than ego trips and even more topping advertising posters.

Climate change and global warming are facts that everyone must deal with equally. Anticipating that ideal vine conditions and vineyards are slowly "moving" north due to warming, Taittinger acquired vineyard land beyond the English Channel, in the English county of Kent. The first harvest yields are expected this year, 2024.[10]

In 2022, 325 million bottles were brought to market, with a global turnover of 6.3 billion EUR. 2021 was the smallest harvest in 35 years due to frost and the vine disease

---

[5] https://www.taittinger.com/de (Accessed 27. November 2023).

[6] https://www.krug.com/ (Accessed 27. November 2023).

[7] https://www.champagne-henriot.com/en/age-gate/ (Accessed 27. November 2023).

[8] https://la-transmission-champagne.com/index.php/de/home-de/ (Accessed 27. November 2023).

[9] https://fabulleusesdechampagne.com/ (Accessed 27. November 2023).

[10] https://www.fr.de/panorama/die-frauen-der-champagne-91209361.html (Accessed 27. November 2023).

Peronospora (Welt am Sonntag 2023). The wine association Comité Champagne conducted an environmental diagnosis in 2001 to calculate the CO_2 footprint and developed its own sustainability certification. Large champagne houses like Vranken-Pommery align all business areas with sustainability. This involves dealing with climate change, reducing energy consumption, plant protection strategy, dealing with employees, as well as crossbreeding for resistance to vine diseases.

Roger Federer ensures game, set and victory as brand ambassador for Moet&Chandon.[11] The market leader in Champagne production was founded in 1743 by Claude Moet. In 2023, it celebrated its 280th anniversary. In 2023, Sibylle Scherer took over the leadership. In 2007, the total sales of the luxury goods conglomerate LVMH, which includes Moet & Chandon, was 62.2 million bottles, accounting for 18% of the total production of Champagne.[12] The 1150 hectares of vineyards in Champagne are spread over about 200 wine-growing communities. The underground chalk cellars measure over 28km and have their own "street names". Dom Pérignon is actually a separate Champagne brand of the Moet & Chandon house.

When Emperor Napoleon Bonaparte visited the legendary estate in Épernay, he awarded Jean-Rémy Moët with the Medal of the Légion d'honneur, the highest award of France at the time. Hence the naming of Moët Impérial as the flagship Champagne. Only in 2023 did two women pass the scepter for the leadership of the Champagne house: Berta de Pablos-Barbier handed over the presidency to Sibylle Scherer.

The future of the diversity of the male-dominated Champagne lies in women's hands. And there are big challenges ahead.

The appeal is based on a study by the American AI startup ClimateAI in 2020,[13] which proves that global warming threatens production. The study warns, among other things, that due to the global temperature increase of 2 degrees Celsius, the Champagne grapes could possibly disappear by 2050. The days are too hot and there are not enough cool nights to build up the necessary acidity.

Passion for the terroir is one thing, climate and environmental responsibility are for all of us.

Reference

Weinland Frankreich Spezial, Welt am Sonntag, No. 46, vom 12. November 2023, p. 53.

[11] https://www.moet.com/de-de (Accessed 27. November 2023).

[12] https://de.wikipedia.org/wiki/Mo%C3%ABt_%26_Chandon (Accessed 27. November 2023).

[13] https://kitogo.de/amerikanisches-ki-startup-warnt-klimakrise-gefaehrdet-champagnerproduktion/ (Accessed 27. November 2023).

Book Recommendations

A.

Bianca Christina Weber-Lewerenz: Value Accents in Construction 4.0
Springer Vieweg, 2022, 517 p., Softcover € 29.99 (D) | € 30.83 (A) | sFr 33.50 (CH)
ISBN 978-3-658-38237-7
https://doi.org/10.1007/978-3-658-38238-4
Also available as an eBook
English version:
Bianca Christina Weber-Lewerenz, Accents of Added Value in Construction 4.0
Springer Vieweg, 2022, 462 p., Softcover € 40.65 (D) | € 41.79 (A) | sFr 45.00 (CH)
ISNB 978-3-658-39406-6
https://doi.org/10.1007/978-3-658-39407-3
Also available as an eBook

The rapid technological development of AI is causing deep cuts, writes Bianca Christina Weber-Lewerenz in the preface of her book "Accents of Added Value in Construction 4.0". Humans, society, and the environment are directly affected. The author demands: "The construction industry, as the sector with the highest economic driving force and at the same time 38% share of global CO_2 emissions—as of 2021—must deal in a special way with how it wants to achieve sustainability goals." The author has published aspects of her book in advance in the construction trade (7–8/2022). "According to national accounts, construction investments in 2019 amounted to around 373 billion EUR, but the gross value added of the construction industry only comes to 171 billion EUR," Weber-Lewerenz continues. Machine Learning is estimated to have a value-added potential of 5.8 trillion US dollars. This means that AI has greater potential than the steam engine. "Forecasts suggest that the market for AI in construction will have a share of around 3.7 billion EUR by 2026. The time is overdue for the construction industry to discover opportunities for itself and to use opportunities to shape the digital transformation and to set up innovative business models in order not to be left behind in the competition," are the clear words of the author. Interdisciplinary cooperation The success depends for Weber-Lewerenz on potentials being recognized and used, risks defined

B. C. Weber-Lewerenz (eds.), *Diversity in Construction—The Gamechangers,*
https://doi.org/10.1007/978-3-658-45308-4

and consequences estimated. Technologies should be developed and used courageously, firmly and sensibly—and not exclusively oriented towards profit maximization: "Digital technologies and AI should make people safer and more efficient in their work and support them. Only when the advantages and opportunities are recognized can people trust these techniques." Morality and values serve as a guideline for the author. Because for the construction culture 4.0, well-considered value accents and ethical principles in dealing with digitization and AI are needed. It is important to consider that humans must be at the center of technical and technological developments and dictate the direction of technological development and its practical application: Interdisciplinary cooperation is necessary for a meaningful examination of the associated interactions. "Human-Technology-Society is essential." A lively dialogue could particularly enrich the construction industry. To reach the international audience, the author has also published the book in English under the title "Accents of added value in construction 4.0". Sustainable use of technologies This perspective is important according to Bianca Christina Weber-Lewerenz, and it is crucial to set the corresponding course now. In the book, the author identifies the key factors and core competencies that are crucial for the sustainable use of technologies and can help to establish a role model and quality seal for the construction industry at home and abroad. Bianca Christina Weber-Lewerenz is a civil engineer and has been following the changes in technologies and value culture in construction since her craft training as the first female bricklayer in Baden-Württemberg. From "learning from scratch" to scientific research, she incorporates her expertise to show practical, value-oriented ways for a sustainable digital transformation and ultimately for a new way of thinking in construction (*From: Book review bauhandwerk/ela, February 23, 2023*).

Bianca Christina Weber-Lewerenz

Accents of Added Value in Construction 4.0

Ethical Observations in Dealing with Digitization and AI

Springer

B.

Ingrid Vasiliu-Feltes: Impact of Women Empowerment on SDGs in the Digital Era
IGI Global Publishing, 2022, 344 p., 215 USD
ISBN 978-1-668-43637-0
https://doi.org/10.4018/978-1-6684-3637-0
Also available as an eBook

The influence of women's empowerment on the Sustainable Development Goals is exponential, as their contributions are essential in all areas relevant to our society and economy. As a society, we face the moral imperative to reshape, redesign, and realign our global approach to women's empowerment. To advance the empowerment movement, a call to action and alternative paths are needed that can address some of the major challenges that exacerbate the global, social, and economic gender gap.

"Impact of Women's Empowerment on SDGs in the Digital Era" discusses global issues related to the gender gap and how empowering women can contribute to each of the 17 Sustainable Development Goals. It highlights opportunities, challenges, success drivers, and the importance of ethical leadership to achieve this. The book creates a legacy for strengthening women's rights for future generations. This reference work is ideal for policymakers, legislators, government officials, researchers, academics, scientists, lecturers, and students. The many academic areas covered in this publication include, among others:

- Development strategies
- Digital identification
- Women's healthcare
- Financial inclusion
- Gender equality
- Global citizenship
- Inclusion
- Leadership
- Legal Authorization
- Strategic Partnership
- Sustainable Development.

(*From: Book Launch and Presentation, May 2022*)

Premier Reference Source
Impact of Women's Empowerment on SDGs in the Digital Era

C.
Shervin Haghsheno, Gerhard Satzger, Svenja Lauble and Michael Vössing: Artificial Intelligence in Construction—Basics and Use Cases
Springer Vieweg, 2024, 464 p., Softcover € 69.99 (D)
ISBN 978-3-658-42795-5
https://doi.org/10.1007/978-3-658-42796-2
Also available as an eBook

This technical book demonstrates the potential of Artificial Intelligence (AI) in construction. The rapidly growing amount of data and the hardware and software used in construction projects form the basis for cross-project and predictive analyses, which are enabled by approaches of machine learning and robotics.

The individual contributions of this book provide a general overview of methods of Artificial Intelligence (AI) and illustrate their use in construction with specific use cases from research and practice. AI methods promise here both efficiency gains and error reduction through automation of repetitive tasks as well as completely new possibilities of decision support in construction planning, execution and operation. The publication is particularly aimed at executives and technical experts in the construction industry who want to achieve a sustainable improvement of their business processes with AI methods.

Shervin Haghsheno
Gerhard Satzger
Svenja Lauble
Michael Vössing *Hrsg.*

Künstliche Intelligenz im Bauwesen

Grundlagen und Anwendungsfälle

Springer Vieweg

D.

Özlem Doger-Herter: Women in Data Science—Female Role Models in AI & ML

Springer Gabler, 2024, 132 p., Softcover € 84.99 (D)

ISBN 978-3-658-42218-9

https://doi.org/10.1007/978-3-658-42219-6

Also available as an eBook

The socially correct handling of data is one of the central challenges of our next decade. Knowledge generation and derived action affects all areas and industries. This requires diversity in perspectives. In this book, leading female experts from science and business write about their fields of activity with data and technologies and address ethical, moral and socio-political aspects. The contributions deal with Data Science in science and policy advice, the analysis of geographical data, gender medicine, data revolution—and the fact that data serves people (and not vice versa).

RESEARCH

Özlem Doger-Herter *Hrsg.*

Women in Data Science

Female Role Models in AI & ML

Springer Gabler

E.
Adolf Opderbecke: The Mason
Nicol Publishing, 2022, 338 p., € 14.95 (D)
ISBN 978-3-868-20691-3
Hardcover

Books on craftsmanship, regardless of the trade, do not lose their modernity. On the contrary, they provide the basic knowledge and understanding that can be built upon.

Anyone dealing with traditional building techniques must inevitably engage with the craftsmanship of past times. This book provides access to traditional expertise. The work provides a comprehensive overview of the standards of the building trade towards the end of the nineteenth century and is an indispensable guide for the creation of contemporary architecture. It was written by practitioners for practitioners. It covers not only all common techniques of processing, but also the material science, properties and possible uses.

In the same series from Nicol Publishing, the following have been published: Timber Construction, Carpentry, Roofing and Plumbing, Wood Constructions, Brick and Stone Construction.

Adolf Opderbecke
DER MAURER
NIKOL

F.

Felix Friedrich and Sharon Heidenreich: German for Architects and Civil Engineers: A language textbook for planning and implementing construction projects with vocabulary, idioms and exercises.

Springer Vieweg, 2021, 221 p. Softcover € 29.99 (D)
ISBN 978-3-658-32299-1
https://doi.org/10.1007/978-3-658-32300-4
Also available as an eBook

A book cover with orange and green and white text Description automatically generated.

A language textbook specifically designed for architects, civil engineers and related professions of all nationalities who want to improve their German communication skills in a professional context. The book can be used both for self-study and as a course companion. After the first half of the book covers many basics of the construction industry, the second half is oriented towards the service phases and the handling of example projects. Each chapter teaches technical vocabulary, idioms, forms of communication and also grammar. Exercises and tasks with solutions round off the work.

According to the Common European Framework of Reference (CEFR) for languages, the textbook corresponds to a level of B2/C1.

Here is a press review: *"... Overall, the textbook, as one of the few with a subject-specific focus (see also the textbook German for Engineers by Steinmetz & Dintera 2014), through the multifaceted topics, the practical exercises and the additional information, makes an important contribution to the basic understanding of architecture and construction and thus also to German as a technical and scientific language." (Stanislav Katanneck, in: Information German as a Foreign Language, Vol. 50, Issue 2–3, 2023).*

Felix Friedrich
Sharon Heidenreich
Deutsch für Architekten und Bauingenieure
Ein Sprachlehrbuch zur Planung und Durchführung von Bauprojekten mit Vokabeln, Redewendungen und Übungen
Springer Vieweg

G.

Also published by Sharon Heidenreich at SpringerVieweg, 2022: "**English for Architects and Civil Engineers—English for Architects and Civil Engineers: A complete project process in English with vocabulary, … expressions, exercises and practical advice**." **(335 pages, € 39.99 (D) ISBN 978-3-658-36029-0**, https://doi.org/10.1007/978-3-658-36030-6**. Also available as an eBook**

Glossary

AI "Artificial Intelligence" AI is the umbrella term for applications, where machines provide human-like intelligence such as learning, judging and problem solving. The development, application and maintenance of such data-driven technologies requires orientation in order to fulfill the associated digital responsibilities. AI is understood as the ability of machines to mimic human intelligence and learn independently, without relying on instructions from a human developer. For this purpose, software evaluates large amounts of data for similarities through machine learning and draws conclusions from it.

Ethical principles provide both the developer of AI methods orientation, as well as in practical application in companies, with the goal of meaningfully supporting people, human work still to make it safer and more efficient and to prepare complex data in a clear and transparent structure for human consumption.

The topic of Artificial Intelligence (AI) is a central component of digitization and touches on a multitude of societal, political and social aspects. The range extends from intelligent robots to learning-capable devices and products. See "Digitization". See "Artificial Intelligence (AI)".

AI "Artificial Intelligence" (English). See "Artificial Intelligence (AI)".

Algorithm Generally speaking, an algorithm provides a procedure to solve a problem. Using this solution plan, input data is transformed into output data in individual steps. Algorithms play a major role in computer science. They form the basis of programming and are independent of a specific programming language. However, algorithms are not only found in computer science or mathematics. Because algorithms are not only executed mechanically by a computer, but can also be formulated and processed by humans in "natural" language. Algorithms are nothing new. Already in his book "On Indian Numerals" from the ninth century, the Arab mathematician Al-Khwarizmi (the namesake of the algorithm) explains the use of Indian numerals. However, the algorithm experiences a new dimension of application in the field of Big Data or Data Mining. Thanks to the increased number of collected data and the increased computing power of computers, large amounts of data can be sifted through and evaluated for

B. C. Weber-Lewerenz (eds.), *Diversity in Construction—The Gamechangers*,
https://doi.org/10.1007/978-3-658-45308-4

patterns and correlations using algorithms. Algorithms not only simplify our everyday life. They can also analyze our behavior. This happens not only retrospectively, but also anticipatively. Algorithms control and influence us, sort out, discriminate and are opaque in the process. Therefore, it is fundamental that humans critically engage with the ethical aspects and consider these in the development and handling of algorithms.

Asylum Seekers Individuals who seek asylum, i.e., admission and protection from political, religious, or other persecution, in a territory or state of which they are not citizens.

Augmented Reality "AR" (English). The computer-aided extension of reality perception. This information can address all human sensory modalities.

BIM "Building Information Modeling" (English). Digital building model, digital method.

BIM2SIM Digital twins of buildings enable the determination of mass and simulation of sustainability aspects.

BPMN "Business Process Modeling Notation" (English)—Modeling. A graphical specification language in business informatics and process management.

CDO "Chief Digital Officer" (English).

CDR "Corporate Digital Responsibility" (English). Entrepreneurially responsible digitization.

Chat GPT "Chatbot Generative Pre-trained Transformer" (English). A chatbot that uses artificial intelligence to communicate with users via text-based messages and images and to understand human language and thus generate a response similar to human language. ChatGPT is the prototype of a dialogue-based chatbot. This was developed by OpenAI in 2022. See also "GPT".

CMS "Compliance Management System" (English).

Cradle to Cradle "from cradle to cradle" (English). Figuratively "from origin to origin"; also abbreviated as C2C. A philosophical or systemic approach for a consistent and consequent circular economy.

Credits The "Credit Points (CP)", also called "Credits", are awarded within the framework of the European Credit Transfer System (ECTS) for services rendered. These include, for example, completed exams, attended lectures, or successfully submitted term papers.

CSR "Corporate Social Responsibility" (English). Corporate Social Responsibility.

Deep Learning "DL" (English). Deep learning, a class of optimization methods of artificial neural networks, which have numerous intermediate layers (English: hidden layers) between input and output layer and thus have an extensive internal structure. In The extension of learning algorithms for network structures with very few or no intermediate layers allows the methods of DL to achieve stable learning success even with numerous intermediate layers.

DEI Diversity, Equality and Inclusion (English). Diversity, justice, and inclusion.

Digital Ethics Digital ethics or data ethics, sometimes also algorithm ethics, as a subfield of ethics and practical philosophy, deals with the moral norms that

should apply to digitization and big data. Digital ethics examines how digital media and technologies are used by individuals, organizations, and in societal contexts, and what solutions can be negotiated to address emerging problems and conflicts. The boundaries to technology ethics are, as with information ethics, fluid.

Digital Technologies They describe technologies based on (computer) hardware, software, and networking, and distinguish themselves from traditional technologies through their flexibility and high availability.

Digital Twin "Digital Twin (DT)" (English). A digital twin is a digital representation of a material or immaterial object from the real world in the digital world. Digital twins enable comprehensive data exchange. They are more than just data and consist of models of the represented object or process and can also contain simulations, algorithms, and services that describe, influence, or offer services about the properties or behavior of the represented object or process.

Digitization There is no clear definition for the term.

Digitization originally refers to the Conversion of analog values into digital formats. These data can be processed informationally. Another meaning of digitization is the digital revolution, also referred to as digital change or digital transformation.

DIN German Industrial Standard.

Ethic Ethics deals with the prerequisites and evaluation of human action and is the methodical reflection on morality. At the center of ethics is specifically moral action, especially with regard to its justifiability and reflection (ethics describes and critically evaluates morality). In ethics, the three questions about the highest good, the right action in certain situations, and the freedom of the will are central.

Engineer of tomorrow The new generation of engineers, who are professionally and personally qualified, shaping the digital transformation.

EU GDPR The "European General Data Protection Regulation (GDPR)" is a regulation of the European Union, which standardizes the rules for processing personal data by most responsible parties, both private and public, as well as the free flow of data within the European internal market. First introduced in 2012, created and came into force in 2016, applicable from 2018.

FM "Facility Management" (English). The totality of instruments and services that support the functionality, safety, and sustainability of buildings, properties, infrastructure, and real estate. Facility Management includes: rental management, including rental administration and accounting.

Foreigner A person who exclusively holds a foreign nationality.

Generative Design 3D CAD function that autonomously generates optimized designs using AI based on a set of system design requirements.

GPT "Generative Pre-Trained Transformer" (English). A multimodal, large, AI-based language model created by OpenAI. See "Chat GPT".

HCBA "Human-Centered Building Automation" (English). The approach of a building automation that is oriented towards people and their usage habits, smart and sustainable.

HI Human. human intelligence

Human-Centered Engineering "human-centered engineering". It is an interdisciplinary science that deals with the technical development and the use of technologies, putting people at the center, to understand the activities of the users, but also the consequences of technical progress on people, to develop a technology that meaningfully supports people.

Human-led AI Human-led AI

iEthics A term that defines the meaningful, human-friendly, sustainable digitization and AI in construction. Developed in 2019 by Bianca Weber-Lewerenz and the goal of the "Excellence Initiative for meaningful and sustainable digital transformation in construction" she founded in 2020.

iMHD "Interactive Module Handbook" (see iMHD: The interactive module handbook for universities to promote procedural, digitized and collaborative work in the construction and real estate industry, Institute for the Management of Digital Processes in the construction and real estate industry at the Bergische Universität Wuppertal.)

Inclusion "Inclusion" (English). Inclusion means inclusion and, that people with disabilities no longer have to adapt their lives to existing structures. Rather, society is called upon, to create structures that enable every person—including people with disabilities—to be a valuable part of society from the start. Lived inclusion means designing the environment so that everyone can participate. Transferred to the corporate context, this means: There is probably at least one person in a team who holds themselves back.

Interdisciplinary collaboration A professional, social collaboration across disciplinary boundaries, with the aim of exchanging knowledge, to gain new knowledge, to further develop approaches and ways of thinking. Each discipline can benefit from this, create a high added value and create value.

IoE "Internet of Energy" (English IoE), not to be confused with Internet of Everything (IoE), It is a concept that goes beyond the boundaries of the Internet of Things (IoT). IoE is an implementation of Internet of Things technology (IoT) in distributed energy systems. IoE is a technology term that refers to the modernization and automation of power infrastructures for energy producers and manufacturers. This can make energy production more efficient and cleaner with the least amount of waste.

IoT "Internet of Things" (English IoT), the term describes that the computer in the digital world is increasingly supplemented by "intelligent objects" up to "AI", artificial intelligence. Instead of, as currently, being the object of human attention, the "Internet of Things" is supposed to support people in their activities.

KPI "Key Performance Indicator" (English). Important performance indicator.

LMS "Learning Management System" (English).

Metaverse "Metaverse" (English). The metaverse is a concept in which a digital space is created through the interaction of virtual, augmented and physical reality. The main aspect is to unite the various action spaces of the internet into one reality.

Migrants People who migrate from one residence/country to other residences/countries.

ML "Machine Learning" (English).

Moodle A free course management system and learning platform. The software offers the possibilities to support cooperative teaching and learning methods.

Morality The term stands for the norms and values, rules and commandments valid in a society. Morality stands for the totality of ethical-moral norms, principles, values, which regulate the interpersonal behavior of a society, which are accepted by it as binding.

New Work In the New Work cosmos, the human being is no longer supposed to perform mere work, but to realize himself through work. The focus is on meaning, freedom, self-determination, independence and personal responsibility in everyday work.

Normative Management The normative corporate management (= Normative Management) determines the identity of a company in values, goals, behaviors as well as organizational norms and summarizes these elements in a mission. It ensures the viability and development capability. Normative management is superior to strategic management.

Open-BIM A uniform, open file format for a software-independent data exchange. The openBIM® initiative represents a modern approach to interdisciplinary collaboration among all members of the AEC industry.

People with a migration background All those who migrated to the current territory of the Federal Republic of Germany after 1949, as well as all foreigners born in Germany and all those born in Germany as Germans with at least one parent who migrated or was born as a foreigner in Germany.

Project-Life-Cycle A comprehensive approach encompassing all phases of the construction process.

R&D "Research and Development" also referred to as R&D. Named as Research and Development (R&D) in English.

Refugee A person who, out of well-founded fear of persecution because of their race, religion, nationality, membership of a particular social group or because of their political beliefs, is outside the country of which they are a national, and cannot or does not want to avail themselves of the protection of this country because of these fears.

Robotics Robotics or robot technology deals with the design, creation, control, production, and operation of robots, e.g., industrial or service robots. In the case of anthropomorphic or humanoid robots, it also involves the production of limbs and skin, facial expressions and gestures, as well as natural language abilities. The focus is on hardware robots with hardware and software. Pure software robots (bots) are primarily developed in computer science, nanorobots in the future in nanotechnology.

SDGs See "Sustainability Goals".

STEM "Science, Technology, Engineering, Mathematics". So-called STEM subjects.

Sustainability Goals 17 political objectives for sustainable development of the United Nations (UN), which worldwide serve to secure a sustainable development on an economic, social and ecological level (English. Sustainable Development Goals, SDG's of the United Nations). They were designed in reference to the development process of the Millennium Development Goals (MDGs) and came into effect on January 1, 2016 with a term of 15 years (until 2030). Unlike the MDGs, which applied particularly to developing countries, the SDGs apply to all states.

The goals include: Health, Energy/Climate/Water, Education, Poverty, Nutrition, Economic Growth, Technology, Inequality, Gender Equality, Hunger, Justice, Governance, Decent Work, Peace, Clean Energy, Land Ecosystems, Oceans/Seas/Marine Resources, Social Inclusion.

The official German title is Transformation of our World: The Agenda 2030 for Sustainable Development (short: Agenda 2030); synonymously Global Sustainability Agenda, Post-2015 Development Agenda, Global Goals of the UN and World Future Contract are used.

The basis for the implementation of the SDGs in Germany is the German Sustainability Strategy adopted by the Federal Government in January 2017. To make it concretely measurable, a catalogue of indicators was developed with the participation of, among others, the German Federal Statistical Office, which was adopted by the UN Statistics Commission in March 2016. See "SDGs".

TGA "Technical Building Equipment".

UIM "Urban Information Modeling" (English). Urban Information Modeling.

Value-based Management Value-oriented management. Value-oriented management (English: value-based management) is a management approach that emerged in the 1980s which defines the goal of corporate management not primarily as the increase in sales or profits, but the company's value. Value-oriented key figures serve both as an internal control instrument and as a basis for management compensation systems.

VDI Association of German Engineers.

VR "Virtual Reality (VR)" (English). Virtual reality, it refers to the representation and simultaneous perception of reality and its physical properties in a real-time computer-generated, interactive virtual environment. A mixture of virtual reality and pure reality is called mixed reality (English. Mixed Reality, also Augmented Reality).

3D models Three-dimensional representations of buildings, areas, cities, infrastructure, built environment.

MIX
Papier aus verantwortungsvollen Quellen
Paper from responsible sources
FSC® C105338

If you have any concerns about our products,
you can contact us on
ProductSafety@springernature.com

In case Publisher is established outside the EU,
the EU authorized representative is:
Springer Nature Customer Service Center GmbH
Europaplatz 3, 69115 Heidelberg, Germany

Printed by Libri Plureos GmbH
in Hamburg, Germany